By Tony Northrup

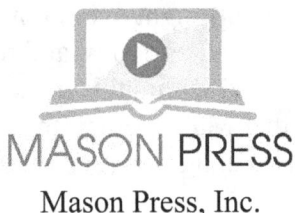

MASON PRESS

Mason Press, Inc.

Published by:
Mason Press Inc.
139 Oswegatchie Rd.
Waterford, CT 06385

ISBN: 978-0-9882634-2-0
Printed and bound in the United States of America.
If you need support related to this book, you can write to the author at tony@northrup.org.
For information on all Mason Press publications, visit our website at *http://www.masonpress.com*.
The trademarks listed in this book are property of their respective owners. Equipment manufacturers retain their full original copyright over photos used in this book.
This book expresses the author's views and opinions, and the contents within should not be treated as fact. The information contained within this book is provided without any express, statutory, or implied warranties. Neither Mason Press, the author, or this book's resellers and distributors will be held liable for any damages caused or alleged to be caused either directly or indirectly by this book.
This book uses Mason Press' Rapid Publishing process to keep it up-to-date. If a new camera is announced, the author can update the book and have the new content available in newly purchased books within a few hours. This allows you, the reader, to always receive an up-to-date book. With a traditional authoring process, a book is always at least four months out of date.
However, you might notice that the Mason Press Rapid Publishing process has several drawbacks. First, the printing quality is not perfect because books are printed individually, after they are ordered. Second, the book has been only lightly edited. Therefore, you will see more mistakes than you would with a traditional printing process. If you find any mistakes, please email them to the author at tony@northrup.org, and he'll fix them for the next customer and add your name to the acknowledgements.

Designer: Chelsea Northrup
Technical Reviewers: Justin Eckert, Kevin Girard, You
Editor: Tanya Egan Gibson, You
Proofreader: You

For Ed & Christine Mercado:

Thanks for making me a part of your family!

Table of Contents

Introduction

PLEASE READ THIS: In order to update this book as frequently as possible, it is very lightly edited. As a result, you'll certainly find typos and mistakes. This book is currently going through a full edit cycle which will fix all typos and grammatical mistakes. If you find a technical error, please email me at *tony@northrup.org* and I'll fix it for the next printing and update everyone's ebook.

This book is about making the best pictures you can for however much money you're comfortable spending. If that's $300 (USD), then Chapter 1 has a recommendation for you. If it's $10,000, following the advice in this book will ensure that every penny is well spent. I've answered thousands of gear questions from readers of my photography techniques book, *Stunning Digital Photography*. As a result, I have a pretty good idea of what most photographers are looking for. If you're looking for quick gear recommendations so you can get started taking pictures as soon as possible, I'll give you those recommendations right away, without any unnecessary technical talk.

I also know that many photographers are interested in the different features of camera equipment, especially when they want to understand whether the features are worth the extra money. Because we all have different budgets and our own styles of photography, there's no one right answer for everyone. For those of you who want to learn more about how your camera gear works and how different features are used, I'll give you all the in-depth information you need to make wise purchases.

According to the people I've already helped by answering questions one-on-one, I've saved my readers hundreds of thousands of dollars. Saving money by carefully choosing your camera gear doesn't make you a cheapskate. In fact, you can spend the money you save on even more camera gear. By spending wisely, you'll pick the camera gear that will make the biggest difference in your photography.

I link directly to most of the camera gear I recommend, and I'd greatly appreciate it if you used the links in the book to buy your gear. I usually link to Amazon.com because I get a small portion of what you spend as an Amazon gift certificate. I'll use the gift certificates to buy more camera gear, which I'll use and review, and then I'll add more information to this book. Of course, you'll get the updates to the book for free. Oh, and you don't just have to buy camera gear through the links—I get a portion of any sale you make after clicking a link. If you're in the US, it would definitely help me if you visited this guide and clicked a link before buying anything on Amazon.

I do get a bit of money if you make a purchase from Amazon using my links, but you'll notice that I often recommend buying used from places such as eBay. I'd rather save you some cash whenever possible, and you can trust that my opinions are unbiased. If there's anything in this book you disagree with, please write me a note at *tony@northrup.org*. As with all of my photography books, I'll update the book based on your comments and send the update to all my readers.

Updating This Book

Please don't be annoyed if my recommendations would be different from yours, or if there's something in the book that is outdated or wrong. Instead, write me an email (_tony@northrup.org_) and tell me how you disagree or what needs to be updated. I'll update the book with your thoughts, so that other readers can benefit, and I'll add your name to the acknowledgements as my way of thanking you for your contribution. Then, I'll send the update out to everyone who has purchased the book for free.

Acknowledgements

This book is FAR too complex for one person to write. First, I'd like to thank the thousands of people who've personally asked me about camera gear in the last several years—your questions taught me what people need to know. I'd also like to thank everyone who told me I was wrong about something gear-related; I am always excited to learn something new. Unfortunately, I haven't kept track of everyone's names. However, since the first release of the book, these folks have provided me important corrections, updates, and outside opinions: Kevin Girard, Mickey Whitlock, Jon Howard, Jayaram Krishnan, Yashar Armaghani, Michael Will, Martin Konrad, Gene Krumenacker, Alma Almanza Cárdenas, Pedro Costa, and you, too, if you contact me (_tony@northrup.org_) with any updates or corrections!

I'd also like to thank the equipment manufacturers who have provided me with loaner equipment for testing, including PocketWizard, Phottix, and Adorama.

How This Book is Organized

First, I will discuss the different types of cameras available: smartphones, fixed-lens cameras, mirrorless cameras, and Digital Single Lens Reflex cameras (DSLRs). Then, I'll help you decide which type of camera you need. Next, I'll provide buying guides for DSLRs and lenses with recommendations for specific models. Finally, I'll recommend specific accessories for you, based on the type of photography you plan to do.

Chapter 1: Camera Gear Basics

I know many people don't care about comparing features or understanding technical details; they just want a simple answer to their questions. If you don't already have a camera, here are

Video: Our Favorite Cameras
11:13 - *sdp.io/favorites*

my recommendations. If you do already have a camera that's a different make than I recommend, feel free to buy the lenses I recommend for your camera.

Do You Really Need More Gear?

Here's something to keep in mind: I've helped review thousands of reader photos, and if the picture can be improved, I tell the reader exactly how to improve it. Only about 2% of the time do I recommend people upgrade their gear, and the vast majority of those times, I'm recommending adding an external flash or using a different lens. Only a couple of times have I recommended someone upgrade their camera body.

The lesson to be learned from this experience is that most people don't need more gear to make better pictures. They just need more practice and some post-processing. For the people who do need to buy gear, most of that money should go to lenses and lighting. The cost of your camera gear is not a reflection of your skill. Many people think that as they improve their photography skills, they should also improve their gear. Only seek out new gear when your current gear is failing you.

Bang for Your Buck

Most of us have limited budgets and need to consider which purchases will do the most to improve our pictures. Let's look at a couple of examples.

First, consider an amateur landscape photographer with the entry-level, 24-megapixel Nikon D3400 and its kit lens. His photography has gotten excellent, so he's considering upgrading to the much more expensive 45-megapixel Z7. The Z7 is amazing, but for his needs, I'd tell him to keep the D3400 and spend the $6,000 he had budgeted for the upgrade (body and lenses) on a trip around the world, visiting all the most beautiful landmarks with his D3400. If you're a landscape photographer, you'll see far more improvement if you spend your budget on travel and practice.

Now, consider a mom with the excellent (but older) Sony a6000. She loves taking pictures of her son's indoor basketball games, but many of the shots are out of focus. I'd tell her that her camera isn't well suited to the task because (in our tests) it doesn't track moving subjects as well as newer cameras. I'd recommend she sell her a6000 and buy a newer (but more expensive) Sony a6400. If she didn't want to spend more money and didn't mind switching to a different brand, she could get better results shooting sports by buying a used Nikon D5300. In this scenario, changing hardware will definitely improve her pictures, even though the camera isn't more expensive.

The landscape photographer would be better off spending his money on travel, while the sports photographer actually needed an upgrade. A portrait photographer might get better portfolio pictures by hiring a model and a makeup artist, or by buying lights. The best way for you to exchange money for improved pictures might not be by buying equipment—but it could be. In this book, I'll try to give you the information you need to decide for yourself.

Your First Camera

I've talked with hundreds of people in the process of buying their first camera. Here's the process they usually follow:

1. Look at a reasonably priced camera kit for sale.

2. Notice a camera that has a few more megapixels and costs slightly more.

3. Start looking at incredibly complex camera reviews, such as those at *dpreview.com*.

4. Discover crazy people who are passionate about nonsense such as high ISO performance, dedicated depth-of-field buttons, or Canon vs. Nikon vs. Sony.

5. Become flustered when they can't decide between two or three different similarly priced bodies and worry that they're going to get the wrong camera and forever be cursed with taking awful photos.

Here's the process that I'm going to walk you through:

1. Determine a total budget that you're comfortable with.

2. Determine how much the accessories are going to cost you.

3. Buy any camera you can afford with the remaining money—used, if possible.

My primary goal is to get you out and taking pictures as quickly as possible. The fact is, your first camera really doesn't matter that much. Even the most basic camera kit is capable of amazing pictures. None of your Facebook friends are going to be able to tell the difference between a photo taken with a $300 camera or a $6,000 camera, but they will notice the difference when you master composition, lighting, and posing.

When searching for Nikon, Canon, Sony and Fujifilm cameras at Amazon, I see kits ranging from under $300 to over $6,000. What do they all have in common? An average review of 4.5 stars—every single camera. They're all amazing cameras, and available at every price point, and the people who use them love them. You literally can't go wrong. If you're afraid you'll buy the wrong camera and regret it, let me comfort you. If you become a casual photographer, any camera will do fine. If you become a serious photographer, you're going to want to upgrade your gear no matter which you buy. You don't know which gear is right for you until you get some practice shooting, however, because you haven't developed a style or discovered which types of photography you love. As your skills develop, there might come a point when your camera equipment can't keep up with you. At that point, you'll be able to sell your existing gear and get most of your money back (especially if you followed my advice and bought used in the first place) and put your money towards gear suited exactly to your specific style.

Here's a quote from Søren Kierkegaard (with the morbid and sexist parts edited out): "Marry, and you will regret it; don't marry, you will also regret it; marry or don't marry, you will regret it either way. Laugh at the world's foolishness, you will regret it; weep over it, you will regret that too; laugh at the world's foolishness or weep over it, you will regret both… This, gentlemen, is the essence of all philosophy."

Spend the money on a camera upgrade, and you will regret it. Don't upgrade, and you will also regret it. Upgrade or don't upgrade, you will regret both. There is no path to perfect happiness and there is no perfect camera.

This book is for people who don't yet own a camera with interchangeable lenses, or who need guidance purchasing accessories. If you already have a camera that you're happy with, skip this book and spend the time taking pictures. I've talked to thousands of readers, and the most common mistake is buying too much camera gear too soon in their photography career. Sometimes, buying new gear is the answer to a problem, but most of the time, getting more experience or spending more time planning is a better answer.

Which Camera Should I Buy?

Which Camera Should You Buy?	
18:19 - *sdp.io/whichcamera*	

If you're asking this question without any additional qualifiers (such as, "Which camera should I buy for portrait/landscape/sports photography?") I'll recommend the Canon EOS RP (about $900 for the body or $1,300 for the kit). It's a great all-around camera that's easy to carry.

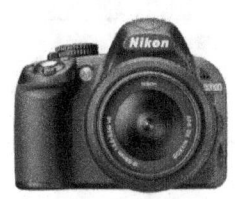 If you'd like to spend less, or you might want to invest in more serious lenses in the future, buy a used Nikon D5300 kit or a Canon T7 kit and don't worry any more about camera gear unless you start to struggle with your equipment's limitations. It should cost you around $400 new or $300 used. I know neither kit is the latest model, but they're great cameras at an amazing price. I've used them both, and they make amazing photos.

If you have an unlimited budget, buy a Canon EOS R5 ($3,900), a Canon 28-70 f/2 ($3,000), a Canon 70-200 RF f/2.8 ($2,700), a Canon RF 15-35 f/2.8 ($2,300) for about $12,000. You can do almost anything with that kit, from video, to sports, to portraits.

These are quick recommendations that work well for most beginners, but there are hundreds of cameras on the market, and if you have specific needs (such as sports or wildlife photography, or if you're a more serious photographer), a different camera might be a better choice. Continue reading!

Which Lens Should I Buy?

This question is a bit trickier to answer, because which specific lens you buy depends on which model of camera you have. I provide specific recommendations for different types of cameras later in this book.

However, your kit lens is good enough for most casual, candid, landscape, and night photography. A great second lens is the 50mm f/1.8, often called the "nifty fifty" or the "fantastic plastic." This lens is cheap and fun, allowing you to get great background blur for nice portraits at a very low cost and it's perfect for many indoor sports. The nifty fifty is available for $100-$400 from Canon, Nikon, and Sony. You can buy less expensive versions from third-party manufacturers, though they don't perform quite as well.

My favorite second lens is a 70-200mm f/2.8 or 70-180 f/2.8. I discuss lenses in-depth in *Chapter 4, "Lens Features"* section of this book.

If the idea of carrying multiple lenses and regularly switching lenses doesn't appeal to you, buy a super-zoom that goes from wide-angle to telephoto, like the Canon RF 24-200. Super-zooms are available for just about every mount.

For more specific recommendations, including lower-priced options, read on!

Which Flash Should I Buy?

All manufacturers produce name-brand flashes. However, after thorough testing, we prefer the Godox flash system (known as Flashpoint in the US, and other brand names in different parts of the world). The Godox V860II is a great starter flash that can be upgraded with a wireless transmitter so you can use it off-camera. I provide recommendations for other flashes and information about flash features that you might need in *Chapter 5, "Flash Features"*.

Which Tripod Should I Buy?

| Video: Tripods |
| **10:06** - *sdp.io/tripodReview* |

Get the *Manfrotto MKC3-H01* (about $60). It's not the right tripod for every situation, but it's inexpensive and easy to carry. If you need a more serious tripod in the future, you'll still use this one as your travel tripod. There are more detailed recommendations and descriptions of the features you should look for in *Chapter 9, "Tripod and Monopod Buying Guide."*

Is Third-Party Gear OK, or Should I Only Buy Name Brands?

You should definitely consider off-brand lenses, and especially flashes.

Generally, however, you usually get what you pay for. Tamron's 70-200 or 70-180 f/2.8 isn't as good as the $2,500+ name-brand 70-200 f/2.8 lenses. It's not as sharp, it doesn't focus as fast and it's not as durable or weather-sealed. However, it will allow you to take professional portraits for less money, and that's a feature the name brands can't offer. Similarly, the $190 Godox V860 II doesn't have all the features of the top-end name-brand flashes, but it has the same flash output and does everything most amateur and even professional photographers require for about one-third of the price. Even if you insisted on spending $600 on flashes, I'd recommend most photographers buy three of the off-brand flashes rather than one of the name-brand flashes.

> **Tip:** Particularly with Chinese manufacturers such as Godox, you can get even lower prices than normal by purchasing them on eBay directly from China.

Throughout this guide, I'll recommend third-party gear when it's a better value than the name-brand gear.

Is It Safe to Buy Used?

Yes, pretty much. You can definitely find horror stories about used sales gone wrong, but eBay (and most used outlets) offer safety measures that protect the consumer from fraud. I often buy equipment shortly after it's released, in which case my only option is to buy new. However, anytime I buy equipment that's been on the market for more than six months, I buy used if I can. On modern equipment, buying used is an easy way to save 10-15% on lenses and 20-30% on bodies.

I've never had a problem, and I've bought about a dozen camera bodies and lenses on eBay. Here are some tips:

- **Know what you're buying**. You typically can return equipment that isn't as described, but you often can't arbitrarily return equipment because you changed your mind, like you might be able to when buying from Amazon or an electronics store. Be sure to read the description of the item for sale.

- **Buy from reputable sellers**. If a seller has more than 100 sales on eBay, and mostly positive feedback, they're serious and honest. You can almost certainly trust them. I generally don't buy from sellers with fewer than 10 sales or with recent negative feedback.

- **Check the return policy**. Many sellers offer returns within 14 days of purchase. If it isn't what you wanted, you can return it and just pay the shipping back to the seller.

- **Pay with PayPal**. PayPal offers important guarantees if something about the deal goes bad. You should read the complete agreement, but basically, PayPal has you covered if someone tries to scam you. Don't pay with Western Union, or a check, or by any other means.

How Can I Get the Best Deal on Used Gear?

Selling prices for used gear vary tremendously. Looking at actual sales of one of my favorite used DSLR recommendations, the Nikon D850, I see the body only selling for anywhere from $1,500 to over $2,500—with no obvious differences between the items. To make sure that you're the guy who got the $1,500 steal instead of overpaying by more than 100%, use these tips:

- **Check the actual sale prices**. Search for your item. Then, in the left page, under **Show only**, select **Sold listings**. This shows you actual sale prices for the item, so you know the least you can hope to pay for the item. With patience, you should expect to get your item for within 10% of that price.

- **Check multiple stores**. In the US, eBay is probably the top market for used camera gear. However, I often find better prices in the Amazon.com marketplace. Simply look up the product you want to buy on Amazon and click the link for used prices.

Also look for used equipment at *keh.com* (an occasional sponsor of our YouTube channel), *cameta.com*, and *adorama.com/c/Used*.

- **Snipe on eBay**. eBay's user interface prompts you to enter a bid and a maximum bid. Your bid takes effect immediately, raising the selling price of the item. If someone outbids you, eBay automatically increases your bid up to your maximum bid. Sniping is the process of using software tools to place a bid at the last few seconds before the auction ends. It sounds evil, but it's perfectly legitimate, and I highly recommend it because it keeps you out of the excitement of outbidding other people (known as a bidding war), which can cause you to pay more than you originally wanted to. I use the *Gixen* service, which is free (though I pay an optional $6/year to help support the service). If you pay for Gixen's service, you can bid on multiple copies of the same item, and if you win one of the auctions, Gixen automatically cancels your other bids. I use this to submit low bids on multiple copies of the same item, increasing my chances of getting a good deal. For example, when I bought a used Canon 7D, the average selling price was $900. I put a $750 bid on about ten different cameras, and got outbid on most of them, but with some patience, my $750 price did eventually win an item.

- **Resell bundle items**. Often, you can find listings that include the item you want, plus some other items that you don't want. Check the sales prices of the items you don't want and consider buying the entire bundle and reselling the unwanted items. Naturally, you need to figure in the cost of the time it will take you to resell them. However, many of the best deals on eBay come from buying bundled items—particularly unusual bundles. For example, if you want a 5D Mark IV, the best deal might be buying a 5D Mark IV bundled with an unusual lens, such as anything other than the kit lens, because many people will overlook the listing. It's downright foolish to sell a camera with multiple lenses and accessories, but it can be very profitable to buy them.

- **Count shipping**. Though it's against eBay's terms of service, some sellers offer low sales prices but overcharge on shipping. Be sure to factor in shipping charges to calculate your total cost for an item.

- **Check the currency**. If you buy gear from another country, the price might be listed in a different currency. I once thought I bid $30 on a part that I won, only to discover that it actually cost me about $45—because the price of the item was in British Pounds rather than US dollars.

- **Don't sweat shutter count or dust**. I've never had a shutter fail in a camera, and I have more than a dozen bodies dating back to the 1940s. While shutters do technically have a limited number of actuations, and they'll eventually break because they are a moving part, it's a very rare occurrence. Broken shutters are also inexpensive to replace. A little dust in the lens doesn't show up in images, either, but many people will avoid listings that show dust—giving you a great opportunity to get a good deal.

- **Go off-brand**. Products from Tamron, Sigma, Rokinon, Yongnuo, and other third-party lens and flash manufacturers tend to have a lower resale value. This makes them particularly strong used bargains (put perhaps weaker long-term investments).

What Is Gray Market (or Grey Market)?

The gray market is the legal sale of products through unintended channels. For camera gear, it's often selling a camera body or lens in the US that was intended for the European market. Often, you can get grey market equipment for a lower price, and it's exactly the same gear.

There's one catch: the camera manufacturer might or might not provide warranty support. If it doesn't provide warranty support and your gear is faulty, you would need to pay for the repair yourself. Of course, if you drop your camera and break it, it wouldn't be covered under warranty anyway. (You'd need insurance for that, instead.)

When equipment is out-of-warranty, the manufacturer might refuse to repair gray market gear; this happened to me recently when I needed to repair a Nikon 70-200 f/2.8 lens. Instead of sending it to Nikon for repair, I sent it to KEH (an occasional sponsor of our YouTube channel), and they did an excellent job.

I happily choose gray market equipment when the price is better. I've never had to send any piece of gear back for warranty repair. I've definitely broken cameras and lenses, but that wouldn't be covered under warranty anyway.

If you buy gray market equipment from _B&H_, they will usually provide their own warranty directly, matching the manufacturer's warranty.

Which Software Should I Buy?

Most camera manufacturers provide some free software with your camera, but it's usually not very good. One free alternative is RawTherapee, though many users will find it technically challenging.

If you get more serious about photography, the _Adobe Creative Cloud Photography plan_ is the best deal: $9.99 per month for both Lightroom and Photoshop (including free updates). Lightroom is for organizing your photos and light editing. Photoshop is for more serious editing. The two applications work best together.

Lightroom and Photoshop are the choice for almost every serious photographer in the world. Simply because of their popularity, there's also an almost unlimited amount of training materials available and a very large number of third-party plugins that extend their capabilities.

For a video overview of Lightroom and my favorite free apps, refer to Chapter 1 of _Stunning Digital Photography_.

Should I use the Adobe Creative Cloud?

Probably! _Creative Cloud_ is a bargain for most US photographers. If you use other Adobe software, such as Premiere Pro (for editing your videos), you might consider Creative Cloud Complete, which is $50 per month, or $30 per month if you have a CS3-CS5.5

product eligible for upgrade. Creative Cloud Complete includes access to most of Adobe's software.

Unfortunately, Adobe charges more for Creative Cloud outside the US (for no good reason). Prices vary, but it's still often the best choice.

If you are a student or teacher with an .edu email address, Adobe will give you Creative Cloud Student and Teacher Edition, with all their software (including Photoshop and Lightroom) for $20 per month.

What Portrait Equipment Should I Buy?

First, your kit lens is fine for group photos. Individual portraits benefit from fast, telephoto lenses that make facial features more attractive and blur the background, as shown in the next picture. Portrait work isn't particularly demanding on a camera's autofocus system, so you don't need to spend much on the body. However, full-frame sensors do blur the background better than compact cameras, as shown in the next picture. For detailed information, refer to Chapter 6 of *Stunning Digital Photography*.

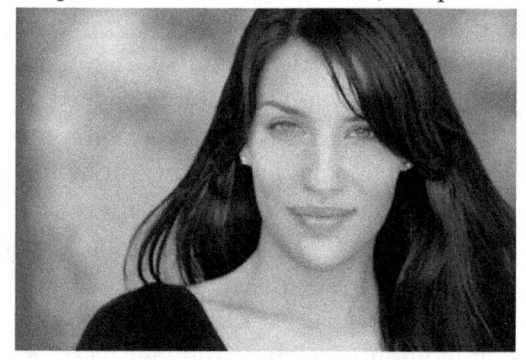

If you're buying your first camera and plan to take portraits, here are recommendations for complete kits at different price points. In addition to the gear listed here, you'll need an inexpensive memory card—but portraits don't require large-capacity or high-speed memory cards, so feel free to get something cheap:

- **$350 budget:** A used *Canon T3 or T5 body ($180)*, a *Canon 50mm f/1.8 II ($100)*, and a *Neewer NW670* flash ($50)

- **$1,000 budget:** A *Canon T6i kit ($750)*, a *Canon 50mm f/1.8 II ($100)*, and a *Neewer NW670* flash ($50)

- **$1,300 budget:** A *Canon T6i kit ($650)*, a *Tamron 70-200 f/2.8 Di LD IF ($570)*, and a *Neewer NW670* flash ($50)

- **$3,000 budget:** A *Nikon D610 kit* ($1,500), a *Tamron 70-200 f/2.8 VC G2 ($1,300)*, and an *Altura AP-N1101 ($75)*

- **$5,000 budget:** A *Nikon D750 kit* ($1,800), a *Tamron 70-200 f/2.8 VC G2 ($1,300)*, a Tamron 24-70 VC G2 ($1,200), and a *Godox TT685N flash kit with remote ($160)*

- **$7,500 budget (mirrorless):** A *Sony a7 III* ($2,000), a *Sony 24-70 f/2.8 GM* ($2,200), a *Sony 70-200 f/2.8 GM* ($2,600), and the *Godox AD600* flash system

- **$10,000 budget (DSLR):** A Nikon D850 ($3,300)*http://amzn.to/YV9PQT*, a *Nikon 24-70 f/2.8* VR ($2,400), a *Nikon 70-200 f/2.8E* ($2,800), and the *Godox AD600* flash system

- **$10,000 budget (mirrorless)**: A _Sony a9_ ($4,500), a _Sony 24-70 f/2.8 GM_ ($2,200), a _Sony 70-200 f/2.8 GM_ ($2,600), and the _Godox AD600_ flash system

Notice that I recommend Canon bodies for the less expensive kits, but Nikon bodies for the mid-range kit. Current Nikon sensors have slightly better image quality, but the Nikon versions of the basic portrait lenses (a 50mm f/1.8 and 85mm f/1.8) cost more than the Canon equivalents. Because the Canon lenses are less expensive, I recommend the camera system for lower budgets.

For the top-end recommendation, the Sony a9's eye-detection autofocus works so well that it has completely changed our portrait workflow; even with shallow depth-of-

Video: Portrait Equipment	
15:28 - _sdp.io/PortraitGear_	

field, moving subjects, and eyeglasses, the a9 nails focus almost every time. It's the first big leap in portrait photography workflow in many years. The ability to shoot silently and to preview the exposure in the viewfinder has made a big difference, too.

These recommendations are for casual portraits with on-camera flash, but you might also need to set aside a budget for multiple lights, props, and software.

Video: 70-200 f/2.8 Shootout	
23:50 - _sdp.io/200test_	

For more about studio and location lighting, refer to the next question.

Many mirrorless cameras less expensive than the Sony a9 are very capable for portraits, too. However, you can get the same effect for significantly less by buying Canon or

Video: 70-200 f/2.8 Discussion	
19:17 - _sdp.io/200talk_	

Nikon DSLR equipment. For example, the _Fujifilm X-T2_ ($1,600) and the 56mm f/1.2 ($1,000) lens will cost you $2,600. You could get similar field-of-view and depth-of-field with a Nikon D610 ($1,500) and the Nikon 85mm f/1.8 ($500), but the full-frame Nikon will give you better image quality and you'll have $600 left over for lighting. You'll also have the option to add the more flexible 70-200 f/2.8 later; nothing equivalent is natively available for any mirrorless system (_Fuji's 50-140 f/2.8_ gives background blur more similar to a full-frame 70-200 f/4 lens).

What Wedding Photography Equipment Should I Buy?

Wedding photography doesn't require a great deal of equipment. Start with a fast normal zoom (such as 24-70mm f/2.8, or your kit lens). The normal zoom will be useful for group shots and photos of indoor receptions in enclosed areas. The Tamron 24-70mm Di VC USD ($1,300) is an ideal choice.

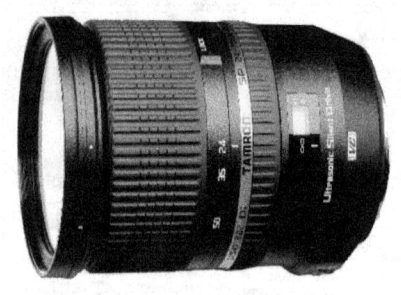

As a second lens, I like to use a telephoto zoom, such as a 70-200mm f/2.8. Follow the recommendations in the portrait photography section of this chapter. The telephoto range of the lens allows you to isolate subjects and simplify the background, allowing a clean composition even when people are crowding around the bride and groom. Using a lens with a low f/stop number (such as f/2.8) allows you to further simplify the composition by blurring the background. When shooting more than one person, however, be sure to use a high enough f/stop number to keep everyone's face in focus.

A powerful external flash is crucial. Outdoors, you'll need it to fill in shadows and add a catch light to your subject's eyes. I average 1,500 shots for a

Video: Tamron 24-70 f/2.8 Review
15:39 - *sdp.io/t70review*

wedding, which is far more than any set of four batteries can provide. To give the flash more staying power and allow it to recycle faster, add an external battery pack. Lower-end flashes don't support connecting an external battery pack, but all higher-end flashes will. The flash recommendations in the portrait sections of this chapter are perfect.

If you have more than one camera body, even if it's an older camera, bring it, attach a different zoom lens to it, and wear them both. This will allow you to quickly switch focal lengths without changing lenses.

If you have enough memory, shoot RAW. Exposure can be very difficult because brides typically wear white, while grooms wear black. Shooting RAW gives you an extra stop or so of leeway, allowing you to recover burnt-out highlights and fill in black shadows on your computer.

You won't get the chance to re-shoot, so it's best to be over-prepared. Bring extra batteries for both the flash and the camera. Bring extra memory cards. If you can, bring an extra camera and lens—even if it's lower quality, it's still better than nothing. Borrow the extra equipment from a friend if you need to.

The wedding photographer's uniform is black: black pants, black shirt. You can wear a *photographer's vest* to carry extra batteries and lenses if you have one. You're going to be moving more than anyone else in the wedding, so choose comfortable (but black) shoes. Ladies, wear pants, because you'll find yourself climbing, leaning, and stretching to get the right angles.

Most professional wedding photographers work in teams of two: a lead and an assistant. Your assistant should have a camera, too. If your assistant is less experienced, give him or her a wide-angle lens, put a bounce flash on it, and set it to automatic. It's good to have two people shooting the ceremony simultaneously; however, only one person at a time should shoot the posed pictures, so that people always know where to look. Also, it helps if at least one of the photographers is a woman, so that the bride will be more comfortable getting ready in front of her.

If you're buying a DSLR specifically to shoot weddings, look for a camera with these characteristics:

- **Dual memory cards**. Memory cards don't often fail, but it can happen. If it happens during a wedding, you wouldn't be able to provide any photos to the happy couple. I've talked to several couples who lost their pictures because their photographer didn't use dual memory cards, and they were crushed. The lost income and bad review can really set the photographer's career back, too.

- **Great continuous autofocus**. You'll be shooting moving people when walking down the aisle and when on the dance floor. Therefore, a good autofocus system is critical. Almost any camera body will be good enough when using the center autofocusing point, so use continuous autofocus with the center autofocusing point on a basic camera, and shoot wide enough to allow you to crop for a nice composition. Cameras with advanced focus systems allow you to use many different focus points for reliable continuous autofocus, providing more flexible composition (at a much higher price).

- **Low noise at high ISOs**. Weddings often require you to shoot in low light, and adding flash could be unpleasant or distracting (especially during a church ceremony). Therefore, you often need to shoot with high ISOs, up to about ISO 6400.

- **Quiet or silent shutter**. Many new cameras support a quiet or silent shutter mode that will make you much more discreet, which is particularly important during the ceremony.

- **High dynamic range**. While less important than autofocus and high ISO, people at weddings tend to wear black and white. A camera with a high dynamic range will allow you to show detail in black tuxedos and white dresses. Be sure to shoot RAW!

If you plan to shoot weddings professionally, you'll need a professional camera. Weddings are hard work; you'll take hundreds of pictures, and you'll need fast focusing in low light conditions. If your camera messes up and causes you to miss a shot, you can't ever recreate it, and your clients will be crushed. Feel free to shoot a wedding with whichever camera you currently have, but be aware of the challenges you'll face. Be sure to read Chapter 7 in *Stunning Digital Photography*.

High-end cameras (typically priced over $2,500) are specifically designed for wedding photographers, and any of them will perform well. My specific recommendations for professional wedding photographers are, in order of preference and price from least to most:

- *Sony a7 III ($2,000)* or *Sony a9 ($4,500)*. Incredibly fast eye-detection autofocus and a silent shutter make the Sony a7 III and a9 the greatest wedding cameras ever. They're very new, however, and not many wedding pros have made the leap to Sony.

- ***Nikon D850 ($3,300)***. The D850 has a fantastic sensor, access to Nikon's incredible lens lineup (as well as inexpensive third-party lenses), and the ability to shoot silently with the rear screen. It also has better dynamic range, which can help with challenging lighting conditions and missed exposures.

- ***Canon 5D Mk IV ($3,400)***. The 5D Mk IV has a a great autofocus system, good-enough dynamic range, and native support for the amazing Canon lens lineup. It's probably the most popular wedding camera in the world.

- ***Nikon D5 ($6,500)*** or ***Canon 1DX II ($6,800).*** The physical size of these beasts will impress many of your clients, and when you trip over a little kid (or drunk lady) while backing up to take a shot, this indestructible camera definitely won't break. The autofocus systems beat everything else out there. Many people will still prefer the D810 or 5D Mark IV, simply because they're much lighter, produce sharper images, and they're still very capable.

Finally, I'll make an unusual recommendation: get a film camera. It sounds crazy to go back to film, I know, but many wedding clients are impressed by a photographer who still uses film. You won't want to do the entire wedding with film, but plan some portraits with a nice film camera.

Film is romantic, whereas digital is quite cold. Many people grew up looking at the prints of their parents' and grandparents' weddings, and dreaming of the day their children would look through their own photo album. Plus, the negative is a true witness; it was there at the wedding, absorbing light that actually touched the happy family. Negatives are undeniably honest, never photoshopped, and survive hard drive crashes.

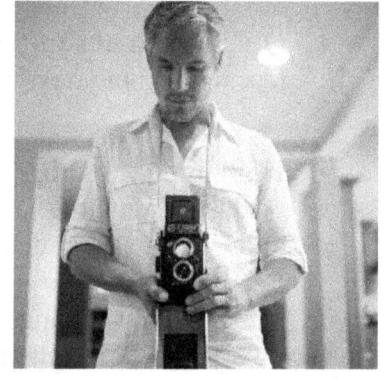

In short, a film camera can make you more marketable as a wedding photographer, even though it's not an efficient addition to your workflow and it won't produce more accurate or detailed images. You can buy 35mm film cameras, such as the Canon 1V or the Nikon F6, that fit your existing lenses. They function exactly like your modern DSLR, but nobody will even notice that you're shooting with film. Instead, I suggest a more romantic film camera, the medium-format TLR (shown in the black-and-white self-portrait). Choose a model that shoots 120 film, use T-Max 400 black-and-white film, and then develop and scan at *thedarkroom.com*. Yashica, Mamiya, and Rolleiflex all make amazing TLRs, which you can find *used for $100-$1000*.

What Landscape Photography Equipment Should I Buy?

Probably none. While great landscape photography often requires intense planning, distant travel,

Video: Perfect Pictures, Inexpensive Gear
10:17 - *sdp.io/perfect*

and long hikes, it doesn't require expensive gear. Any camera and a kit lens will work fine for most landscapes. Honestly, your smartphone can produce results that are good enough for most people. For examples of how to get more out of an inexpensive camera, watch our Perfect Pictures video at *sdp.io/perfect*.

While almost nobody needs to buy anything more than an inexpensive kit for landscapes, if you're planning to make large prints and you have a camera budget, or you want the best equipment possible, here are my ideal landscape recommendations for different budgets:

- **$400**: A Nikon D3200 kit

- **$500**: A Sony a5100 kit

- **$1,500**: A Nikon D5500 with the Sigma 18-35 f/1.8 lens

- **$2,400**: A Nikon D610 ($1,500) and a Sigma 24-105 f/4 ($900). Or, if you're willing to buy used gear, choose a used Nikon D800e for about the same price.

- **$3,500**: An Olympus E-M1 II ($2,000), the Metabones .71X speed booster ($700), and a Sigma 18-35 f/1.8 lens ($800)*http://amzn.to/12OQvHD*

You'll notice that I'm recommending full-frame cameras for those with the budget; they offer significantly less noise than compact cameras at their base ISO. The Nikon D810's ISO 64 support provides unbeatable detail and noise.

Canon cannot currently match the image quality of Sony and Nikon, and that's a key factor for landscape photography. However, if you're already invested in the Canon world, the 6D has a built-in GPS, which is useful for keeping track of the locations of your landscape photos.

The Olympus E-M1 II as our top choice might surprise you. After all, the sensor is ¼ the size of full-frame cameras like the Pentax K-1 (our previous top choice). The E-M1 II's High-res mode absolutely blows away full-frame cameras, including the 50 megapixel Canon 5DS-R. There's one big catch, however: you must use a tripod. That's usually not an issue for landscape photography, however. I'm also suggesting adapting a Sigma lens with a Canon mount, because that lens is sharper than native micro four-thirds lenses. It also gathers much more light, which will create much cleaner astrophotography pictures. You'll also notice that I recommend the Sigma 24-105 f/4 lens instead of the wider *Canon 16-35mm* or *Nikon 14-24mm* zooms. Those are awesome lenses, but very few landscapes allow shooting at wider than 24mm. In fact, many landscapes are more telephoto, because few scenes left in the world allow you to shoot so wide without including distractions. Additionally, at super wide angles, distant objects such as mountains become tiny. For that reason, I often bring only a single 24-105 when I'm shooting landscapes. If I have the opportunity to shoot wider, I shoot a panorama, as discussed in Chapter 2 of *Stunning Digital Photography*.

Many landscape photographers, most notably Ansel Adams, used tilt-shift techniques to provide greater depth of field. Using these techniques, they could focus on flowers or rocks in the foreground (shown in the next picture), while keeping distant mountains sharp. Canon offers several tilt-shift lenses (*17mm*, *24mm*, *45mm*, and *90mm*) and Nikon has a *24mm* and an *85mm*, each costing between $1,200 and $2,400. They're necessary for film work, but for digital landscapes, I instead recommend using focus stacking techniques. For detailed information about focus stacking, refer to Chapter 12 of *Stunning Digital Photography*.

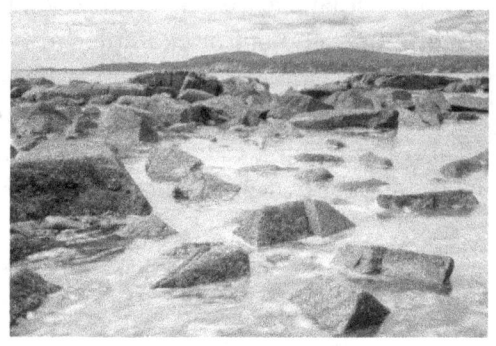

If you plan on working in low light or at night, you will also need a tripod, as discussed earlier in this chapter.

What Wildlife Photography Equipment Should I Buy?

If you don't yet have a camera, here are some recommendations at different price points:

- **$500:** A *Canon T3 body ($320)* and a *Canon 75-300mm ($170)*. Check eBay

  Video: Wildlife Photography Equipment
  14:36 - *sdp.io/BirdGear*

  or other used outlets to find gear even less expensive.

- **$750:** A *Canon T3 body ($320)* and a *Canon 70-300mm IS ($400)*. Adding image stabilization to the lens will allow you to use lower shutter speeds, reducing the noise in images of still animals such as perched birds.

  Video: Tamron 150-600 Review
  16:27 - *sdp.io/t600review*

- **$1,400:** A used *Canon 7D ($400)* and a used *Canon 400mm f/5.6 ($1,000)* prime. The Canon 7D is an excellent camera and offers far better autofocusing than the T3, but it's only a good value when bought used.

- **$3,250:** A *Nikon D500 ($1,850)* and a *Nikon 200-500mm ($1,400)*. The D500's focusing system and high frame rate make it the greatest wildlife camera ever.

- **$15,000:** Buy a Nikon D850, Nikon's vertical grip for the D850 with the EN-EL15b battery (to reach 10 FPS), a Nikon 600mm f/4 lens, and a Nikon 1.4X teleconverter. The D850's 45 megapixels create the most detailed wildlife images currently possible, allowing you to crop heavily and retain great detail. It has a higher frame rate, a better focusing system, and a bigger buffer than the Canon 5DS-R, too.

The reason I recommend Canon gear for less expensive wildlife setups is the amazing _Canon 400mm f/5.6 ($1,300)_ prime lens, which provides the sharpness of a $10,000 lens but is light enough to carry on hikes. I used it to take the following picture of an osprey flying, using my Canon 7D. Even though I have a much more expensive Canon 500mm f/4, I often choose the 400mm f/5.6 just because it's easier to carry. Nikon simply has no equivalent, and it drives many Nikon wildlife photographers insane.

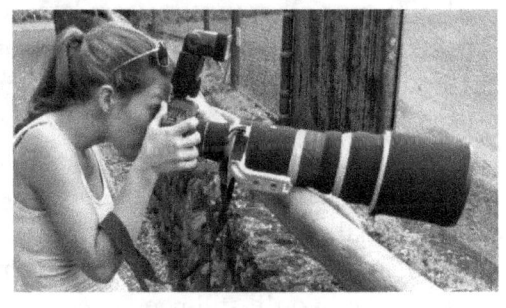

If you just want a wildlife lens for your existing camera body, here are the available lenses at different price points:

- 70-300, 75-300, or 100-300 f/4-f5.6: $125-$1,500, 1.5 lbs. (for all cameras, including Micro Four-Thirds)

- Tamron 200-500mm f/5.0-6.3: $950, 2.7 lbs.

- Sigma 150-500mm f/5.0-6.3: $1,070, 3.2 lbs.

- Pentax 300mm f/4: $1,120, 3 lbs.

- 70-400mm f/4-f/5.6: $2,000, 5.6 lbs.

- _400mm f/5.6: $1,200, 2.8 lbs._

- 300mm f/4: $1,400, 4.5 lbs.

- _400mm f/4: $6,000, 4.3 lbs._

- 300mm f/2.8: $3,400-$7,500, 5 lbs.

- 400mm f/2.8: $7,000-$11,500, 11.8 lbs.

- 500mm f/4.5: $5,000, 6.9 lbs.

- 500mm f/4: $7,000-$10,000, 8.5 lbs.

- 600mm f/4: $10,000-$13,000, 11.8 lbs.

- 800mm f/5.6: $6,500-$13,000, 26 lbs.

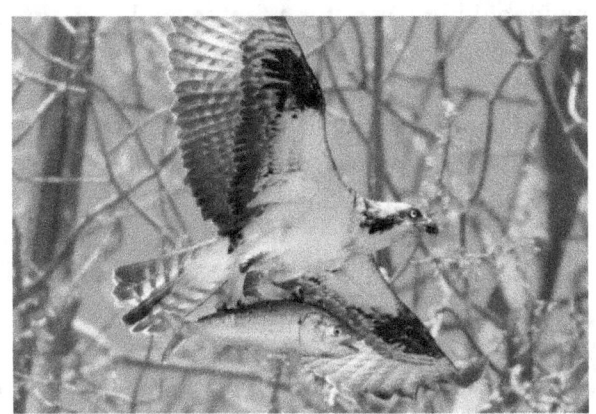

Avoid telephoto lenses without autofocus, including Canon FD lenses, the Nikon AI-S lenses, and mirror lenses, unless you're only photographing still animals. You just won't be able to keep up with flying birds, and you'll miss a lot of still animals.

As you can see, bird photography can be a very expensive hobby. You can definitely find cheaper telephoto lenses, such as mirror lenses, but they won't give you as sharp pictures. One way to offset the high cost is to plan to sell your lens. Used lenses on eBay tend to get about 85-95% of the original purchase price. I bought a _100-400mm telephoto zoom_ for

$1,500 and sold it twelve years later for $1,250, so it only cost me a few cents per day to own. I've owned my 500mm for about ten years, and it has actually increased in value 12% because the manufacturer discontinued it. In fact, it's not usually worthwhile to buy a used lens because it doesn't save you much money. While it's usually cheaper to buy and then sell a lens, you also have the option of renting them from a local camera shop or an online service (such as *http://borrowlenses.com*).

Notice that most of the suggested lenses are not zooms. Zoom lenses are heavier and less sharp, and you won't need to zoom anyway—you'll spend all your time zoomed all the way in, and you'll still need to crop. For that reason, I typically recommend primes. However, at the entry level, wildlife photographers have gotten some amazing photos with the Tamron and Sigma zooms.

Unless you plan to only use your lens in your backyard, weight is an important factor. The heavier lenses take much sharper pictures but traveling with a massive 9 lb. lens is difficult, and hiking any distance with that much weight is tiring. Most people can only handhold big lenses for a few seconds at a time. The big, heavy lenses definitely take better pictures, but you must be willing to sacrifice convenience, or you'll find the lens sitting at home unused.

When choosing a camera body for bird photography, choose a body with a good autofocus system and a compact (cropped) sensor. While full frame cameras are overall superior (but more costly), the smaller sensor extends your telephoto lens even further. Essentially, it performs some of the cropping in-camera. Cameras with smaller sensors typically have a higher pixel density, meaning your pictures will actually have more resolution after cropping.

You can use a teleconverter to further increase your lens' focal length. If your lens has an aperture of f/2.8, you can add a *2X teleconverter* to it to double the focal length and increase the minimum f/stop number to f/5.6. If your lens has an aperture of f/4, you can add a *1.4X teleconverter* to it to multiply the focal length by 1.4 and increase the minimum f/stop number to f/5.6.

Teleconverters always reduce autofocusing speed, so I don't recommend using them for moving subjects. If you use a teleconverter on a lens with an aperture of f/5.6, you will lose the ability to autofocus. While you could manually focus, you generally can't do it quickly or accurately enough. Additionally, a 1.4X teleconverter slows your shutter speed by half, and a 2X teleconverter slows your shutter speed by four times. Those slower shutter speeds will force you to use higher ISOs, resulting in lower quality pictures. For those reasons, I recommend just cropping pictures instead of using a teleconverter with a lens that has an aperture of f/5.6.

> **Tip:** If you have a 70-200mm f/2.8 lens, use it with a 2x teleconverter to get sharp pictures of larger birds.

Image stabilization is helpful when birding for two reasons: it decreases camera shake and it stabilizes the viewfinder. If you don't have image stabilization, you can eliminate camera shake by using a higher ISO and a fast shutter speed. In the case of flying birds, you will need a fast shutter speed to freeze motion anyway, which might seem to completely eliminate the benefit of image stabilization. However, I find that image stabilization makes it much more pleasant to look through the viewfinder—without it, looking through the lens is shaky and can even be nauseating if you do it for long enough.

For more information about wildlife techniques and equipment, refer to Chapter 8 of *Stunning Digital Photography*.

What Sports Photography Equipment Should I Buy?

Sports are challenging to photograph, and it's one of the few types of photography where a basic camera simply might not do the job. For action sports, you need a camera with a fast autofocus system, high frames-per-second, and a fairly fast lens. For indoor action sports, you'll also need a camera that can support high ISOs with relatively low noise and the fastest lens you can find.

Here are some recommendations for specific camera bodies:

- ***Canon 80D ($1,200)* or *Canon 7D ($used for $500-$800)*.** These two cameras have the same amazing autofocus system and similar image quality. Additionally, their compact sensors bring you closer to the action. However, their compact sensors also show a great deal of noise at higher ISOs, making pictures of indoor sports unbearably noisy. They're therefore perfect for outdoor sports, but you might consider upgrading to a full-frame camera for indoor sports. The fast 7 and 8 frames per second allow you to take more photos during the most exciting moments, improving the odds that you'll catch the perfect shot.

- ***Nikon D7500 ($1,250).*** The Nikon equivalent of the 7D II, the D7500 has a fantastic autofocus system. The 8 frames per second is great for sports.

- ***Nikon D500 ($1,500).*** With 10 frames per second and an unbeatable focusing system, the D500 tracks fast-moving subjects like a professional camera. It's literally my favorite sports camera of all time.

- ***Sony a9 II ($4,500).*** The Sony a9 and a9 II are my favorite sports camera of all time, despite being less expensive than the Nikon D5 or Canon 1DX II. 20 frames per second really increases your odds of catching the perfect moment, and a viewfinder with no blackout is game-changing. The ability to have a silent shutter makes it viable for quieter events like tennis and golf, and silence is nice when you're shooting on the sidelines of your kid's sporting events. The only reason I would recommend the more expensive D5 or 1DX II is if you need a telephoto lens that Sony currently lacks. While you can adapt Canon lenses to the a9 with autofocus, the autofocus isn't fast enough for professional sports work.

- ***Nikon D6 ($6,500)* or *Canon 1DX III ($6,500).*** These monsters are clearly overkill for your kid's soccer game, but I felt like I had to mention them, since they're the choice for every professional sports photographer. These cameras have the best autofocus systems human technology has developed, and they each shoot over 10 frames per second. And, if an angry soccer dad attacks you, they double as weapons.

For lenses, I typically recommend the same lenses I do for portraits, though you might want to add a 1.4x teleconverter, *as discussed in the Lens Buying Guide chapter*. The Nikon recommendations are:

- $200: *Nikon 50mm f/1.8G AF-S* (sufficient only for sports you can get close to, such as basketball and volleyball)

- $500: *Nikon 85mm f/1.8G AF-S*

- $770: *Tamron 70-200 f/2.8*

- $1,500: *Tamron 70-200 f/2.8 VC*

- $2,400: *Nikon 70-200 f/2.8 ED VR II*

The Canon recommendations are:

- $100: *Canon 50mm f/1.8* (sufficient only for sports you can get close to, such as basketball and volleyball)

- $420: *Canon 85mm f/1.8*

- $770: *Tamron 70-200 f/2.8*

- $1,500: *Tamron 70-200 f/2.8 VC*

- $2,500: *Canon 70-200mm L IS II*

The Sony E-mount recommendations are:

- $170: *Sony 50mm f/1.8* (sufficient only for sports you can get close to, such as basketball and volleyball)

- $2,600: *Sony 70-200 f/2.8 GM*

I don't typically recommend using a flash for sports. Even though the light would help, the flash can disturb the players.

What Macro Photography Equipment Should I Buy?

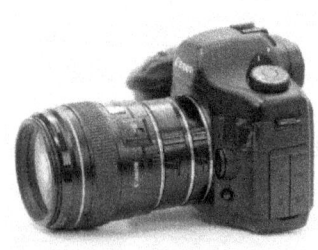 First, I recommend everyone start with a set of extension tubes. Extension tubes are literally just empty tubes; they have no optical elements at all. They simply move your lens farther from the sensor, increasing the size of the image circle in the same way that moving a projector farther from the wall increases the image size. The following picture shows two extension tubes connected between a lens and a camera body.

Like diopters, extension tubes prevent you from focusing on subjects in the distance—the more extension you add, the shorter your maximum focusing distance becomes. Extension tubes also reduce light to the sensor, increasing your shutter speed and making it more difficult to focus. My unscientific experiments show that each 36mm of extension cuts the light by about half—requiring you to double your shutter speed. Therefore, you should use the least amount of extension possible to get the focusing distance that you need. Because it is different for every lens, the best way to find the right length of extension tube is trial-and-error.

Even if you later upgrade to a full macro lens, you'll still use your extension tubes for wildlife and portraits of babies. You can buy two different types of extension tubes:

- **Without autofocus ($10-$15)**. These extension tubes don't have wiring to carry electronic messages between the camera body and lens, which causes you to lose autofocus and aperture control. Losing autofocus isn't a problem for still-life macro photography because you will usually manually focus anyway. Losing aperture control is a problem, however, because you frequently need to choose a high f/stop number to get the depth-of-field you need. There's a work-around described in the next section. I don't recommend extension tubes without autofocus, but if that's all that's in your budget, the Fotodiox models for _Canon_, _Nikon_, and _Sony_ are under $15.

- **With autofocus ($50-$180)**. These extension tubes cost more, but they allow you to autofocus (which is important for moving subjects) and they give you complete control over your aperture. Don't waste your money on the more expensive name-brand tubes; those from Kenko are the best, but many users report success with the Vivitar, CowboyStudio, and Zeikos models. They're really just hollow tubes, and the build quality won't impact your image quality. I recommend the Xit XTETS autofocus extension tubes for _Canon_, _Nikon_, and _Sony_ ($57).

Macro lenses cost about $500-$1,000, which makes them far more expensive than using diopters or extension tubes with an existing lens. However, a true macro lens offers several benefits:

- **1:1 magnification**. This means that the image on the sensor is the same size as the subject itself. Basically, it means you can get very close to the subject. Some specialty lenses, such as the Canon MP-E macro lens ($1,000), which focuses from 1:1 to 5:1, allow even greater magnification. It's more like a microscope, however, than a traditional lens, and it can be very difficult to use.

- **Small minimum apertures**. Depth-of-field gets _very_ small when taking macro pictures. To allow the greatest depth-of-field possible, macro lenses provide maximum f/stop numbers of f/22 or higher.

- **Restricted autofocusing**. It can take several seconds to autofocus a macro lens. To reduce that time, macro lenses often provide a switch to restrict focusing to specific ranges, such as between one and two feet, and between two feet and infinity.

- **Precision manual focusing**. Macro lenses tend to have finely adjustable focus rings that make it easier to focus on close-up subjects, but would require a great deal of spinning to focus on a distant subject.

- **Ring flash availability**. Depending on the size of your lens, the height of your flash, and the distance to your subject, your lens might cast a shadow on your subject when you are using a flash mounted to your camera. A ring flash mounted to the front of the macro lens eliminates shadows and provides a more even light, as shown in Figure 12-3. The ring flash must be matched to the front of the macro lens. Also consider using off-camera flash, as discussed in Chapter 3 and Chapter 6 of *Stunning Digital Photography*.

- **Infinity focus**. When you add extension tubes or diopters to a standard lens, you lose the ability to focus on subjects in the distance. A macro lens can always focus from extreme close-up to infinity.

You can add extension tubes to a macro lens to get even closer focusing. Add too much to the macro lens, and you'll literally focus inside the lens.

When shopping for a macro lens, one aspect that you need to pay particular attention to is dust resistance. While all lenses get some dust, you don't generally notice it. However, because you tend to use macro lenses to shoot close-up subjects, the dust inside the lens can become more in focus, ruining your pictures. Before you buy a used macro lens, look for dust inside of it by looking through the lens at a light from both ends. If, and when, you get dust inside your macro lens, be prepared to spend $150-$250 to have it disassembled and professionally cleaned. While you can clean the front and rear elements yourself, disassembling a lens to clean the internal elements is a task best left to optics professionals. The macro lens you choose depends on your subject:

- **Still life**. You can get as close as you need to with still subjects such as flowers, so I recommend the Sigma 50mm f/2.8 macro lens ($370). The 50mm focal length provides more depth-of-field than lenses with longer focal lengths, and depth-of-field is a real challenge with macro photography. You'll also want a tripod for still life macro.

- **Insects**. Living subjects will run or fly away if you get too close, so you need more working distance. Often, they move so quickly that you need to handhold the lens, so image stabilization is an important feature. There's one lens that meets both these requirements: the Sigma 150mm f/2.8 OS macro lens ($1,100).

For information on using extension tubes and macro lenses, and information about other macro equipment (including reversing rings, diopters, focusing rails, and the amazing *Canon MP-E* lens), refer to Chapter 12 of *Stunning Digital Photography*.

What Equipment Should I Buy for Stars and Star Trails?

Note that I'm addressing wide-angle photos of the night sky in this section, but not close-up pictures of distant planets or moons.

First, I recommend starting with your kit lens. Master the technique, and make sure you have the stamina to stay out in the cold late at night. For detailed information about taking pictures of stars, refer to Chapter 10 of *Stunning Digital Photography*.

While your kit lens can get usable pictures of stars, it's not ideal. The ideal gear for

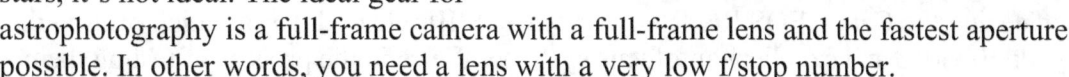

astrophotography is a full-frame camera with a full-frame lens and the fastest aperture possible. In other words, you need a lens with a very low f/stop number.

On the other hand, autofocus is generally useless with stars; you'll always need to manually focus. Autoexposure is useless, too.

Rokinon, a third-party lens manufacturer you've never heard of, makes a series of all manual wide-angle lenses that are perfect for stars. Because Rokinon doesn't bother making the lens communicate with any of your camera's electronics, it's easy for Rokinon to attach different mounts to their lenses. As a result, these lenses are available for just about every camera mount ever made, including Canon, Nikon, Sony A and E, Pentax, Four-thirds, Micro Four-thirds, Sony, Samsung NX, and Fujifilm X.

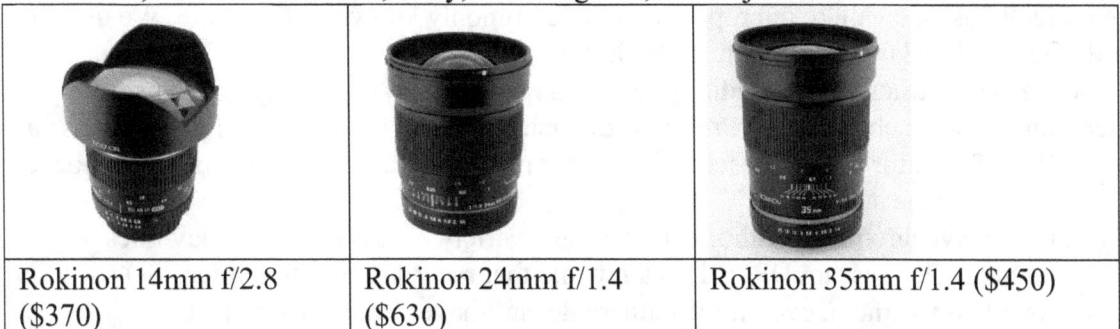

Rokinon 14mm f/2.8 ($370)	Rokinon 24mm f/1.4 ($630)	Rokinon 35mm f/1.4 ($450)

Of these three lenses, I find the 24mm the most useful for stars; it provides sweeping views, and you can always crop if you need to. If you find you want to show even more of the sky, the 14mm is a good, but specialized, lens. Unfortunately, it's two stops slower than the other lenses, so you'll need an ISO that's two stops higher when you use it—significantly increasing the noise in your photos.

May absolute favorite lens for stars is the *$900 Sigma 20mm f/1.4* (for full-frame cameras).

What Equipment Should I Buy for Video?

While many of my recommendations have been DSLRs, mirrorless cameras shine for video because they tend to autofocus better while recording video and they have electronic viewfinders that you can use while recording.

When choosing a camera for video, choose a camera with a tilting or articulating screen, a touchscreen, and an electronic viewfinder. Though very few cameras currently support 4k resolution, I highly recommend it. 4k cameras provide sharper video, even if you view them at standard HD (1080p). If you plan to shoot handheld, you need a lens or body with image stabilization. Power zoom helps, too.

At different price points, my recommendations are:

- **$500 kit: _Canon EOS M50_**. This little camera has excellent autofocus and great controls. It's perfect for vlogging. It doesn't have 4k video, however, just 1080/60. Still, it's good enough for our YouTube channel.

- **$1,200: <u>Canon EOS RP kit</u>**. This full-frame mirrorless camera focuses reliably and accepts a wide variety of Canon RF lenses and cheaper EF DSLR lenses (when using an inexpensive adapter).

For higher budgets, you'll have to wait for an updated version of this book. We're on the verge of seeing the Canon EOS R6, R5, and Sony a7S III released, all of which offer high-quality 4k/60 video and excellent autofocus systems. We're in the process of testing them now.

Audio is extremely important. All modern cameras have a built-in microphone, but that sound will usually be awful. A full discussion of audio is outside the scope of this book, but you can get much better audio by attaching an external shotgun microphone and connecting it to your camera's mic jack (if your camera has one). Rode makes several excellent shotgun microphones.

If you're recording someone talking and they're not within a couple of feet of the camera, you'll get the best audio by placing a microphone physically on the person. For our videos, we typically use a lavaliere microphone, more commonly known as a lav mic. We use the Rode Go wireless lav mics, which are $200 for a set.

If you learn the basics of video editing, you can record audio separately and synchronize it later using a tool such as Adobe Premiere Elements or Final Cut Pro. We frequently use a Zoom H1n ($130) for this purpose; we'll simply place it on a table or in someone's pocket and synchronize the audio later.

If you master synchronizing audio, you can use a variety of less expensive lav mics to get quality results. The Tascam DR-10L is a lav mic that records audio to a micro SD card, so it works with any camera, even if the camera doesn't have a microphone jack.

Chapter 2: Choosing a Camera Type

Today, popular cameras fall into four main categories: smartphones, fixed-lens cameras (also known as point & shoots), mirrorless interchangeable lens cameras, and Digital SLRs (DSLRs).

Each has its own advantages, as the following sections discuss.

Smartphones

An interest in taking photos can start with a smartphone. You simply can't beat the convenience of a camera that can send a picture to your friends in 10 seconds. As much of Instagram demonstrates, it's possible to make amazing pictures with a smartphone.

- **Convenience**. I carry my phone with me everywhere, so I always have it. As commercial photographer Chase Jarvis is credited with saying, "The best camera is the one that's with you."

- **Discretion**. Nobody notices a smartphone because they're everywhere. That makes them great for street photography and candids.

- **Detail**. Modern smartphones take pictures with more than enough detail for Facebook or even moderately large prints.

- **Low-light handheld**. The latest generation of smartphones, including the iPhone 11, Samsung Galaxy S20, and Google Pixel 3 phones have a special night mode that automatically stacks dozens of images over 3+ seconds to produce incredibly detailed photos at night. They even work for astrophotography and aurora borealis. In our testing, they outperform even the greatest full-frame professional mirrorless cameras when shooting handheld.

- **Speed**. Older smartphones often waited more than a second to take a picture after pressing the shutter. However, many modern smartphones take a picture the exact instant you press a shutter. They do this by constantly capturing photos and then going back in time to the photo it took when you pressed the shutter, completely eliminating delay.

- **Software**. Because modern smartphones have a great deal of processing power, the camera apps are often feature-rich, supporting High Dynamic Range (HDR, as discussed in Chapter 11 of *Stunning Digital Photography*), panoramas (as discussed in Chapter 2), and powerful image editing capabilities such as those provided by *Instagram* and similar apps.

- **Connectivity**. Smartphones are constantly connected to the Internet, allowing you to take a picture, edit it (optionally), and then message it to someone directly or post it on Facebook or Twitter. Recency is one of the most important factors in photography—a picture of your son hitting a home run last week will be interesting to his grandparents, but a picture of him hitting a home run 20 seconds ago can make them feel like they're standing next to you. It also allows you to get immediate feedback from friends anywhere in the world.

While smartphones provide great detail, they fail in many other ways:

- **Limited zoom**. While some modern smartphones, such as the iPhone 7+, offer multiple lenses, the range is still very limited. Digital zooms aren't useful because they make everything blurry; you're better off just cropping the picture after you take it. The Samsung Galaxy S20 brags about a 100X zoom but in our testing it produced results that are less sharp than an inexpensive camera with a zoom lens. You can buy *lenses that attach on the outside of some smartphones*, but they're inconvenient and the picture quality isn't great.

- **Awful flash**. Smartphones feature a bright LED located very close to the lens. Typically, you have no control over the flash except to turn it on or off. The flash produces red-eye (which might be fixed automatically in software), blows out nearby subjects, and leaves anything farther than a few feet away in darkness. Once you use a bounce flash (discussed in Chapter 3), you'll never be able to tolerate your smartphone's flash.

- **Limited or difficult controls**. Changing the settings is a slow process of tapping on the touch screen. Full cameras typically have physical dials that you can spin to immediately get the settings you want.

- **No real background blur**. While smartphones like the iPhone 7+ offer portrait mode that creates fake (but convincing) bokeh, you'll never get a nicely blurred background from a smartphone picture unless you download an app that blurs it after you take it. Background blur in video is available on some smartphones, such as the Samsung Galaxy S20, but it's ugly and unusable. I expect future smartphones to improve on this.

Because of these drawbacks, smartphones are still frustrating to use for all but the most casual snapshots. No matter which serious camera you buy, though, your smartphone will always get used because of its convenience.

Fixed-Lens Cameras

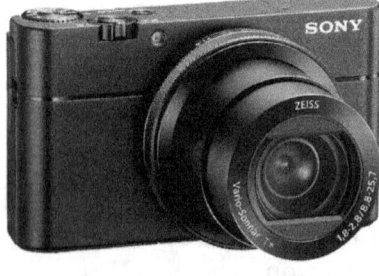

Fixed-lens cameras come in every size, shape, quality, and cost, and the only trait that brings them together is their lack of an interchangeable lens.

Most casual photographers only ever use their kit lens. If you want a high-quality camera for general use, and you don't plan to shoot serious sports, wildlife, weddings, portraits, or astrophotography, a fixed lens camera is the right choice for you. Basically, I recommend a fixed-lens camera to everyone who wants to take great pictures but doesn't want photography to become a hobby or profession.

The smallest fixed-lens cameras (P&S), fit in your pocket and require very little understanding of photography to take in-focus and well-exposed pictures. P&S cameras were the most popular camera type from about 1980 until about 2010. Before 1980, the technology simply didn't exist to make cameras that were easy enough to use to be considered point-and-shoot.

P&S cameras are a dying breed, however. While having a zoom lens is a big advantage over smartphones, the cost of buying a separate camera and the inconvenience of carrying it around just isn't worth it for most people. As a result, I recommend people use their smartphones for snapshots and a mirrorless camera or DSLR for more serious photos, and completely avoid P&S cameras.

There are also a growing number of fixed-lens cameras that bridge the gap between P&S cameras and DSLRs by offering bigger sensors, powerful lenses, manual controls, and external flash support. Their picture quality can be similar to DSLRs, giving you all the capabilities in a smaller, (sometimes) less-expensive package. Here are some quick recommendations:

- **Casual**. The Sony RX100 series of cameras are an excellent all-around camera and a big step up from a smartphone in some ways. However, the user interface is confusing to all but the most technical users, and sharing photos requires using either an unreliable smartphone app or a memory card reader. As a result, we rarely recommend these cameras.

- **Full camera companion**. In the past, photographers often bought two cameras: a small, pocketable camera and a full-sized camera for great results. Today, we don't find that to be necessary because the quality of images produced by smartphones has become amazing. We recommend spending the money you would have spent on a pocket camera and upgrading your smartphone to the latest generation.

- **Sports and wildlife**. Many fixed-lens cameras brag about "40X" f/2.8 lenses, but they're being deceptive. You can find photographers who've taken great sports and wildlife photos with a fixed-lens camera, but it's much more difficult than it would be with a full camera. If you plan to shoot sports and wildlife, your money is simply better spent on a full camera.

- **Video**. The Panasonic FZ1000 ($500) is the cheapest way to get into high-quality 4k video. Panasonic is misleading when they advertise the lens as "25-400mm f/2.8-f/4.0", however. The lens is physically 9.1-146mm f/2.8-f/4.0, and in full-frame 35mm terms, it behaves like a 25-400 f/8-f/11 lens. That's lousy for a still camera, but fairly useful for a video camera.

- **Serious, planned photography ($1,400)**. The Fujifilm X100V ($1,400 new) has an APS-C sensor and a prime 23mm f/2 lens that's equivalent to a 35mm f/3 lens. It feels like an old-fashioned viewfinder camera in your hands. It lacks a zoom, or the ability to change lenses, making it very specialized.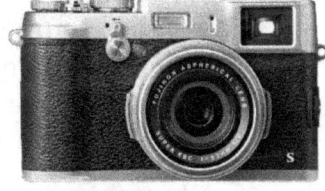

- **Serious, planned photography ($3,300)**. The Sony RX-1 RII ($3,300 new, or $2,700 used) has a full-frame 42 megapixel sensor and a prime 35mm f/2.0 lens with amazing image quality and low-light capabilities unmatched at this size. Like the X100V, the prime lens limits your options, but when the conditions are right, this camera takes sellable, professional-grade images.

As discussed later in this book, f/stop numbers are meaningless unless you factor in the camera's sensor size. Most people don't know this, however, and many fixed-lens camera manufacturers take advantage of this ignorance to mislead customers into believing a lens performs better than it actually does.

For example, consider the Sony RX-10, which advertises a 24-200mm f/2.8 lens. Many photographers mistakenly believe that it will perform like a DSLR with a 24-70 f/2.8 and a 70-200 f/2.8 lens attached. However, if you calculate the crop factor (about 2.7X), you discover that the lens gathers the same amount of total light and provides the same background blur as a 24-200 f/7.6 full-frame 35mm lens. That's substantially worse than a full-frame f/2.8 lens. For more information, refer to the *Sensor Size and Crop Factor* section.

Mirrorless Interchangeable Lens Cameras

Mirrorless cameras (formally known as Mirrorless Interchangeable Lens Cameras, or MILCs) are the right choice for most photographers who don't need access to the wide variety of Canon and Nikon DSLR accessories.

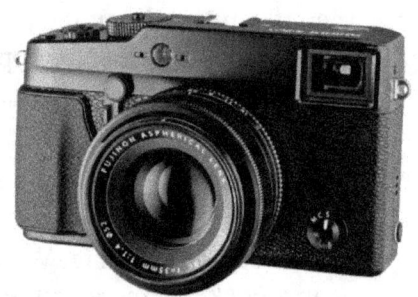

Mirrorless cameras combine some of the best qualities of fixed-lens cameras and DSLRs:

- **Interchangeable lenses**. Whereas fixed-lens cameras have a lens built-in, mirrorless cameras support interchangeable lenses. This makes them substantially larger than a typical fixed-lens camera when the lens is attached; however, being able to change lenses makes the cameras infinitely far more versatile than a fixed-lens camera.

- **Small size and light weight**. Besides being more portable, small cameras are less intimidating. That makes them better for candid and street photography, where the photographer doesn't want to be noticed. Though the size difference disappears when you use bigger lenses, you always have the option to use a small pancake lens on your mirrorless camera. Wide-angle lenses (such as a 12-24mm) tend to be much smaller than their DSLR equivalents.

- **Manual controls and advanced features**. Mirrorless cameras typically allow for manual control of shutter speed, aperture, ISO, and exposure compensation. Additionally, through the power of software, they provide advanced features such as bracketing, panoramas, and HDR.

- **Electronic viewfinders (EVFs)**. EVFs preview your exact exposure and allow advanced information in the viewfinder, such as depth-of-field preview (without darkening the viewfinder), highlighting the parts of the frame that are in focus, and showing a histogram. If you're shooting black-and-white, you can see the world in

B&W in realtime, which makes a huge difference in your shots. DSLRs simply bounce light to your eye, without any added information.

- **Wider focusing points**. Mirrorless cameras can focus almost anywhere in the frame, even when using the viewfinder.

Mirrorless camera technology is advancing far faster than DSLRs. If you used a mirrorless camera in 2012, you might have been annoyed by the limited lens selection, slow performance, poor viewfinder, and noisy images. In 2018, many mirrorless cameras match DSLRs in many different aspects of photography, including image quality and focusing speed. By 2020, I'm confident that even professionals will be buying more mirrorless cameras than DSLRs.

If you're a DSLR user and you have your doubts about mirrorless cameras, spend a full day shooting with a Panasonic GH5, Fujifilm X-T2, Olympus E-M1 II, or the Sony a9. The next time you pick up a DSLR, it will seem clunky, primitive, loud, and impossibly heavy. The optical viewfinder will seem outdated compared to the electronic viewfinders, which are bright even at night, show you your exposure before you take a picture, and provide histograms and focus peaking.

But I don't recommend mirrorless cameras to everyone. No mirrorless system can match the variety of Canon and Nikon DSLR lenses and flashes. Mirrorless lenses and flashes cost more than comparable DSLR lenses and flashes, too. Despite manufacturer's claims, only the Sony a9 can track moving subjects as fast as a comparably priced DSLR.

There are no bad mirrorless cameras on the market now—you literally can't go wrong. However, to make your shopping a little easier, here are some suggestions, depending on your priority (prices include lens):

- **Cheapest**. The Olympus E-m10 ($400 used for a kit) isn't the newest model, it's a little slow, and Olympus is leaving the camera business. However, it's small and light, it has a tilt screen, amazing stabilization, it accepts the massive variety of Micro Four-thirds lenses, and the image quality is more than enough for beautiful 8x10" prints.

- **Smallest**. The Fujifilm X-T30 ($1,000 new for a kti) has fantastic still and video image quality in a very small size. The touchscreen, wireless capabilities, and small size make it a great family camera.

- **Best handling**. The Fujifilm X-T4*http://amzn.to/17F6rKN* ($2,200 with a lens) is expensive, but old-school photographers will love the real shutter speed, exposure compensation, and ISO dials with the number written right on them. For best results, use a Fujifilm prime lens or fixed aperture zoom with an aperture dial.

- **Best vlogging**. If you're going to be in front of the camera, try the Canon M50 ($500 for a kit). The face-detect autofocus is the best in its class. You can buy a handheld tripod (HG-100TBR, $130) with a built-in remote control. Consider upgrading to the Canon M 11-22mm zoom lens ($400) for walking with the camera, or the Sigma 16mm f/1.4 lens ($400) for stationary video in low light or with a nice background blur.

- **Best video**. If you're going to be behind the camera, you can't beat the amazing Panasonic GH5 ($1,400) with 4k/60p video. Combine it with the Panasonic 14-140 ($630) lens for amazing results in a tiny, discreet package. The Fujifilm X-T4, described previously, is also an amazing option.

- **Most DSLR-like**. If you like the feel and functionality of a DSLR, but you want something smaller and lighter, you have many options. The Sony a7 III ($1,800 new) or Sony a7R IV ($3,200 new) has become the choice of many professionals. Using the Sigma MC-11 adapter ($200), you can adapt Canon EF DSLR lenses with satisfactory autofocus. The Canon EOS R mirrorless system is starting to mature, and the Canon EOS R6 and R5 (which we are currently in the process of testing) look promising. If you have existing Nikon lenses, the next-generation of Nikon Z bodies should be worth waiting for, though we currently recommend Nikon users continue to use their Nikon DSLRs instead of upgrading to the Z6 or Z7.

- **Best image quality.** The Sony a7R IV ($3,200) uses a 60 megapixel full-frame stabilized*http://en.wikipedia.org/wiki/APS-C* sensor, more than professional DSLRs but at a much smaller size. The Sony 24-70 f/2.8 GM ($2,200) is the best walking-around lens for it. That kit is expensive at, but you'll get image quality similar to the best DSLRs at a lower price and weight. If you have the budget, you might also consider the 100 megapixel Fujifilm GFX 100 ($10,000), though the lens selection is limited compared to the Sony E-mount.

Beyond those recommendations, I would steer you towards either the Sony full-frame E-mount lineup or Fujifilm X-mount cameras because of the wider range of lenses and accessories available to you. The Canon R and Nikon Z mounts show promise, but as of 2020 their camera and lens lineups are still limited in comparison to competitors. Micro four-thirds has a fantastic lineup of lenses and cameras from both Panasonic and Olympus, but the future of the lens mount is uncertain because Olympus announced they are exiting the camera business and Panasonic seems to be dedicating most of their energy to the full-frame L-mount. Speaking of the L-mount, it definitely has the potential to be successful, but we would like for it to mature further before recommending it.

Digital Single Lens Reflex (DSLRs)

DSLR cameras are both the least and most expensive types of cameras. If you want to get the most camera for your budget, buy a used DSLR. If you are a professional photographer, or you want to shoot like one, a DSLR is probably the best way to spend your budget.

Note, however, that Canon and Nikon (the most popular DLSR manufacturers) have shifted their development resources from DSLRs to their mirrorless lineups. We still expect new DSLR cameras and bodies from both companies, however, most of their new lenses will be for their mirrorless platforms. Therefore, if you feel like you need to always have the latest-and-greatest, a mirrorless camera might be the right choice. A DSLR, however, still provides better value for your money.

Like with mirrorless cameras, you can change the lenses and flashes on DSLRs. DSLRs have a few key advantages over mirrorless cameras:

- **Lens selection**. Canon and Nikon DSLR systems are the oldest, and thus they have the widest selection of lenses. This doesn't make any difference if your choice of camera system has all the lenses you need. However, serious portrait, wedding, sports, and wildlife photographers often need lenses only available to Canon and Nikon DSLR users.

- **Lens cost**. DSLR lenses tend to be less expensive than equivalent mirrorless lenses. For example, the Sony E-mount 50mm f1.8 is $250, while the Canon equivalent is $125 and the Nikon equivalent is $215. Kit lenses aside, mirrorless lenses tend to be 50%-300% more expensive than the DSLR equivalents. DSLR systems also have a wider variety of used lenses and less-expensive third-party lenses. If you plan to buy a wide variety of lenses, a DSLR might simply be less expensive than a mirrorless camera.

- **Focusing on moving subjects**. DSLRs are defined by the mirror that bounces light from the lens to the viewfinder. This same mirror also reflects light to a dedicated phase-detect focusing system. While mirrorless cameras focusing systems have improved to the point that they focus just as well as DSLRs on still subjects, our tests show they are substantially less reliable for tracking moving subjects. The one exception to this is the latest generation of Sony mirrorless cameras, including the Sony a9, the Sony a9 II, and the Sony a7R IV, which have the best focusing of any camera we've ever tested.

If you're in the market for a DSLR, I have good news: there are no bad DSLRs on the market now—you literally can't go wrong. However, to make your shopping a little easier, here are some suggestions for different budgets, depending on your priority (prices include lens):

- **As inexpensive as possible**. The _Canon T6 kit_ or _Nikon D3500_ (both about $400-500). These basic cameras do almost everything the more expensive cameras do, but they might require just a bit more patience. Remember, every dollar you save is a dollar more you can spend on lenses, flashes, and tripods.

- **Up to $800**. The _Canon 80D_ or _Nikon D5600_. A solid step up from their basic counterparts, these models offer improved usability and image quality. Notice that I'm not recommending the latest-generation Nikon cameras; the older models are equally capable and lower cost.

- **Up to $1,500**. The _Canon 6D Mark II_ or _Nikon D750_. These models provide full-frame sensors with image quality that leaps ahead of the less expensive models. Full-frame sensors are ideal for landscapes and portraits, but less useful for sports and wildlife.

- **Up to $4,000**. The *Canon 5D Mark IV* or *Nikon D850*. These full-frame models offer the greatest possible image quality with dramatically improved autofocus systems.

Whenever possible, save yourself some money and buy refurbished or used equipment. Bodies, in particular, are often available used much cheaper than new, and they work just as well as their new counterparts. Newer bodies are rarely available used, requiring you to pay full price for them. Be sure to buy from a reputable buyer and choose buyers that offer a return period whenever possible.

I generally have no strong preference between Canon and Nikon. Each of the bodies at a similar price point creates similar pictures, and you shouldn't worry too much about minor features. If you're the type who wants to get started shooting as soon as possible, just buy the camera from the previous list that fits your budget and start shooting.

If you're the type who wants to understand every aspect of a camera you're buying, or if you're considering upgrading your existing body and you don't know if it's worth it, continue reading this section. My goal is to inform you about the different factors that might influence your buying decision so that you can determine the factors that are most important to you and make an educated decision.

If you just want to see an overview of the different camera bodies available, skip to the buying guides for *Nikon*, *Canon*, and *Sony*.

Chapter 3: Camera Body Features

This chapter marks the end of quick recommendations until you get to the brand-specific buying guides at the end of the book. We'll dig deep into the factors that influence camera-body buying decisions, including lens availability, image quality, and speed.

Lens Mount

There's one camera trait you need to choose before thinking about features: lens mount. Lens mount determines the availability of accessories such as add-on lenses and flashes. There are currently more than 20 different popular lens mounts, which makes camera shopping incredibly complex. For most people, I recommend choosing one of the five most popular lens mounts:

- **Canon EF (APS-C and full-frame DSLR)**. The most popular lens mount in history, Canon's DSLR system has the biggest selection of bodies and lenses for both tight budgets and professionals.

- **Canon R (full-frame mirrorless)**. Canon's newer full-frame mirrorless system supports EF lenses with an adapter, as well as a variety of high-quality (but expensive) RF mirrorless lenses. Currently, though, native R bodies and lenses are extremely limited and rather expensive, so R remains a choice for forward-looking people with a bigger budget.

- **Fujifilm X (APS-C mirrorless)**. The perfect system for enthusiast photographers, Fujifilm X offers fun, compact cameras and lenses, but it lacks the variety and third-party support of the other systems here.

- **Nikon DX & FX (APS-C and full-frame DSLR)**. Dating back to the 1950s, the Nikon DSLR system includes some of our favorite cameras of all time, including the Nikon D500 and D850. In the future, we hope to recommend Nikon DSLR users an upgrade path to the mirrorless Z system.

- **Sony E-Mount (full-frame mirrorless)**. Sony's has the largest variety of full-frame mirrorless cameras and lenses because they got almost a decade head-start on the Canon and Nikon mirrorless systems. Though Canon and Nikon are quickly catching up, Sony is currently the only manufacturer creating full-frame mirrorless cameras with a professional-grade focusing system and dual-card slots.

Though you'll no doubt find horror stories about each of them, your odds of getting good prices, support, and accessories are roughly equal. Canon and Nikon are constantly playing catch-up with each other. At times, Nikon has had superior technology. Within six months, though, Canon will release a camera or lens that very slightly surpasses Nikon's. This is free-market competition at its finest, and it works so well that you can be confident with your purchase from either brand.

With that said, some photographers might appreciate the subtle differences between brands:

- **Sony**. Sony was the first manufacturer to popularize full-frame mirrorless cameras, now the best-selling camera segment. As a result, they have the widest variety of cameras and lenses, and they also have superior autofocus technology and high frame rates. Sony manufactures their own sensors (as well as sensors for other manufacturers, including Nikon and Apple), so they often get the newest technology first. While Sony is the tech leader in many aspects of photography, their cameras can be annoying, with many people hating their menu system, user interface, and physical controls.

- **Canon**. Canon has been the #1 camera manufacturer since the 1980s, and that brings many benefits, such as the largest market of used equipment and the best support from third-party lenses and flashes. While Canon's technology fell behind Sony for many years, the release of the Canon R5 in 2020 promises to put Canon back in the competition. Where Canon has always led is in lens quality; consistently their lenses have outperformed every other manufacturer. This is particularly true in the realm of full-frame mirrorless, where amazing lenses like the Canon 85mm f/1.2 have single-handedly drawn portrait shooters into the Canon realm.

- **Nikon**. Nikon DSLRs such as the D500 and D850 have always been our favorite cameras to shoot with; they are reliable and feel great in the hands. Nikon spent the last 40 years in the #2 spot behind Canon, but they always seemed to work a little harder, staying one step ahead of Canon in both image quality and usability. However, the decline of camera sales in the mid-2010's has impacted Nikon harder than Canon and Sony, resulting in Nikon falling behind in important technology such as mirrorless autofocus. Nikon sales have fallen to a distant third behind Canon and Sony. In the last several years, Nikon has continued to cut their research and development budget, which makes us worry that Nikon might no longer be a technology leader in the years to come. Still, for those looking for the best DSLR, we recommend Nikon.

- **Fujifilm X-mount**. Fuji X-mount cameras are perfect for old school photographers who love buttons, dials, and manual settings, and who aren't afraid to read a camera manual. If you like Leica cameras, I'd probably steer you to Fujifilm instead, because they have the same soul at less cost. Fuji's not popular enough to get much lens support from Tamron and Sigma, which is a big drawback—you're limited to the handful of lenses Fujifilm sells. You can adapt lenses with limited autofocus capabilities, but we've found adapted results to be less-than-ideal.

- **Micro four-thirds**. Previous editions of this book heavily recommended Micro Four-Thirds, and you can still create great photos and video with the system. However, With Panasonic dedicating the bulk of their research and development to the (still young) full-frame L-mount, and Olympus selling their camera division to a company known for creating low-quality products, the future of the Micro Four-Thirds platform is uncertain.

I hesitate to recommend many other lens mounts, such as the Pentax K mount or the Panasonic/Sigma/Leica L-mount because they have limited lens support or I'm uncertain that the manufacturer will continue making new lenses and cameras in the future. Nonetheless, I have dedicated chapters later in this book to every single major lens mount, so you can explore every manufacturer's offerings.

Don't take those comments too seriously; the differences don't matter too much. These camera manufacturers are 99% the same. Plus, they're constantly releasing new gear to fill in any gaps, so any advantages or disadvantages are temporary (provided your camera company stays in business).

When you buy an interchangeable lens camera, you'll get the best results if you use the lenses designed for that system. Therefore, it's important to consider the body and lens variety for each system. The following table lists the most popular mirrorless systems in rough order of popularity, along with the approximate number of camera bodies and lenses designed for the system. The highlighted rows are my most commonly recommended systems.

System	Current Bodies (approx.)	Native Lenses (approx.)
Nikon Full-frame DSLR (FX)	4	100+
Nikon APS-C DSLR (DX)	5	25
Nikon Full-frame Mirrorless (Z)	2	10
Nikon APS-C Mirrorless (Z)	1	2
Canon Full-frame DSLR (EF)	3	100+
Canon APS-C DSLR (EF-S)	7	15
Canon Full-frame Mirrorless (R)	2	6
Sony APS-C Mirrorless	3	20+
Sony Full-frame Mirrorless	4	33
Micro four-thirds Mirrorless	25	50+
Canon EOS M Mirrorless	3	5
Pentax K	5	40
Fujifilm X/XF	2	27

Note that Canon, Nikon, and Sony APS-C camera bodies can use same brand's full-frame lenses, albeit with a drop in image quality. With a few exceptions, I generally recommend upgrading to a full-frame body before investing in full-frame lenses.

Note that I'm not listing the number of DSLR lenses that can be connected to the system using adapters; I discourage you from factoring these lenses into your decision-making process because their size defeats the purpose of using a mirrorless system in the first place, and handling and autofocus tend to be clumsy. I do understand that DSLR lens compatibility is an important factor for people with existing equipment, so if you have several Pentax K-Mount lenses that you want to use on a modern, digital, mirrorless camera, you should buy the Pentax K-01. If you want to use your Canon DSLR lenses on a mirrorless body, you should get the Canon EOS M.

Do think through this choice carefully, however, because lenses designed for mirrorless systems are much smaller and lighter, and those are the biggest reasons to get a mirrorless system in the first place. If you want to use DSLR lenses, I advise you simply to buy a DSLR. If you don't want to lose your investment in DSLR lenses but you do want to use a mirrorless body, I advise you to sell your DSLR lenses and put the proceeds towards native lenses.

Display Articulation

Cameras have four types of displays:

- **Non-articulating**. The screen is fixed to the back of the camera and can't move. This is the most durable design, and all high-end DSLRs have a non-articulating screen.

- **Tilt screen**. Tilt screens tilt up or down 90 degrees. This is very useful for holding the camera low to the ground (for kids and flowers) or high in the air (for shooting over a crowd).

- **Selfie screen**. Selfie screens tilt up 180 degrees, allowing you to see the display while being in front of the camera.

- **Fully articulating ("flip screen")**. Fully articulating screens flip 180 degrees from the side, providing the greatest flexibility. They're also the least durable, though I've never broken an articulating display.

Whenever possible, I choose a camera with a fully articulating display, especially if I plan to take video. However, many high-end cameras lack the option. The following product images show a tilt screen, a selfie screen, and a fully articulating screen.

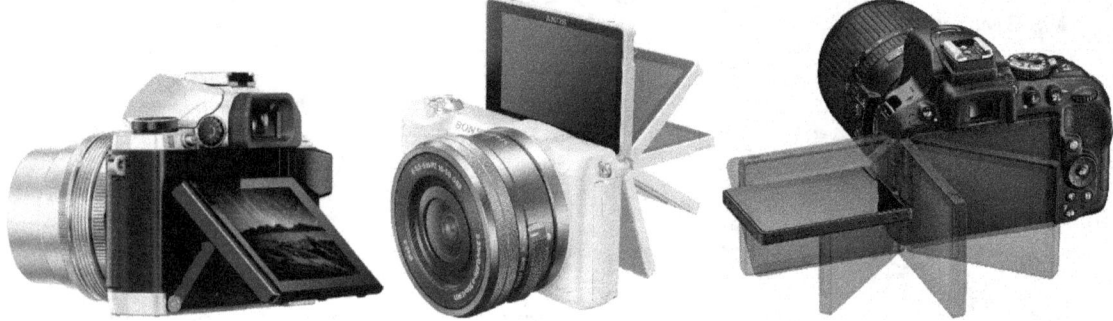

Viewfinder

The viewfinder is the window you put your eye to before taking a picture. Many modern cameras don't have a proper viewfinder at all, instead requiring you to use the large LCD screen on the back of the camera to compose your shot. That's fine; you don't need a viewfinder for casual photography, as demonstrated by the popularity of smartphone photography.

Do You Need a Viewfinder?

Viewfinders are useful for serious photographers and anyone shooting sports or wildlife, however. First, they block out sunlight and distractions, making it easier to concentrate on your composition. If you're using a telephoto lens (as you would be for sports and wildlife), they make it much easier to hold your camera steady and pinpoint a distant subject.

This picture shows the Sony a5100 (which has an LCD display but not a viewfinder) on the left and the slightly more expensive Sony a6000, which has a viewfinder in the upper-left corner.

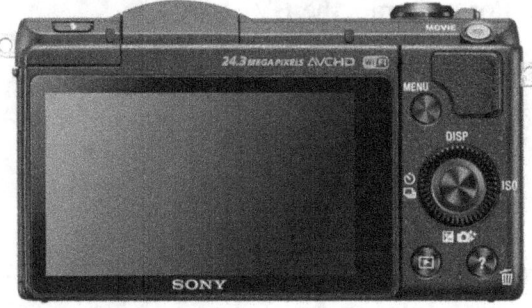 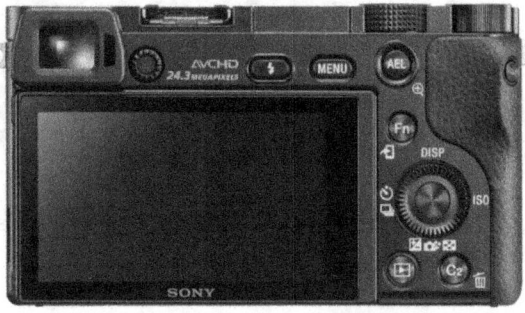

Optical vs. Electronic Viewfinders (EVFs)

Modern cameras have one of two types of viewfinders, and this can have a significant impact on your photography:

- **Optical viewfinders**. All DSLRs (except the Sony A-mount SLT cameras) have an optical viewfinder, the basic design of which dates back to the late 1800s. It really is a primitive design; the camera inserts a mirror between your lens and sensor and it bounces the light into a prism which then bounces the light into your eye. This guarantees that the viewfinder shows what your sensor will see, because it's exactly the same light.

- **Electronic viewfinders (EVFs)**. Mirrorless cameras process the data from the digital sensor and display an image on either a large display on the back of the camera or a small, simulated viewfinder. This means that you can see the effects of exposure compensation. The camera can also supplement the image with histograms, focusing peaking, and other useful data. Because the camera has to process the signal, there is always a very short delay of a few milliseconds.

The following three figures show the Canon 70D's optical viewfinder (first) followed by two different views of the Fujifilm X-T1's EVF. As you can see, the X-T1's EVF is bigger and brighter, shows far more information, and overall looks far more modern and polished.

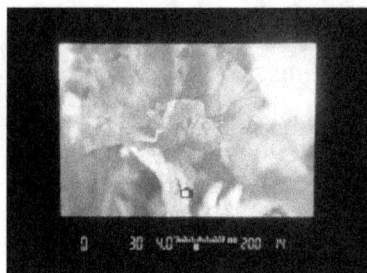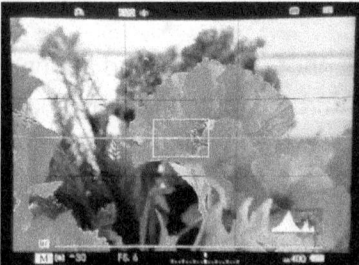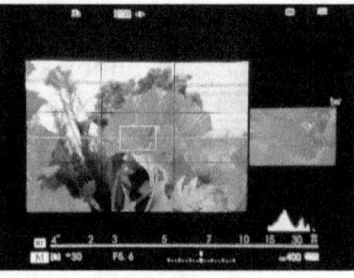

EVFs allow for a much more complex display while looking through the viewfinder, because the camera can add anything it wants over your view, almost like the heads-up display from the *Terminator* or *Robocop* movies. For example, the EVF can:

- Display a histogram while you look through the lens.

- Show you the actual exposure that the picture will have, including any exposure compensation you've dialed in.

- Allow you to monitor video recording using the viewfinder.

- Allow you to review the last photo without removing your eye from the viewfinder.

- Reduce or eliminate the viewfinder blackout period DSLRs experience when taking a picture.

- Provide a depth-of-field preview that doesn't darken the viewfinder. Depth-of-field preview on an optical viewfinder is less useful because the viewfinder gets very dark, making it difficult to see.

- Provide 100% viewfinder coverage. Optical viewfinders on less expensive cameras tend to hide very small portions of the edges of the frame.

High-quality electronic viewfinders (such as those in the Sony a7S III, Fujifilm X-T4, and Canon EOS R5) are superior to optical viewfinders for most types of photography. They're better than real life, allowing you to see in the dark, view the exact effect of any exposure compensation, preview filters (such as seeing in black-and-white), and zoom in for precise focusing. They don't go dark when you take a picture, and you don't have to take your eye away from the viewfinder to review your last shot. You can even use the viewfinder while recording video.

Optical viewfinders seem outdated in comparison (and they are), but they operate at the speed of light, so there's no lag. That makes them superior for sports and wildlife. They also don't require any battery power, so while a DSLR battery can last a week on vacation, my mirrorless cameras with electronic viewfinders are out of batteries in 4-6 hours.

For true DSLRs with optical viewfinders, you can realize most of these benefits by switching your camera to live view mode and looking at the display on the back of the camera. Of course, this requires you to hold the camera away from your face, so it is not a perfect substitute for an EVF. DSLR live view displays also tend to be slower and more laggy than a mirrorless camera's EVF.

In practical use, EVFs have some disadvantages when compared to an optical viewfinder. Specifically, the lag, refresh rate, and sharpness (discussed in the next section) are never as good as those of an optical viewfinder. They also consume far more batteries than an optical viewfinder does.

In my opinion, EVFs are so great that they're definitely the future. Whether they're right for you, right now, though, depends on your subject. You certainly don't *need* an EVF, though they can be nice to have.

EVF Lag, Refresh Rate, and Sharpness

All EVFs are not created equal. Better quality EVFs look nicer, allow for more precise manual focusing, and enable you to better track moving subjects.

The three most important criteria for evaluating an EVF are:

- **Lag**. This is the delay between when an event occurs in the real world and the time you see it on the viewfinder. All EVFs have some lag because the camera has to process the data from the sensor, but a shorter lag is better. Longer lags can be quite frustrating to use with moving subjects. As an example of the variation in lag, our recent test found the Fujifilm X-T1 had a lag of .005 seconds, the Olympus E-M10 had a lag of .025 seconds, and the Sony a6000 had a lag of 0.046 seconds. In low light, lag can increase substantially. To put that lag into perspective, humans have a reaction time of about 0.5 seconds.

- **Refresh rate**. Higher refresh rates cause movements in the EVF to be smoother. For example, if you pan the camera sideways with a slow viewfinder, the pan will seem jagged instead of smooth. EVFs measure their refresh rate in frames per second (fps). Fast EVFs, like the Olympus E-M10, have a refresh rate of 120 fps. Slower EVFs might have a 30 fps refresh rate, similar to that of a television.

- **Resolution**. The more pixels in the viewfinder, the sharper the screen. Sharper screens look more like the real world, and can help make manual focusing easier. For example, Sony gave the high-end a7S III a 9 million-dot display, whereas Olympus gave the lower-end E-M10 only a 1.44 million-dot display.

A few years ago, lag, refresh rate, and resolution problems could often make EVFs almost unusable on mirrorless cameras. However, all modern mirrorless cameras have EVFs that won't interfere with the picture-taking process. If you shoot action, however, it might be worth investing in a camera with a higher quality EVF.

Viewfinder Placement

All DSLRs have the viewfinder in the upper-middle of the back of the camera because it must be physically aligned with the lens. Mirrorless cameras could theoretically put the viewfinder anywhere, but they generally position it in the upper-middle (like an SLR) or the upper-left (like a viewfinder camera), as shown in the following picture. If you shoot with your right eye, having the viewfinder in the upper-left is more comfortable because it doesn't require you to press your nose against the camera back. If you shoot with your left eye, the center viewfinder is more comfortable.

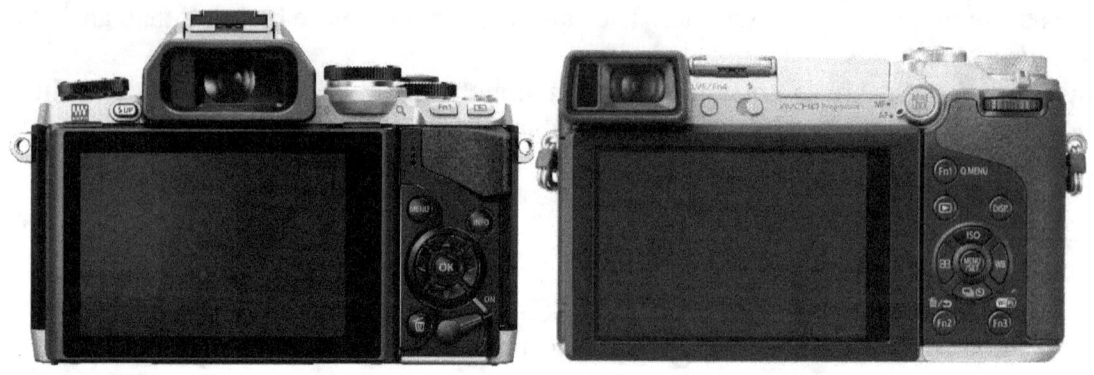

Touch Screen

Touch screens allow you to navigate menus by touching the screen, just like you would on your smartphone. Touch screens are particularly useful for choosing an off-center focus point and for quickly zooming in to check focus while reviewing pictures.

A touch screen is a nice feature, and it can definitely improve your workflow. Hand a camera without a touchscreen to anyone under 25, and you'll see him or her poke at the screen; younger people assume everything should have a touch screen. And they should—touch screens are wonderful.

You definitely can live without a touchscreen, however, and if you want a professional-level camera, you'll have to. Unfortunately, only mid-range cameras currently have touch screens available. For most casual camera buyers, though, a touch screen should be a requirement.

Autofocusing

Almost any camera will focus on well-lit still subjects with the kit lens, so the casual photographer doesn't need to worry about it.

Focusing becomes a challenge when tracking moving subjects (such as sports and wildlife), when shooting in low-light, and when working with shallow depth-of-field. Telephoto lenses and fast lenses (with f/stop numbers such as f/1.8) have a shallow depth-of-field, which means the eyes of your subject can be in focus, but the background will be very blurry. For more information about depth-of-field, refer to Chapter 4 of *Stunning Digital Photography*.

Mirrorless vs. DSLR

Generally, DSLRs autofocus better than mirrorless cameras (when using the viewfinder). This difference in focusing speed means a low-end mirrorless camera won't be useful for candid portraits, sports, or moving animals. If you're interested in those types of photography, check out DSLRs instead.

Note that mirrorless cameras support both single shot and continuous autofocus modes. However, focusing tends to be so slow that continuous autofocus on moving subjects doesn't work very well. In practice, you'll generally get better results with single shot autofocus.

Notable exceptions include the most recent generation of Sony E-mount cameras, including the Sony a6600, Sony a9, Sony a1 and Sony a7 IV. It also includes recent Canon RF cameras, such as the R10, R7, R5, R6, and R3.

Different Focusing Technologies

Most modern DSLRs and many higher-end mirrorless cameras feature two different focusing mechanisms:

- **Phase detection**. The quickest way to autofocus, phase detection uses a pair of sensors for each focusing point. With a DSLR, light travels through your DSLR's partially translucent mirror. Behind that mirror, at the exact same distance from the lens as your camera's sensors, are one or more pairs of focusing sensors for each focus point on your camera. Those sensors see a small part of your picture from two slightly different angles. When the view from the two sensors lines up, that part of the picture is in focus and your camera can stop the autofocus process.

- **Contrast detection**. Contrast detection autofocus is much slower than phase detection autofocus; however, contrast detection is more flexible because it can focus on any part of the picture, not just where a focusing point exists. DSLRs can use contrast detection in live view mode and some support contrast detection autofocus while recording video. Contrast detection examines data from the camera's sensor while adjusting the lens' focus, and by comparing subsequent frames captured from the sensor, can determine whether a focus movement causes the captured image to get more contrasty (indicating more focused) or less contrasty (indicating less focused) at any point on the frame. Therefore, with contrast detection, a focus point can be anywhere in the frame. Most DSLR sensors only provide data at 30 frames per second or 60 frames per second, limiting how quickly the camera can capture subsequent views of the scene while focusing, and thus the overall focusing speed. Therefore, mirrorless cameras tend to have much faster contrast detection autofocus than DSLRs.

DSLRs, and some mirrorless cameras, focus primarily using phase detection autofocus. With a DSLR, any time you're looking through an optical viewfinder, you're using phase detection autofocus.

All modern digital cameras also support contrast detection autofocus. If you're looking at the live view display on the back of a DSLR, that means the camera's sensor is receiving all the light from the scene and the camera has moved the mirrors out of the way. Without the mirrors, your camera can't redirect light to the phase detection autofocus sensor. Therefore, it must use the slower contrast detection autofocus instead.

The Canon 80D, 5D Mk IV, and other dual-pixel AF cameras are exceptions to this. Dual-pixel AF technology provides phase detection autofocus in live view, greatly improving focusing speed when the mirror is up—such as when you're recording video.

That's more technical detail than you need to remember, so here are a few key points:

- Phase detection is generally much better than contrast detection, especially for moving subjects.

- DSLRs support phase detection autofocus only when you're looking through the optical viewfinder.

- All digital cameras support contrast detection. DSLRs use contrast detection when you're viewing the live view display.

- Phase detection autofocus is limited to specific focusing points, while contrast detection allows you to focus anywhere in the frame.

How Phase Detect Focusing Works

With most modern lenses, the focusing motor (which physically turns the lens elements to change the focus) is built into the lens. Every other part of the focusing is controlled by the camera body.

All DSLRs, and some high-end mirrorless cameras, use phase detect focusing. When you focus your camera, it looks at the image coming through the lens and then tries to focus a little closer or farther away. Just like you have two eyes in your head, each focusing point in your camera is two separate sensors. When the focus is correct, the image from those two sensors lines up. At that point, the camera stops focusing.

Of course, that's an oversimplified explanation of a really complicated process. Focusing speed and precision are extremely important to portrait, wildlife, and sports photographers, so camera makers do everything they can to make the process as fast as possible. Unless you become a serious sports or wildlife photographer, you really don't need to understand all the nitty-gritty about how different camera bodies autofocus. However, keep reading if you're interested or you just want to understand the lingo.

Every camera body has multiple focus points. There's always one focus point in the center of the image and several others spread around the frame. Using the focus-and-recompose technique (described in Chapter 4 of *Stunning Digital Photography*), you only really need the center focusing point. However, the other focus points are useful for action shots where you don't have time to focus and recompose.

Not all focus points on a camera are created equal. Typically, the center focus point is the fastest, and the farther you get from the center, the less powerful the focus points are. Common types of focusing points include:

- **Vertical**. Vertical focus points detect contrast by looking up and down an image. Therefore, a vertical focus point would be able to focus very quickly on a shirt with horizontal stripes, but it might not be able to focus at all on vertical stripes.

- **Horizontal**. Horizontal focus points detect contrast by looking left and right through an image. They're good at detecting vertical stripes (or any type of vertical contrast), but not great at focusing on horizontal subjects.

- **Cross-type**. These use both vertical and horizontal contrast detection. They can focus on just about anything, as long as the subject is well lit and isn't a solid color.

It's common for the center focusing point to be cross-type, and focusing points towards the edges of the frame to be horizontal or vertical. However, some cameras, such as the Canon 7D and 70D, use only cross-type sensors.

Sometimes, a focusing point's capabilities depend upon the minimum f/stop number of the lens you're using. For example, the center focusing point might be cross-type with f/2.8 lenses, but only horizontal with lenses of f/4 or higher. There's no practical application for this knowledge; it's not likely to be a significant enough factor to justify spending hundreds or thousands on a different body or faster lens. In practice, you'll use the camera and lenses you have and do the best you can with the focusing it provides.

Most camera bodies support autofocusing when the lens' minimum f/stop number is f/5.6 or lower. However, some Canon bodies (specifically, the Nikon D500, Nikon D5, Nikon D810, Canon 1DX II, and Canon 5D Mark IV) support focusing with f/8 lenses on some of their focusing points. This capability is important to wildlife photographers who often use teleconverters to extend the reach of their telephoto lenses because teleconverters also increase the lens' minimum f/stop number.

Focus Point Positioning

DSLR manufacturers often advertise the number of focusing points a camera offers, but the number of focusing points is not nearly as important as how they're positioned in the frame. For example, some cameras have all their focusing points clustered in the middle of the frame, which is only useful if the subject will be centered in the frame. If you use the rule of thirds, as described in Chapter 2 of *Stunning Digital Photography*, you'll want your focusing points to be 1/3 of the way through the frame. Only higher-end DSLRs provide that.

Therefore, don't consider a camera with a large number of focus points to be automatically superior to another camera. Instead, pick the camera with the focusing points closer to the edge of the frame.

For example, the previous-generation Sony Alpha a99 bragged about having 102 autofocus points. This is mostly marketing fluff, because it has only 19 autofocus points that will allow you to attain initial focus. The remaining 83 points are only useful for continuing to track moving subjects after you initially focus, and then, only with specific lenses. 19 AF points are plenty, but there's a bigger problem: those 19 AF points are clustered around the center of the frame.

The Canon 5D Mark III had 61 autofocus points, and at a glance, would seem to have an inferior autofocus system compared to the a99. However, comparing the autofocus points that can be used for initial focus on a rule-of-thirds grid, as shown in the following figure, tells a different story. The a99's focus points, on the left, are clustered around the center, requiring you to use the focus-and-recompose technique for every single shot that follows the rule of thirds—and that will be most of your shots. The 5D Mark III's autofocus points reach much farther, allowing you to autofocus on the left or right third of the frame (but still not all the way to the corners, unfortunately).

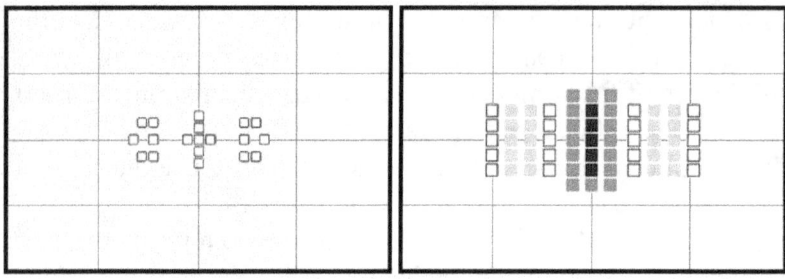

The Sony Alpha a99's autofocus points (left) are clustered around the center of the frame, which is less useful than the Canon 5D Mark III's autofocus points (right), which are distributed farther across the frame.

I spent the first decade of my photography career only using the center autofocus point (combined with the focus-and-recompose technique), so even just a single autofocus point will get the job done. I'm only showing this example because, as you assess different camera bodies, I want you to put less emphasis on the number of autofocus points and more emphasis on how they're distributed across the frame.

Nikon Focusing Motors

For Nikon camera bodies, there's one more factor to consider: whether the body has a focusing motor. Whereas most modern lenses have the focusing motor built into the lens, older Nikon lenses relied on a special motor built into the body to adjust the lens focus. Higher-end Nikon DSLRs still include focusing motors to allow autofocus with these older lenses. Specifically, the following recent Nikon DSLRs have a focusing motor: D90, D300, D7100, D7200, D600, D610, D800, D810, D3x, D4, D4S, D500, D750, D780, D5, and D6.

All other new Nikon DSLRs don't have a focusing motor. Therefore, autofocus will work fine with all newer AF-S and AF-I lenses, but you won't be able to autofocus with older AF lenses that require a focusing motor (though you can still use them with manual focus). Fortunately, most of Nikon's current lineup is AF-S. All you'd really be missing out on is autofocusing with their fisheye lenses, but the wide depth-of-field with fisheye lenses makes focusing easy.

Sensor Size and Crop Factor

To watch the crop factor videos, visit *sdp.io/crop*.

Cameras can be divided into several categories by their sensor size. Starting with the smallest, they are:

- **CX (2.7X)**. These tiny mirrorless cameras, such as the Nikon 1, have the smallest common sensors.

- **Micro Four-Thirds (2X)**. These small mirrorless cameras have relatively small 16-megapixel sensors that are capable of producing excellent images when paired with the right lenses.

- **APS-C (1.5X for most, 1.6X for Canon)**. The smallest type of DSLR is also a common mid-range mirrorless format, and the right choice for most non-professional photographers. Lenses designed for their smaller sensors are lighter and less expensive than those designed for bigger, full-frame cameras. You can also connect full-frame lenses to Canon APS-C, Nikon DX, and Sony Alpha bodies, but when you take a picture, the camera will crop out a smaller section from the center of the lens image. This is known as the crop factor, and it's actually helpful when using telephoto lenses with wildlife or sports. In fact, many wildlife photographers prefer an APS-C or DX camera over their full frame counterparts. Crop factor is discussed in more detail later in this section.

- **Full-frame 35mm (1X)**. Matching the sensor size of 35mm film, full-frame DSLRs require bigger, more expensive lenses. Full-frame DSLRs are the right choice for most professional photographers, but not simply because the sensor is bigger. Instead, I recommend full-frame cameras to photographers because they tend to have more features completely unrelated to the sensor. Additionally, full-frame Canon and Nikon cameras have access to the widest variety of native lenses simply because the formats have been used by professionals for decades. Given the same variety of native lenses, you could get the same photos with smaller sensors, but other formats simply don't have the same variety available, and adapting the lenses always comes with significant penalties.

- **Medium format (less than 1X)**. Medium-format DSLRs provide professional studio photographers the resolution they need for shooting magazine covers and posters. The 60 megapixel Hasselblad H4D-60 retails for about $42,000, yet it can't take decent indoor pictures without flash, it's too slow for wildlife or sports, and it's too big for most people to carry around (though I often travel with a medium format film camera). For those reasons, most professional photographers use full-frame 35mm DSLRs instead.

The following figure compares a compact camera sensor to a full-frame camera sensor.

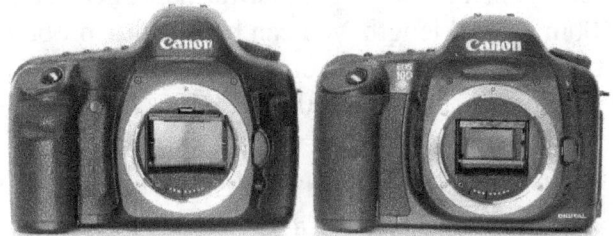

The following figure shows the relative sizes of different sensor types along with the crop factor of each. The crop factor is very important to understand when purchasing lenses or even just reading this book.

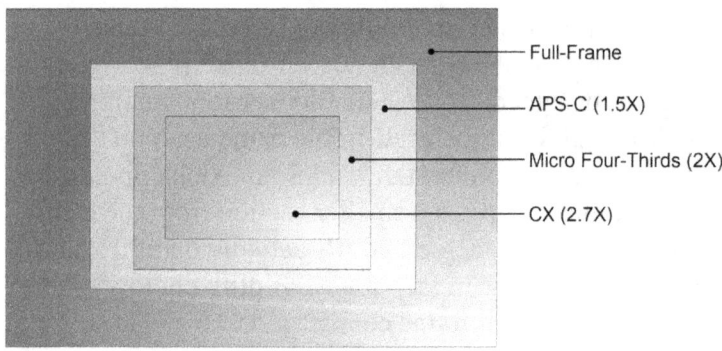

Crop Factor and Focal Length

Throughout this book and *Stunning Digital Photography*, I list focal lengths in 35mm equivalent. Therefore, if you want to calculate the equivalent focal length for a compact DSLR, you would divide the 35mm focal length by 1.6 for Canon or 1.5 for Nikon. If you want to calculate the equivalent focal length for a Micro Four-Thirds camera, you would divide the focal length by 2. If you want to calculate the equivalent focal length for a medium format camera, you would multiply it by 2.

For example, a "normal" or "standard" lens sees about the same angle of view as our eyes. On a full-frame camera, the normal view is 50mm. On a compact camera, the normal view is about 31mm, or 50 / 1.6. On a Micro Four-Thirds camera, the normal view is about 25mm, or 50 / 2.

Given this conversion, these three lenses provide similar zoom ranges when attached to the type of camera they were designed for:

- **Micro Four-Thirds**: the Olympus 14-42mm.

- **Compact**: the Canon 18-55mm.

- **Full frame**: The Sigma 24-105mm.

Most of the photography world, including myself, uses 35mm equivalents when discussing focal length just because 35mm has historically been the most popular format. Therefore, if you see an example picture that shows a 200mm focal length, you can bet that it's probably 200mm using the full-frame 35mm format. If you were using a compact DSLR and wanted the same field of view, you would use a 125mm lens, because 200 / 1.6 = 125.

> **Tip:** Big and small sensors can both have very high megapixel counts. However, smaller sensors capture less light because they have less surface area. So, the larger full-frame format will always be better in low-light and produce less noise—but the difference might never be important to you.

Crop Factor and Aperture

Different sensor sizes change the field of view provided by a focal length. We refer to this as the "crop factor," and it allows us to quickly determine that a 45mm micro four-thirds lens is equivalent to a 90mm full frame lens.

While the crop factor works for determining the field of view, it does not work for determining the depth of field and background blur that you'll get from any given lens. While camera manufacturers often provide a "35mm equivalent" when describing a lens, they don't tell you that you won't get the same background blur when using smaller lenses, which has led to many frustrated portrait photographers.

For example, consider two portrait lenses that seem very similar:

- The full-frame Canon 85mm f/1.8 ($400)

- The Micro Four-Thirds Olympus 45mm f/1.8 ($400)

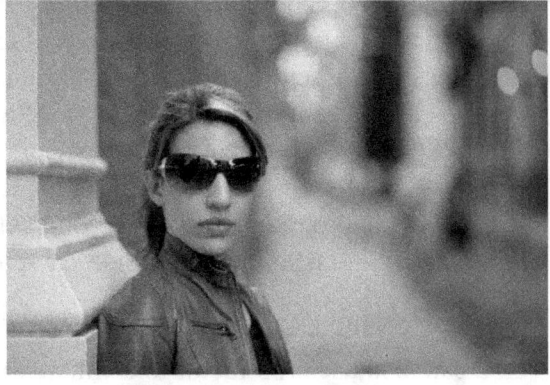

Knowing that you double the focal length of micro four-thirds lenses to determine the 35mm equivalent, the Olympus seems to compare favorably to the Canon. You might see portraits taken with the Canon 85mm f/1.8 (such as the following), and assume that you'd be able to achieve similar background blur. The Olympus can't achieve the same background blur, however, because you must apply the crop factor to the aperture to calculate depth-of-field (and thus background blur). In this example, the Olympus 45mm f/1.8 lens is equivalent to a full-frame 90mm f/3.6 lens when considering both field of view and background blur.

You can also multiply the depth-of-field by the crop factor. Therefore, a micro four-thirds camera with a crop factor of 2x has about twice the depth-of-field (and thus half the background blur) of a full frame camera, even after you multiply the focal length by the crop factor. An APS-C sized DSLR has 1.5X to 1.6X more depth of field, or 50-60% less background blur than a full frame camera.

For calculating the shutter speed you'd need in any given lighting scenario, you wouldn't need to multiply the aperture—the Olympus would still have the same shutter speed as a full-frame 90mm f/1.8 lens, or any f/1.8 lens, for that matter. However, for portrait work, lenses for smaller sensors have far less background blur. To achieve similar background blur to the Canon 85mm f/1.8, you would need a 45mm f/0.9 lens, and nothing like that is currently available.

Currently, the best micro four-thirds autofocusing lens for achieving a nice background blur is the Olympus M Zuiko ED 75mm f/1.8 ($830). For calculating background blur, this lens is equivalent to a full-frame 150mm f/3.6 lens. Unfortunately, that doesn't compare favorably to traditional 35mm portrait lenses. My budget full-frame portrait recommendation, the Tamron 70-200 f/2.8 ($750), is just as sharp, less expensive, offers much better background blur, and provides a very useful zoom range.

There is one micro four-thirds lens that provides full-frame SLR like background blur in a portrait focal length, but it lacks autofocus, and autofocus is critical when working with short depth-of-field portraiture because people move too much to manually focus on the eye. The *SLR Magic Noktor 50mm f/0.95* is equivalent to a full-frame 100mm f/1.9, and is $1,100.

Let's consider that lens on crop- and full-frame DSLRs. On a Nikon DSLR with a compact sensor, it becomes equivalent to a 105-300 f/4.2. On a Canon DSLR with a compact sensor, it becomes equivalent to a 112-320 f/4.5. Only on a full-frame body will you be able to achieve the full potential of the lens' ability to blur the background.

I don't want you to feel bad about purchasing a Micro Four-Thirds or APS-C camera; they're very capable cameras, and background blur is only one aspect of photography. Smaller sensors, and their large depth of field, actually show you much more of a scene, making them ideal for landscapes. There are also other ways to control background blur, including moving your subject farther from the background. For detailed information, refer to Chapters 4 and 6 of *Stunning Digital Photography*.

I hope that highlighting this weakness of smaller sensor designs for portrait work will help push lens manufacturers to offer faster lenses for the smaller sensors. The first manufacturer to give fair treatment to APS-C sized sensors is Sigma, with the Sigma 18-35mm f1.8 lens. When calculating background blur, this lens is roughly equivalent to a 27-52mm f/2.8 lens. It's still too wide angle to make a viable portrait lens, but it's an amazing lens for general photography with compact sensors, and I'm glad to see Sigma manufacturing faster lenses for smaller sensors.

Crop Factor and ISO

You can also use crop factor to estimate the total image noise different sensors will have at a specific ISO. Simply multiply the ISO of the smaller sensor by the crop factor twice:

Smaller Sensor ISO * Crop Factor * Crop Factor = Full Frame ISO

Or, to write it another way:

Small Sensor ISO * (Crop Factor)^2 = Full Frame ISO

For example, you can expect ISO 200 on a Micro Four-Thirds camera (which has a 2x crop factor) to have similar total image noise as ISO 800 on a full frame camera, because 200 * 2 * 2 = 800. You can expect ISO 100 on a Nikon APS-C camera (which has a 1.5x crop factor) to have similar total image noise as ISO 225 on a full frame camera. The following table gives you estimates of the amount of total noise you can expect from different ISOs and different sensor sizes, given similar sensor technology. That last clause, "given similar sensor technology," is very important, and I'll discuss it further.

Full-frame	APS-C (1.5X)	Canon APS-C (1.6X)	MFT (2x)	CX (2.7X)
64				
100				
200				
225	100			
256	114	100		
400	178	156		
640	284	250	160	
729	324	285	182	100
800	356	313	200	110

1,600	711	625		400	219
3,200	1,422	1,250		800	439
6,400	2,844	2,500		1,600	878
12,800	5,689	5,000		3,200	1,756
25,600	11,378	10,000		6,400	3,512

Some of the table cells are blank because those ISOs are not natively available on modern cameras. For example, the full-frame Nikon D810 supports a native ISO of 64.

Theoretically, a Micro Four-Thirds camera that supported an ISO of 16 would provide the same total image noise as a full-frame camera at ISO 64. However, no Micro Four-Thirds camera supports a native ISO lower than 160.

Therefore, in bright light or studio conditions that allow you to use your cameras lowest native ISO, bigger sensors will provide lower total noise. This isn't because bigger sensors are inherently better, but simply because camera manufacturers haven't yet designed smaller sensors to work with lower ISOs. In the future, a Micro Four-Thirds camera with a native ISO of 16 could theoretically compete with the Nikon D810 for studio work.

My calculations are estimates based on the total light gathered by each sensor size at a specific ISO. That means the crop factor math isn't quite perfect, because some sensors are more efficient than others. However, the math is accurate to around $1/10^{th}$ of a stop for most cameras (excluding Canon cameras, which are not as efficient).

The following table compares *www.DxOMark.com*'s measured scores for the top-end bodies from popular camera lines. Below the DxOMark score, I show the ISO you would use to achieve the same level of noise as ISO 800 on the Nikon D810 (based on the DxOMark measurements). The ISO 800 Equivalent (total light) row shows the equivalent ISO based on total light gathered (as estimated by the crop factor^2 formula).

	Nikon D810	Sony a7R	Sony A7S	Canon 5D Mk3	Nikon D7100	Sony a6000	Canon 70D	Olympus E-M1	Panasonic GH4	Nikon 1 V3
DXOMark Score	2853	2746	3702	2293	1256	1347	926	757	791	384
ISO 800 Equivalent (measured)	800	770		643	352	378	260	212	222	108
ISO 800 Equivalent (total light)	800	800	800	800	341	341	309	201	201	108
Efficiency vs D810 (stops)	N/A	-0.06	+0.38	-0.32	+0.05	+0.15	-0.25	+0.08	+0.14	0.00

The last row shows the number of stops the camera over- or underperforms the D810 while gathering the same total light. As you can see, differences in sensor technology make very little difference compared to differences in sensor size; even professionals would not notice a difference of less than .2 stops. You'll also notice that Canon's sensors don't perform as well given the same light as sensors from other camera manufacturers.

It's interesting to compare the 36-megapixel Sony A7R to the 12-megapixel A7S. Obviously the A7R can capture more detail, but the A7S gathers light about 40% more efficiently.

Estimating total light using the crop factor^2 formula is useful for quickly estimating the noise performance of different cameras. To more precisely calculate the noise performance of specific cameras, check the Sports (Low-Light ISO) rating of the camera on DxOMark.com.

Crop factor for far away subjects

In some sports and most wildlife scenarios, you can't get close enough to your subject to fill the frame. For example, with bird photography, even with a very expensive 500mm f/4 lens on a full-frame body, most photographers will need to crop significantly.

In the Canon, Nikon, and Sony SLR lens-mounts, all super-telephoto lenses are designed for full-frame bodies. In most scenarios, I recommend using full-frame lenses with full-frame bodies. However, if you would need to crop significantly, you can get much sharper, more detailed pictures by using an APS-C body.

A Nikon example

In the Nikon world, many wildlife photographers debate the APS-C D7100 (with a 1.5x crop) and the full-frame D810. The D7200 is "only" 24 megapixels, while the D810 is an astounding 36 megapixels. If you can fill the frame with your subject and the Nikon 600mm f/4 lens, the D810 provides about 24 megapixels of visible detail (according to DxOMark's perceptual megapixel ratings). It doesn't show the entire 36 megapixels it's theoretically capable of, because the lens optics are not perfect.

The D7200, when paired with the Nikon 600mm f/4, produces only 10 megapixels of visible detail. Therefore, the D810's 240% greater detail makes it a far superior choice. Now, imaging you're photographing eagles at a distance, and you need to crop 1.5X, anyway. The D7200 ($1,000) has a native 1.5X crop, so it still produces 10 megapixels of detail. The D810 ($3,300) in DX crop mode will also produce about 10 megapixels of detail with that lens. If your primary purpose was wildlife, you might be better off buying the D7200 and spending the extra $2,300 on a safari.

Why isn't the math perfect?

The D810's DX mode is 16 megapixels, and the D7200 (which is always in DX mode) is 24 megapixels. So why doesn't the D7100 produce 50% more detail than the D810?

The two cameras use different sensor technologies. The D810 has the anti-aliasing and optical low-pass filters removed, so each pixel is sharper. This demonstrates that you can't simply count megapixels; you need to factor in the lens sharpness and the technology.

A Canon example

In the Canon world, the differences are more extreme. Comparing the 7D and the 5D Mark III with the Canon 600mm f/4 lens, the Canon's full-frame 22 megapixel sensor produces 20 megapixels of visible detail. The 7D's APS-C (1.6X crop) 18 megapixel sensor produces only 12 megapixels of visible detail.

For faraway subjects where you would need to crop 1.6X anyway, the 7D still produces 12 megapixels of visible detail. The 5D Mark III produces only 7.8 megapixels of visible detail. Therefore, the 7D produces 53% more detail than the 5D Mark III, at about 30% the cost.

Again, why isn't the math perfect?

It might seem odd that the Canon 600mm f/4 produces 20 megapixels of visible detail with the 5D Mark III—91% of the sensor's physical resolution. Yet, that lens on the 7D only produces visible detail at 66% of the sensor's physical resolution. Are the 7D's pixels less sharp?

Maybe, but not by much—the sensor technologies are similar. The more likely explanation for the lower percentage is that the 7D's higher pixel density more greatly exceeds the lens' physical sharpness. In other words, by cramming more pixels into the same area, the 7D is putting more pressure on the lens sharpness. It extracts more detail, but an even sharper lens could extract even more detail.

So, should I get APS-C for the higher pixel density, or go full-frame?

Get APS-C for the higher pixel density if you plan to crop anyway. Realistically, almost all wildlife photography is heavily cropped. Even with super telephotos, even with an APS-C body, most wildlife photographers need to crop almost all of their photos.

Outside of captivity, the only times you don't need to crop wildlife photos are when you're shooting large mostly tame animals, such as deer, or when you've camouflaged yourself and spent hours getting close to your subjects.

In other words, if you're masking your scent and wearing a ghillie suit, a full frame body for wildlife might be worth the extra money. Otherwise, an APS-C body is probably a better overall value.

Megapixels

Megapixels describe the picture size a digital camera produces. Usually, the more megapixels a camera has, the better. However, if you're using an unsharp lens (such as a consumer kit lens), megapixels might not make any difference.

> A *megapixel* is a million *pixels* (picture elements, or more simply, colored dots) that make up a picture. So, a one megapixel picture (about the size you'd see on the web) is made up of a million dots, and an eight megapixel picture (which would make a nice 8x10" print) is made up of eight million dots. To calculate the dots, multiply the width and height of the picture. So, a 1600x1200 picture is about 1.9 megapixels, and a 4000x3000 picture is about 12 megapixels.

The following table puts megapixels and photo resolution in perspective by showing the maximum size you can make a camera's photos while still maintaining clarity. The cameras used range from a smartphone to a professional medium-format studio camera that costs more than a new BMW.

Megapixels	Resolution	Print Size (300 dpi)	Sample Camera
8	3264x2448	8x10"	iPhone 5
12	4272x2848	9x12"	Canon T3
16	4928x3264	11x14"	Nikon D7000

21	5616x3744	12x19"	Canon 6D, 5D Mark III
33	6726x5040	17x22"	Mamiya DM33
36	7360x4912	18x24"	Nikon D810
60	8956x6708	22x30"	Hasselblad H4D-60
200	16352x12264	40x54"	Hasselblad H4D-200MS (still subjects)

You can make a nice 8x10" print with a smartphone. So, why would you ever want more?

- **Bigger prints**. 300 dots per inch (DPI) are ideal for printing. 200 DPI look good, too; the picture won't be quite as sharp, but you won't notice when viewing the picture from a distance. The farther away you are, the lower the DPI you need; I've made billboards with a 6 megapixel camera, and they looked fine because of the large viewing distance. At 300 DPI, you'll need a 15 megapixel camera to make an 11x14" print, a 30 megapixel camera to make a 16x20" print, and a 50 megapixel camera to make a 20x30" print.

- **Cropping**. Those print size estimates assume you use every last pixel. Yet, if you print an 8x10", you'll crop off about 7% of the image because your camera's sensor is a bit wider than an 8x10" print. If you want to rotate an image 90° to crop a horizontal shot to a vertical portrait, you'll lose more than half your megapixels. Cropping a full-body picture of someone to a headshot can reduce your pixel-count by 75%. Having more pixels means having more cropping flexibility.

- **Noise**. All camera sensors have noise visible when you view a picture at full resolution. Generally, pictures aren't viewed at full resolution, however, the more pixels you start with, the less noise there will be when you share pictures on the Web or view smaller prints.

- **Selling pictures**. Microstock agencies charge more for larger pictures, so higher resolution images can make more money.

There is a downside to larger megapixel counts—bigger file sizes. Bigger files take more space on your memory card and on your computer, and editing your pictures takes longer. However, memory cards and disk drives are cheap nowadays, and most cameras allow you to shoot at a lower megapixel count when you need to.

Many people choose Sony and Nikon cameras over Canon and Micro Four-Thirds cameras because Sony sensors (which Nikon uses in their cameras) have higher megapixel counts. For example, many new photographers must choose between the 18 megapixel Canon T5 and the 24 megapixel Nikon D3300.

With 33% more pixels, the D3300 must produce sharper images, right? It can, but the sharpness will be the same for most photographers, because there's another factor: the lens. Both cameras come with a kit lens that's not as sharp as it could be: DxOMark rates them both at about 9 megapixels. Therefore, if you just use the kit lens, both cameras will produce similar details, and you won't even be making use of the T5's full 18 megapixels. So, if you plan to use only the kit lens, you can safely disregard the difference in megapixels and instead base your decision on how the camera feels in your hand, focusing speed, price, and other factors.

However, if you plan to replace your kit lens with the incredibly sharp Sigma 18-35 f/1.8 lens (the sharpest APS-C zoom lens available), you will get sharper pictures from either camera. DxOMark rates the Canon T5 with the Sigma 18-35 at 15 megapixels, while it rates the D3300 with the same lens at 17 megapixels. In this example, the D3300's 33% more megapixels produced about 13% more detail.

Here's another comparison: the 24-megapixel Canon 5D Mark III and the 36-megapixel Nikon D800E. When connected to the excellent Tamron 24-70 f/2.8 lens (the sharpest zoom lens available in that range), DxOMark rates the Canon combination at 18 megapixels of detail. The Nikon combination rates 23 megapixels of detail. In this example, Nikon's 50% more pixels results in 28% more sharpness.

These tests are conducted in ideal conditions. Any sort of camera shake, subject movement, or atmospheric conditions (such as humid air) will further decrease the importance of additional megapixels. However, in our testing with sharp lenses, Nikon's higher megapixel sensors do have noticeably more detail when you use a sharp lens.

Before you decide to buy a high megapixel camera and an expensive, sharp lens, ask yourself whether sharpness and detail will have a noticeable impact on your photography. For most casual photographers, sharpness doesn't matter at all. Most casual photographers only ever share their images online, where people view images at no more than 2 megapixels, and nobody would ever be able to tell the difference between a high megapixel camera with a sharp lens and a low megapixel camera with a cheap kit lens. Even if you have a huge 4k monitor and you view the image full-screen, it will only have a resolution of 3840 x 2160, which is about 8 megapixels.

In other words, even the most basic cameras capture far more megapixels than you could ever view online. Rather than spending their budget on expensive cameras and lenses just to get more sharpness, most photographers would be better served by choosing less sharp gear and setting aside more of their budget for lighting and training.

Consider a portrait photographer with a total budget of about $5,500. He could spend his budget in two ways:

- Buying a 36-megapixel D810 ($3,300) and a Nikon 70-200 f/2.8 ($2,200).

- Buying a 24-megapixel D3300 ($400), a Tamron 70-200 f2.8 ($1,300), a set of four Paul C. Buff Einstein lights with modifiers (about $3,000), and light stands, backdrops, stools, and fans ($800).

Sure, the D810 would produce sharper images, but the extra lights will have more impact on the final picture quality.

Similarly, a landscape photographer with an extra $1,000 in his budget will do more to improve his portfolio by spending the money on a trip to Glacier National Park than he would buying more megapixels and sharper glass.

Megapixels are the last thing most photographers should worry about. Wildlife photographers, however, are an important exception. Wildlife photographers almost always crop their pictures because they can't get close enough to the animals to fill the frame. Sharpness makes a huge difference on image quality for wildlife photos, too. Therefore, high megapixels and a sharp lens are more important to wildlife photographers than they are to portrait, landscape, or even sports photographers.

High Megapixels and Noise

Many photographers believe higher megapixel sensors produce noisier images. Often, the exact opposite is true: higher megapixel sensors in Nikon and Sony cameras actually produce less image noise than the lower megapixel sensors in Canon cameras. So, why the confusion?

The lower the megapixel count, the larger the individual pixels (known as photosites). Just like a larger bucket gathers more rain, a larger photosite gathers more light. The more light a photosite gathers, the less noise it will produce in that pixel.

Therefore, since pixel noise is primarily determined by the amount of light a single photosite gathered, lower megapixel sensors have less pixel noise. When photographers compare the noise produced by two sensors, they often zoom in and view images on a pixel-by-pixel basis (1:1). When they do that, they're comparing pixel noise, and lower megapixel sensors will seem to have less noise.

But there's an important difference between pixel noise and total image noise. We only see pixel noise when we compare photos on a pixel-by-pixel basis. In the real world, images are scaled to the size of our display or print, and multiple pixels are blended together. In the real world, we don't notice pixel noise. Instead, we see total image noise.

Total image noise is determined primarily by the total light gathered by the sensor, and the light gathered by each individual pixel makes no measurable difference.

In practice, the higher megapixel sensors from Sony and Nikon actually have less total image noise than the lower megapixel sensors from Canon. In other words, if you take pictures with a 42-megapixel Sony a7R II and a 24-megapixel Canon 5D Mark III and make a 20x30" print, the a7R's print will have less visible noise (and it will be sharper, too).

Maximum ISO Speed and Noise

Like film, digital cameras need a specific amount of light to create an image. Just about any camera can take great, noise-free pictures in bright sunlight. When the sun goes behind a cloud, or you move indoors, you'll begin to notice the difference between sensors.

Digital cameras measure their sensitivity to light using ISO. The following table shows common ISO speeds and when you might use them (without flash).

ISO speed	Use
50-200	Outdoors, full sunlight
400	Outdoors, cloudy days
800	Indoors, brightly lit
1600	Indoors, normally lit
3200	Indoors, poorly lit
6400	Indoors, poorly lit
12,800+	Indoors, dark (such as a bar at night)

Higher ISO speeds are one of the greatest features of professional cameras because they allow you to take photos in dimly lit environments without using a flash or a tripod. At

ISO 12,800 and higher, you can handhold pictures at night—imagine being able to take pictures of your friends in a restaurant without a flash, or doing night photography without a tripod. Unfortunately, current digital cameras produce so much noise at ISO 6400 and above that you're still better off using a tripod or a flash and a lower ISO setting whenever possible.

The more serious you are about your photography, the less important high ISO speeds will be. Only the most casual photographer will be happy with photos taken at ISO 12,800 on any camera; the noise in the image would simply be too high to sell the picture for any purpose other than, perhaps, photojournalism. However, high ISO speeds typically produce fine results for online use, such as posting a picture to Facebook. In other words, if you want to be able to take snapshots of your friends in a dim bar and put them online, pick a camera with a high maximum ISO. If you want clean, professional results, even in dark environments, you'll carry a tripod with you, set your camera to ISO 100, and use a long exposure, so the high ISO won't matter.

Stunning Digital Photography will tell you everything you need to know about choosing the right ISO setting for different situations. For now, understand that one of the most important factors to consider when buying a camera is the noise levels at different ISO settings and the maximum ISO speed. The less noise and higher the maximum ISO speed, the less you'll need that annoying flash.

> A lower-megapixel sensor with low noise can create a picture of similar quality to a higher-megapixel sensor with high noise. For a detailed explanation, read "Contrary to conventional wisdom, higher resolution actually compensates for noise" at http://dxomark.com/index.php/eng/Insights/More-pixels-offsets-noise!.

If you really want to nitpick the noise your camera's sensor makes, visit http://DxOMark.com. DxO Labs measures the signal-to-noise ratio (SNR) of different camera sensors at different ISO speeds and allows you to compare them. As you'll see if you examine a camera at the site, the SNR is very high at ISO 100 and 200. At ISO 1600 and above, the SNR is very low—reflecting the noisier pictures you'll see at higher ISO speeds.

Minimum ISO Speed and Noise

Landscape and still life photographers have the opportunity to use a tripod and long exposures, and thus always have the option of using their camera's lowest native ISO. Studio photographers have the opportunity to add as much light as they want and also use their camera's base ISO. For these types of photographers, high ISO performance doesn't matter, but optimal ISO performance can be very important. You can get the lowest possible noise by using a full-frame camera at its base ISO.

Most cameras have a base ISO of 100, and you should use that base ISO whenever possible to take the cleanest pictures. One camera in particular, the Nikon D810, offers an unusually low native base ISO of 64, providing lower noise than is possible on any other camera.

Olympus Micro Four-Thirds cameras have a base ISO of 200, and Panasonic Micro Four-Thirds cameras have a base ISO of 160. Because they have higher base ISOs, their very best image quality is noisier than other cameras. Later, this chapter discusses how the Micro Four-Thirds sensor size further reduces the total light gathered at the base ISO, meaning Panasonic's base ISO of 200 produces images with about the same noise as a full-frame camera (with similar sensor technology) at ISO 800. For many professionals, that means even the most well-lit shots will be too noisy to use commercially.

Disregard extended ISOs. For example, many cameras with a base ISO of 100 offer an extended ISO of 50. Choosing ISO 50 will not provide better image quality than ISO 100; your image quality will be exactly the same as taking a raw image at ISO 100 and overexposing it by one stop.

Lens and Body Compatibility

Lenses should be matched to your sensor size. If you have a Canon or Nikon DSLR, use the following table to determine which type of lens to buy for your DSLR.

Sensor Type	Lens Type	Sample Lens
Canon APS-C	EF-S	EF-S 17-85mm f/4-5.6 IS
Canon full-frame	EF	EF 100-400mm f/4-5.6 L IS
Nikon DX	DX	18-105mm f/3.5-5.6G ED VR AF-S DX
Nikon FX	(not indicated)	AF-S Nikon 24-70mm f/2.8G ED

If you have a Canon APS-C DSLR, you can attach a full-frame Canon EF lens. However, the picture will be cropped by 1.6X, which is known as the focal length *multiplier*. Everything will look fine through the viewfinder, but you'll be zoomed in 60% more than you would be with a Canon full-frame camera body. If you're using a wide-angle lens to shoot at 24mm, and the camera uses an APS-C sensor with a 1.6x crop factor, your picture will be at the not-so-wide-angle 38mm (24 x 1.6 = 38).

You cannot use Canon EF-S lenses with a Canon full-frame body. The following table shows Canon body and lens compatibility.

	Canon EF-S	Canon EF
Canon APS-C	Ideal	1.6X crop
Canon Full-frame	Not functional	Ideal

If you have a Nikon DX DSLR, you can attach a full-frame Nikon lens. The picture will be cropped by 1.5X, effectively zooming everything in by 60%. Unlike Canon, you can use a Nikon DX lens on a full-frame Nikon FX camera body—however, because the lens was designed for smaller sensors, your pictures will be at a lower resolution. The following table shows Nikon body and lens compatibility.

	Nikon DX	Nikon Full-frame
Nikon DX	Ideal	1.5X crop

| Nikon FX | 62% smaller pictures | Ideal |

For most people, crop factor isn't a concern at all. Just buy lenses designed for your camera, or use the lens that came with your camera. You can still get ultra-wide-angle lenses—for example, Canon makes a 10-22mm lens that is equivalent to a 16-35mm full-frame ultra-wide-angle zoom. Sites such as bhphotovideo.com have separate categories for DSLRs with *Canon APS-C* or *Nikon DX* sensors and *full-frame DSLRs*, so as long as you're aware of your sensor type, you can get the right lens.

The only types of lenses that Canon and Nikon don't create for compact cameras are the professional super-telephotos. Canon and Nikon both make 600mm f/4 super-telephoto lenses for full-frame cameras only because if you're willing to spend $8,000 on a lens, you're probably also willing to spend a few more bucks on a full-frame DSLR. You can still use these super-telephotos or any full-frame lens on a DSLR with a crop factor, but you'll be losing part of the image projected by the lens, meaning you're paying for something you aren't using and carrying around unnecessary glass.

Frames Per Second and Buffer

Frames per second (FPS) measures how fast your camera can take pictures when it is in continuous shutter mode. The slowest DSLRs can take shots at about 3 FPS, whereas professional models designed for sports can shoot at up to 14 FPS.

If you plan to shoot sports or wildlife, FPS will be very important to you, and it might be worth spending more to get a camera with a higher FPS. Moving from 3 FPS to 6 FPS doubles the number of shots you take when the action is occurring. In sports, that increases the chances of you capturing a stunning shot with a fleeting expression, or recording that split-second where the baseball meets the bat. In wildlife, it means you're more likely to get a shot of the flying bird with its wings just at the top of a flap.

Basically, higher FPS increases the overall quality of your action shots by increasing the chances that you capture that split-second where the light, expression, and pose are perfect. However, even 3 FPS is enough for the casual sports and wildlife photographer. If you aren't serious about shooting sports or wildlife, FPS really doesn't matter at all, and it shouldn't be a factor in your buying.

Buffer size is another factor that is closely related to FPS. When you take a picture, your camera first copies the picture from the sensor to a temporary, but high-speed buffer. From the buffer, the picture is then copied to your permanent, but much slower, memory card. Therefore, your camera's advertised FPS can only be achieved while there is room in the buffer. When shooting RAW, the buffer can fill up quite quickly. For example, the Canon 5D Mark III shoots at about 6 FPS. However, the buffer can be completely filled after about 2 seconds of continuous shooting, and future shots will be much slower to capture. Using a faster memory card can extend your continuous shooting by allowing the buffer to unload images faster, allowing you to take many more shots before the FPS slows down. With the Canon 5D Mark III and a high-speed UDMA 7 CF memory card, the camera can shoot continuously for about 6 seconds (instead of the usual 2) before the buffer is full.

If you plan to shoot long bursts of action and use RAW, research two factors:

- The number of continuous shots the camera can capture before the buffer is full.

- The performance of different memory cards with your camera.

Buffering isn't typically a problem with JPG files, because JPG files are much smaller, and therefore consume much less of the buffer.

Maximum Shutter Speed

Different camera bodies have different maximum shutter speeds. For example, the base model Nikon D3100 has a maximum shutter speed of $1/4000^{th}$ of a second, whereas the top-end Nikon D5 has a maximum shutter speed of $1/8000^{th}$ of a second.

Very few people will ever need to shoot above $1/4000^{th}$ of a second. $1/4000^{th}$ is sufficient for any sports or wildlife, even hummingbirds. Even if you have the need to go faster, it's difficult to get enough light to allow your camera to exposure at that shutter speed without using a very high ISO. Therefore, the maximum shutter speed isn't a factor for most photographers.

Controls and Ergonomics

One of the most underrated features of a camera body are the buttons and controls. Lower-end cameras tend to have fewer buttons. Instead, they allow you to change settings by navigating a menu. Higher-end cameras have more buttons dedicated to specific settings, allowing you to more quickly change a setting.

For example, the Canon T3 lacks an exposure compensation dial. If you want to adjust the exposure compensation, you need to navigate the menu system. Higher-end cameras have a dedicated dial that you can quickly adjust with your thumb.

Even though adding controls can make a photographer much more efficient, very few people spend extra money on a higher-end camera just for the additional buttons. Instead, you can think of more advanced controls as a free bonus included with the higher-end cameras.

Display

Like the controls, the display is a very important part of a camera—yet not a feature that's likely to push you to spend hundreds or thousands of extra dollars on a camera.

In a nutshell, larger, brighter, higher-resolution displays are better because they give you a better preview of your picture. However, you're not likely to choose between two cameras just based on the display quality.

Some cameras have articulating displays which flip up, down, out, or some combination of those. Flip up and flip down displays are quite useful; they allow you to hold the camera up high over your head or down low to the ground, making unusual compositions much easier. Flip out displays can be rotated towards the front of the camera so you can see yourself when you're taking a self-portrait, which helps you frame yourself in the picture. However, articulating displays take up space and reduce the durability of a camera because they're difficult to weather seal. For that reason, articulating displays are features on lower-end cameras that are missing from higher-end cameras. If an articulating display is important to you, you might be forced to choose a less expensive camera to get that feature.

Built-In Flash

Some expensive DSLRs have a built-in flash above the lens that pops up when needed. DSLRs designed for professionals don't have a built-in flash. This seems counter-intuitive, and like articulating displays, it's one of the few examples of a feature that's cut from higher-end cameras.

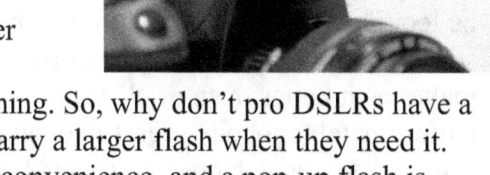

I like having a built-in flash on a DSLR. Sure, it never looks as nice as an external flash, but it's convenient because it's always with you, and it's better than nothing. So, why don't pro DSLRs have a built-in flash? Because pros tend to plan ahead and carry a larger flash when they need it. Also, weather sealing is more important to pros than convenience, and a pop-up flash is difficult to make weather proof.

If you get a pro-level DSLR without a built-in flash, be prepared to also buy an external flash and carry it with you. If you get a consumer-level DSLR with a built-in flash, you should also be prepared to buy an external flash, because the built-in flash is quite ugly.

USB Charging

For those of us planning to travel with our camera, USB charging is a helpful feature. With USB charging, you can charge your camera's battery using a USB cable, exactly like you charge your smartphone.

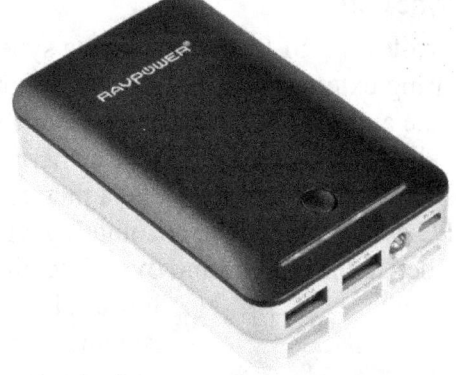

This means you don't have to travel with a battery charger. It also means that you can connect your camera to a USB battery charger, such as the RavPower model shown below, and charge your camera when you don't have access to a power outlet. Look for a model that supports 2A (2 amp) charging; this will charge your devices faster. I always keep a battery charger and several USB cables in my backpack while traveling.

Of course, USB charging is more of a concern for cameras with shorter battery life. My DSLRs can often go an entire week with a fresh battery. On the other hand, my mirrorless cameras typically run out of batteries around 2:30pm on a day of casual travel photography.

Currently, only some Sony, Fujifilm, and Samsung mirrorless cameras support USB charging.

Sound

Lower-end cameras, such as all smartphones and most P&S cameras, can be completely silent when they take a picture because they don't have a physical shutter. DSLRs have both a mirror and a shutter that needs to open and close with every picture you take, and that process makes a distinct thud or clanking sound.

If you're shooting portraits or landscapes, the sound will make no difference to you whatsoever. However, if you're a photojournalist or a wedding photographer, or if you shoot candids or wildlife, the mirror/shutter noise can be extremely important to you. Imagine needing to photograph a funeral without disturbing the grieving, or waiting hours for a fox to come out of its den, only to have it startled by the noise from your first shot. Some newer DSLRs feature quiet or "silent" modes (which typically aren't really silent). On some bodies, the mode is only available when shooting in live view mode, so you can't be looking through the viewfinder when you use it. However, many new bodies reduce sound when using the viewfinder by moving the mirror a bit slower, reducing the noise it makes as it bangs open or closed.

Unfortunately, nobody seems to document the sound levels of different cameras in any standardized way. However, if you search the web for specific models and the words shutter and sound, you can often find users who have done testing on their own.

If low sound levels are really important to you, you might look instead to mirrorless cameras. If your DSLR really must be silent, consider buying a camera muzzle or a sound blimp from _http://www.soundblimp.com/_.

Sync Cord

Pro-level DSLRs tend to have a sync cord connector, which is a very old standard for firing external flashes and studio lighting when you take a picture. Basically, you plug a special cable called a sync cord into your camera and your flash system, and when you take a picture, the flash fires. PC sync cords are an absolutely awful connector type. They're unreliable, and over time, they tend to come

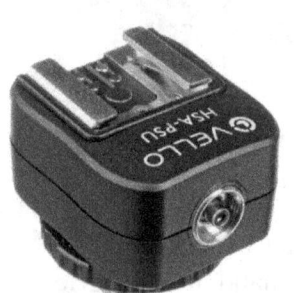

loose. There's always a cable running from your camera to your lights, and at some point, you're going to trip over it and pull the cord out (if you're lucky) or pull your light or tripod down (if you're not so lucky).

If your camera doesn't have a sync cord and you really want one, don't fret. you can get an inexpensive flash hot shoe to PC sync cord adapter (shown next). Better yet, you can always trigger external lights by using a wireless remote attached to your flash shoe. For more information, refer to the Wireless Flash Trigger Buying Guide in this book.

Weatherproofing and Durability

Another factor that distinguishes consumer-level DSLRs and lenses from pro-level DSLRs and lenses is weather sealing. Every button and opening on your camera is an opportunity for dust, sand, and moisture to enter the camera. Pro-level DSLRs are designed to better keep these elements out, resist rain, and survive more serious bangs and drops.

Most consumers don't need that extra weatherproofing and durability. If it's raining, the consumer will probably just leave their camera at home. If you're a photojournalist or a sports photographer, however, you don't have that option; you need to take pictures regardless of the conditions.

Camera designers often have to choose between convenience and weatherproofing. For consumer cameras, the designers choose convenience. For professional cameras, the designers choose weatherproofing. That's why consumer cameras have some cool features that professional cameras lack, such as articulating LCD screens and built-in flashes. Pro cameras also need to have every button and opening weather sealed, which adds weight and can make the buttons a little less friendly to use.

Overall, I feel like most people overestimate the value of weatherproofing. A few drops of water on your camera are fine; it's only a good soaking (such as leaving it out in the rain or dropping it in water) that will kill a camera. Carry a plastic bag with you, and if it starts to rain, put the camera in the bag.

I should also add that weatherproofed or weather-sealed cameras are far from indestructible. Chelsea & I were caught in a flash storm while hiking a mountain in Glacier National Park. We had five cameras with us: a waterproof video camera, two weather-sealed Canon 5Ds, and two non-weather-sealed Panasonic GH4s. The waterproof camera survived, and the GH4s survived, despite not having weather sealing. Both the Canons were totaled beyond repair, despite their weather sealing. We were using weather-sealed lenses on the Canons.

Two companies have decided to use weather sealing as a significant part of their marketing: Olympus and Pentax. However, these weather sealing marketing claims can be a bit sketchy. There's no way for us to objectively test the weather sealing, so we can't validate their claims. If your camera is ruined by weather, it won't be covered by your warranty, so there's not necessarily a financial penalty for the manufacturer if they can't fulfill their claims.

Metering

Metering is the process your camera uses to determine how bright or dark to auto expose a scene. In the days of film, metering was exceptionally important because you wouldn't know if you exposed a shot properly until after you developed your film. With digital cameras, you can instantly glance at your photo and immediately know whether the camera over- or under-exposed the shot, and make any adjustments necessary. Additionally, if you shoot raw, you can often adjust the exposure one or two stops in either direction and still get excellent results. So, even if your camera does expose the picture incorrectly and you can't adjust exposure compensation and re-shoot, you can still get a great result.

For those reasons, metering systems are much less important than they were in the film days. Nonetheless, higher-end cameras tend to have more advanced metering systems that will more accurately expose complex scenes, such as heavily backlit scenes. The differences between metering systems aren't significant enough to influence your camera choice, however.

Memory Card Slots

Most modern digital cameras use two different memory card standards:

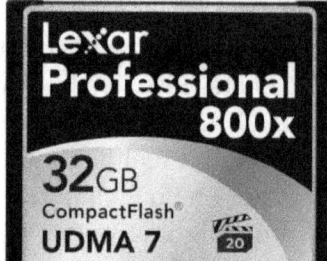

- **SD**. SD cards (shown right) are small and inexpensive. They also tend to take longer to write pictures to, though some cards are faster than others. Newer UHS-2 cards can be faster than most CF cards.

- **CF**. CF cards (shown next) are larger and more expensive than SD cards. They tend to be faster, allowing you to take more pictures in a short amount of time when using continuous shooting.

- **CFast**. Currently only available on the Canon 1DX II, the CFast format is very fast but the cards are also very expensive.

- **XQD**. Currently only available on the Nikon D500, D5, Z6, and Z7, XQD cards are expensive but incredibly fast.

- **CFExpress type B**. Very similar to XQD, these cards are fast and expensive.

- **CFExpress type A**. The newest format, CFExpress type A is about the size of an SD card but with much better performance (and higher cost).

Lower-end cameras tend to use SD cards, whereas higher-end cameras use CF cards. Higher-end cameras often have multiple slots for memory cards.

In the Canon world, the mid-range Canon 6D has one SD card slot. The higher-end Canon 5D Mark IV has two slots: one SD and one CF. The top-end Canon 1D X II also has two memory card slots, both CFast.

Having two cards slots is useful; you can have twice the capacity, and if one card fails, you have a backup. Having a second slot can also allow you to keep an Eye-Fi card in one of the slots, providing Wi-Fi connectivity for quickly previewing pictures on a computer or mobile device.

GPS

Some newer cameras have built-in GPS tagging capabilities, automatically adding location information when you take pictures. While not particularly useful for professional purposes, GPS tagging is both useful and fun for personal pictures. Applications such as Lightroom (shown next) can display your pictures on a map, allowing you to see exactly where they were taken and allowing you to browse pictures by the location.

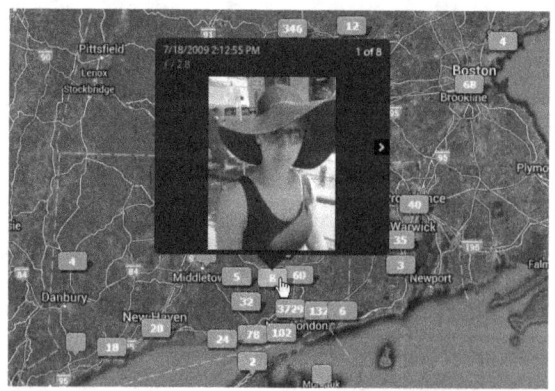

Only a handful of DSLRs, including the Canon 6D, have built-in GPS tagging. If your camera does not have GPS capabilities, you can often purchase an overpriced accessory to add GPS tagging to your camera, such as the Canon GP-E2 ($250) or Nikon GP-1 ($200). Third parties might have a unit compatible with your camera, such as the Marrex MX-G10M.

A less-expensive alternative is to use a smartphone app to record your location and then synchronize your location on your PC. Lightroom supports automatically synchronizing GPS data from your phone. Search your smartphone app store for geotagging to find supported apps.

Wi-Fi and Ethernet

Wi-Fi is the wireless network technology that laptops and tablets use to connect to the Internet. Most new cameras support Wi-Fi. Ethernet accomplishes the same thing as Wi-Fi, but uses a cable and transfers images quite a bit faster. Top-end cameras such as the Canon 1D X II allow you to connect them to a network using an Ethernet cable. Additionally, you can buy overpriced accessories to connect other camera bodies to wired and wireless networks.

Connecting your camera to a network is useful for tethering, which is the process of instantly transferring images from your camera to a PC or mobile device. Tethering is useful in several different scenarios:

- A casual photographer could transfer a picture to his or her smartphone so he or she could post it on Facebook without going back to his or her computer.

- A photojournalist could transfer a picture of a crime scene directly to his or her editor.

- A portrait photographer could transfer pictures to a PC so the customer can preview them immediately.

- A fashion photographer could allow an art director to examine the pictures as they're taken on a computer display, so the art director could provide immediate feedback.

- A commercial photographer could examine the pictures on a PC to verify that the images meet the stringent quality requirements.

Most casual photographers won't ever need to tether their camera, but many professional photographers do need to do so. If your camera has Wi-Fi or wired Ethernet built in, you can install an app on your mobile device, PC, or Mac to receive and preview pictures you take in real time. If your camera does not have tethering built-in, you can probably add an Eye-Fi or Transcend SD Wi-Fi card and accomplish the same thing. If your camera takes CF memory cards, you can use the SD Wi-Fi cards with an inexpensive adapter.

Wi-Fi can also provide remote viewing and remote control of your camera. While cool, I have yet to find a practical application for this.

X-sync Speed

As described in "Flash Sync Problems" in Chapter 5 of *Stunning Digital Photography*, flash synchronization problems can lead to uneven lighting in a flash picture (shown next).

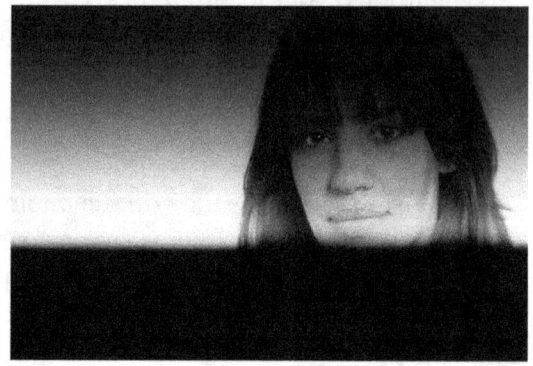

Most name-brand flashes support high-speed sync with the manufacturer's camera bodies, allowing you to use flash at any shutter speed. Thus, if you use a name-brand flash, you can usually shoot at any shutter speed with anybody (but check your flash's manual to be sure).

When using a flash that does not support high-speed sync, such as a generic flash, you cannot use shutter speeds faster than your camera's X-sync speed. The X-sync speed is the fastest shutter speed at which the camera's shutter fully exposes the entire sensor at once, giving the flash the opportunity to fire and illuminate the entire picture evenly.

I don't know that anyone has ever chosen one body over another based on the X-sync speed. It's only important if you want to use a generic flash with faster shutter speeds, and even then, X-sync speed on camera bodies varies very little. For example, most of the Canon lineup has an X-sync speed of 1/200th, but the top-end camera, the Canon 1DX, has an X-sync speed of 1/250th. The Sony NEX-6 has an X-sync speed of 1/160th. The Nikon D40 has a remarkably fast X-sync speed of 1/500th, due to a rather special shutter mechanism, but the more expensive D4 has an X-sync speed of only 1/250th.

While I don't expect you to choose a camera body based on its X-sync speed, portrait photographers using studio lighting or generic flashes in bright sunlight should be familiar with their camera's X-sync speed. In bright sunlight, you often need to use shutter speeds that might be faster than your camera's X-sync speed. If that's the case, and your flashes don't support high-speed sync, you will need to watch your shutter speed closely and verify that your flash is evenly illuminating the frame.

The PocketWizard wireless camera triggers have a clever feature called HyperSync that can actually increase a camera's X-Sync speed. For example, the Canon 50D normally has an X-sync speed of 1/250th, but when using a PocketWizard FlexTT5 and the HyperSync feature, can achieve X-sync speeds of 1/400th. For more information about wireless flash triggers, refer to the Flash Buying Guide later in this book.

Image Stabilization

Image stabilization, as discussed in *Chapter 4, "Lens Features,"* allows you to hand hold pictures with slower shutter speeds. Many Sony, Olympus, and Pentax have it built into the body. Other manufacturers put it into the lenses.

Having stabilization built into the body, known as In-Body Image Stabilization (IBIS), has no benefit for users who use only kit lenses. Since all manufacturers offer image stabilization with their kit lenses, your results will be similar.

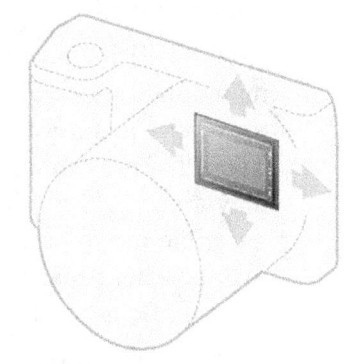

However, IBIS has massive benefits to those of us who use fast prime lenses in low-light conditions. Many photographers can get sharp hand-held pictures with a 50mm lens at 1/10th of a second. However, Canon and Nikon photographers can't buy a stabilized 50mm lens, so they'd have to use a zoom lens with a higher f/stop, requiring them to use a much higher ISO. A stabilized body, such as the Sony a7 II, allows you to use a fast prime lens at f/1.4 or f/1.8 with stabilization.

In the real-world, this has allowed me to take handheld night pictures in cities using ISO 100 and shutter speeds of 1/5th or 1/10th. With a Canon or Nikon body and a zoom lens, I would have needed to use an f/4 zoom and an ISO of 400 or 640.

All Sony SLT/DSLR bodies have sensor shift image stabilization built in, and any lens you use it with will be automatically stabilized. In theory, this would allow you to save money on lenses, because you wouldn't need to buy lenses with the image stabilization feature. In practice, Sony lenses aren't any less expensive than Nikon and Canon lenses. For example, the image-stabilized Sigma 70-200 f/2.8 is the same price for all three systems, about $1,250. Sony's 70-200 f/2.8 costs about the same as Canon's image-stabilized lens ($2,000), despite being significantly less sharp.

Most Pentax, Panasonic, Sony and Olympus mirrorless cameras have IBIS, too.

Anti-Aliasing filter/Optical Low-Pass Filter/Blur Filter

Most camera sensors have an anti-aliasing (AA) filter, also known as an optical low-pass filter or blur filter. Basically, this filter blurs the image just a bit before your sensor captures it, helping to reduce aliasing and moiré.

Aliasing and moiré are odd-looking artifacts in images, and they're definitely something you want less of in your photos. However, most photos wouldn't have any of those artifacts, anyway. You see aliasing and moiré primarily in tight patterns, such as brick walls and checkered or striped clothing.

Most of the time you don't need the AA filter, but it's always in front of your sensor, very slightly reducing sharpness. A few cameras, including the Nikon D800, provide the option of removing the AA filter by choosing an alternate model (the D800E). A few other cameras, including the Nikon D5500, D7200, and D810, simply do not include an AA filter.

Recent Pentax cameras don't use an AA filter, but they do offer to shake the sensor a bit to simulate the filter.

Most photographers should never bother thinking about whether or not they want an AA filter. Simply choose the camera that's right for you based on more important features and use the sensor as-is. Technical photographers who are looking for the ultimate in sharpness can seek out one of the few bodies without an AA filter, such as the Canon 5DS-R, Nikon Z7, Nikon D850, or Sony a7R IV.

Image Processing

Newer DSLRs have very powerful processors capable of performing really complex photo manipulation in-camera. This has allowed camera manufacturers to add HDR and panorama features to cameras.

These features are convenient for the casual user, but they're not for enthusiasts because they give you very little control over how the processing is done, and often they simply do a terrible job of processing the pictures. Therefore, I would never recommend one model over another because of the presence of HDR, panoramas, or other image-processing special effects. Instead, I would recommend the photographer do the processing on his or her computer.

Video

Starting with the Nikon D90 and the Canon 5D Mark II, video capabilities in interchangeable lens stills cameras have been an important factor to many users. In fact, many commercial TV shows and movies use such cameras, rather than traditional video cameras, for some tasks because of the relatively low cost, their great low-light capabilities, and the shallow depth-of-field they can achieve. I shot all of the videos for this book using stills-centric cameras.

Some cameras are better than others for video. If video is an important part of your buying decision, and you want to make professional-level videos, here are the features to look for:

- **Video quality**. Most new cameras can record 1080/60p video, which provides incredible sharpness and a smooth frame rate. However, there are subtle differences in the quality of the video that professional videographers will notice, such as chromatic aberration (odd colors at the edges of objects), tearing (a strange artifact that occurs when you pan a DSLR), moiré (bizarre effects in tight patterns and grids), and aliasing (jagged edges that should be smooth).

- **Frames per second**. Most video is filmed at 30 frames per second (fps), and 30 fps is standard for all video playback. Films are typically shot at 24 fps. However, cameras that support higher fps allow you to use video editing software to create slow motion video without dropping the frame rate. If a camera supports 60 fps, you could play it back at half speed using the standard 30 fps rate. Some cameras support very fast frame rates, such as the Canon EOS R5, which can record 1080/240 video.

- **Maximum video length**. Due to technical limitations, such as overheating, some cameras will only record for a limited amount of time, such as 12 minutes. After that time, you have to manually restart recording. This is important to anyone who plans to record a long event, especially if you plan to leave the camera on a tripod.

- **Low light recording**. One of the advantages of using a stills camera for video is that their sensors are better at recording in low light than typical video cameras. Some stills cameras are better in low light than others, and every camera's video gets very noisy in dark environments.

- **Autofocus while recording**. Currently, Canon and Sony cameras have the best autofocus while recording. It smoothly changes focus when you touch the LCD, and it looks great when being played back. No other DSLR is particularly good at autofocusing while recording; the focusing will be slow, jerky, and disturbing to watch. You're better off stopping recording, refocusing, and then restarting. Serious videographers buy expensive focus pulling equipment and hire a second person just to control the cameras focus. Nonetheless, for casual shooting, you might want to choose a camera that supports autofocusing while recording.

- **Articulating display**. An articulating display is very useful for video because it allows you to hold the camera high or low while still watching the display. If a camera does not have an articulating display, you can probably attach an external monitor, as described in the "HDMI out" bullet point below.

- **Audio input**. DSLRs have terrible microphones. For all but the most casual video, you will want to use an external mic. While it's possible to record your audio to an external device, such as a Zoom H4n, it's much more convenient to record the audio from an external mic directly to your camera. Look for a camera with a mic jack. If stereo audio is important, verify that the camera supports stereo input; some only support mono input.

- **Manual audio levels**. If a camera does have a mic jack, make sure you can manually control the audio levels. Some cameras automatically adjust the mic levels, which can lead to unpredictable and difficult-to-edit sound.

- **Headphone jack.** If sound is important to you, your cameraman should be monitoring the audio using headphones to ensure the mics are working properly. Most cameras do not offer a headphone jack, but some do.

- **HDMI out.** The display on the back of your camera is too small for serious videographers. Additionally, it might not be easily visible when holding the camera high or low. Therefore, many videographers attach a larger, external LCD camera monitor to the camera.

- **Uncompressed/clean HDMI output**. Modern DSLRs cannot record uncompressed or raw video. Instead, the camera processes the video before saving it to the memory card. Serious videographers working in studio environments often want to record uncompressed video to an external computer, both for image quality and to make the workflow more convenient; if you record video directly to a computer, a technician can edit the video as it records and doesn't have to copy video files from the memory card later.

- **Availability of video-friendly lenses.** Video-friendly lenses tend to support image stabilization, and smooth, silent focusing and aperture adjustments. Of particular note are the Canon STM lenses, which are great for video, but your options are currently very limited: the only DSLR zoom lens is the EF-S 18-135mm f/3.5-5.6, which only works with Canon APS-C DSLRs and is only suited for amateur video.

Chapter 4: Lens Features

> If you're buying your first camera and you're buying a kit with a lens, skip this section and come back when you've run into the limits of the current lens. Your kit lens is perfect while you're still getting comfortable using your camera.

A lens shapes incoming light and focuses it on your camera's sensor. The lens is the single biggest factor in image quality—even more important than the sensor. The lens helps determine how close you can zoom to your subject, how sharp and contrasting your pictures are, how fast your camera focuses, how nicely the background is blurred, and when you need to use flash.

Lens Mounts and Sensor Size

Make sure you buy lenses that match your lens mount and, usually, your sensor size. A common mistake is that buyers choose a lens with the right brand (such as Canon or Sony), but the buyer doesn't realize the camera manufacturer makes multiple different incompatible lens mounts. Canon, Sony, Nikon, and Pentax all make cameras with compatible lens mounts but different sensor sizes, and choosing a lens designed for a different sensor size will usually work, but produces less-than-optimal results.

For example, Canon currently offers four lens mounts: EF (full-frame DSLR), EF-S (APS-C DSLR), M (APS-C mirrorless), and RF (full-frame mirrorless). You can attach EF lenses to EF or EF-S bodies, but if you attach an EF lens to an EF-S body, there's a 1.6X crop factor and lower image quality (discussed later in this book). You cannot attach EF-S lenses to an EF body, unless it's a third-party lens such as the Sigma 18-35 f/1.8, in which case you produce pictures with dark corners. You can attach EF or EF-S lenses to an M body using an adapter, but you cannot attach M lenses to any other lens mount. You can attach EF (but not EF-S) lenses to an RF body using an adapter, but RF lenses cannot be attached to other lens mounts.

It's incredibly confusing and many buyers choose incompatible or less-than ideal lenses. For example, consider this one-star Amazon review for a Canon RF lens that the buyer purchased to use on a Canon EF-S body. My first instinct was that the review was unfair, and perhaps it is, but the understandable confusion around Canon lens and camera compatibility led to both a frustrated camera buyer and a 1-star review.

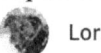 Lori

☆☆☆☆☆ **Is not compatible with Canon EOS 80D or 90D**

Reviewed in the United States on July 31, 2020

Verified Purchase

I was so excited to get a 600ml lens at this price. I thought any Canon lens would fit any Canon camera. I was sadly mistaken. It also is not compatible with the Canon Extender EF 1.4x.So my low rating is only due to that, as I do not know how it would perform, hopefully someone else has better luck than I did.

The sections that follow will discuss compatibility in more detail.

You can buy third-party lenses from Sigma and Tamron, but just be sure they're made for your camera type. The description for third-party lenses will say something like, "for Canon SLRs" or "for Nikon SLRs."

Canon Lens Varieties

Originally introduced in 1987, the Canon EOS name identifies all their cameras and lenses. However, that doesn't mean all EOS cameras and lenses are compatible.

Within the EOS family, Canon has three lens mounts and four lens varieties:

- **EF**. Lenses designed for full-frame Canon DSLRs, such as the 5D, 6D, and 1D families. Like the full-frame camera bodies, EF lenses are designed for more professional purposes. Canon uses the "L" designation for professional lenses, and all L lenses are also EF lenses.

- **EF-S**. Lenses designed for compact APS-C Canon DSLRs, which are everything except the 5D, 6D, and 1D families. EF-S cameras include the T5, 7D, and 90D.

- **M or EF-M.** Lenses designed for compact APS-C Canon mirrorless cameras, such as the M50 and M6.

- **R or RF.** Lenses designed for full frame Canon mirrorless cameras, such as the R, RP, R5, and R6.

The following tables show camera body and lens compatibility for the most popular camera systems.

Canon		Lenses			
		EF-S	EF	M	RF
Cameras	EF-S	✓	1.6X Crop	X	X
	EF	X	✓	X	X
	M	With adapter	With adapter 1.6X crop*	✓	X
	RF	X	With adapter	X	✓

* You can use the "Metabones T Speed Booster Ultra 0.71x Adapter for Canon Full-Frame EF-Mount Lens to Canon EF-M Mount Camera" ($500) to reduce the crop factor to 1.1X.

If you have a full-frame Canon DSLR, such as a 5D, 6D, or 1D, you can only use EF lenses. You cannot use EF-S lenses, M, or R lenses.

If you have a Canon compact DSLR, such as a T2, T3, T4, 7D, Rebel, Kiss, or anything with more than two numbers in the name (like 80D or 600D), you *can* use either EF or EF-S lenses. However, you *should* choose Canon EF-S lenses whenever possible. Canon EF-S lenses are optimized for the compact sensor.

However, if you have a Canon compact DSLR, you can use EF lenses. However, it's rarely the right choice. Because EF lenses are designed for use on a full-frame sensor, your compact DSLR sensor only records the middle of the image. Basically, using an EF lens on a compact DSLR is like zooming in 1.6 times (for Canon) or 1.5 times (for other manufacturers). It also means you're spending more money and carrying around a heavier lens than you can take advantage of.

For those reasons, I don't recommend using EF lenses with Canon compact DSLRs to casual users. More serious compact DSLR users will probably need to buy full-frame EF lenses for portraits, sports, and wildlife, however, and the $100 Canon 50mm f/1.8 lens, also known as the "fantastic plastic" or "nifty fifty" is a great choice for any camera.

Lens descriptions are filled with numbers and acronyms that are baffling to all but the most hardened photographers. Here's a typical Canon APS-C lens description diagrammed:

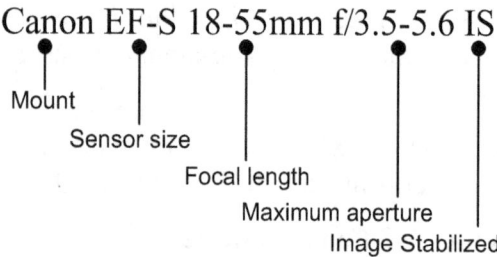

Canon EF-S 18-55mm f/3.5-5.6 IS

Mount
Sensor size
Focal length
Maximum aperture
Image Stabilized

For this lens, every part of the description is important:

- "Canon" means it's designed for Canon SLRs. You can't use it on a Nikon SLR.

- "EF-S" indicates a lens designed for Canon APS-C DSLRs with a smaller sensor size. You cannot use EF-S lenses on full-frame Canon cameras.

- "18-55mm" is the effective focal length. Because EF-S lenses have a 1.6X crop factor, the zoom range is equivalent to 29mm-88mm on a full-frame camera.

- "f/3.5-5.6" is the maximum aperture. Because the description lists a range of apertures, the lens is a variable aperture lens that changes the maximum aperture throughout the zoom range. Therefore, the maximum aperture is f/3.5 when zoomed out to 18mm, and f/5.6 when zoomed in to 55mm.

- "IS" is an acronym that means Image Stabilization, which reduces camera shake.

A typical Canon full-frame professional lens adds a couple of elements to the description:

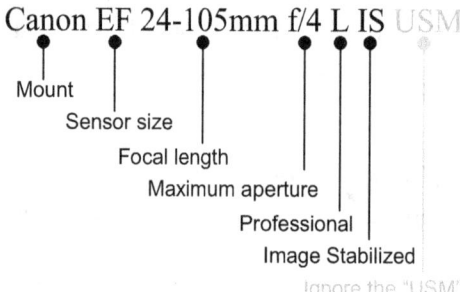

Canon EF 24-105mm f/4 L IS USM

Mount
Sensor size
Focal length
Maximum aperture
Professional
Image Stabilized
Ignore the "USM"

Because the sensor size is "EF," rather than "EF-S," the lens is designed for full-frame cameras. If you use it on an APS-C camera, multiply the focal length by 1.6X to create an effective focal length of 38-168mm. The "L" means it's an expensive professional lens.

The "USM" can be ignored; it's just an acronym for Ultrasonic Motor—Canon's quiet and fast-focusing motors from the late 1980s.

Nikon Lens Varieties

Nikon's current lens system, the F-mount, was introduced in 1959. Because it was introduced so much earlier than Canon's, the original design relied on mechanical linkages between the body and lens to support focusing the lens and changing the lens aperture. These 50-year-old design elements are still present on many Nikon lenses.

Considering how much has changed about photography in the last 50 years, it's fairly remarkable that you can use many of the 400 F-mount lens from the last 50 years on your modern DSLRs (though you probably won't want to). With that said, there are countless compatibility issues between old lens and new bodies, and between new lenses and old bodies, so you should refer to the manuals included with your newer equipment before trying anything.

Like Canon's, Nikon's F-Mount system has separate lens varieties for compact and full-frame cameras:

- **FX**. These lenses are designed for full-frame Nikon DSLRs, such as the D610, D810, Df, and D5. Like the full-frame camera bodies, FX lenses are designed for more professional purposes.

- **DX**. These lenses are designed for compact Nikon DSLRs, which are the D3x00, D5x00, D7x00 and D500 families.

Either type of lens can be used on any Nikon body. However, if you have a full-frame Nikon, you will definitely want to choose FX lenses. If you use a DX lens on a full-frame camera, you will only be using a small part of the camera's sensor, losing a great deal of detail.

If you have a compact Nikon DSLR, you can use either type of lens without serious penalty, however, you should choose DX lenses whenever possible. FX lenses will be unnecessarily large, heavy, and expensive because their design is optimized for a full-frame sensor, and your compact sensor will only capture the center of the image from the lens. Using an FX lens on a DX camera is like zooming in 1.5 times, so a wide-angle lens becomes a standard lens, and a standard lens becomes a telephoto lens.

For those reasons, I don't recommend using FX lenses with Nikon compact DSLRs to casual users. More serious compact DSLR users will probably need to buy full-frame FX lenses for portraits, sports, and wildlife, however, and the $100 Nikon 50mm f/1.8 lens, also known as the "fantastic plastic" or "nifty fifty" is a great choice for any camera.

In addition to understanding DX and FX lenses, you also need to understand AF and AF-S lenses:

- **AF-S**. Most new Nikon lenses are the AF-S variety, which includes a focusing motor in the lens. AF-S lenses autofocus with all modern Nikon DSLR cameras.

- **AF**. Many older Nikon autofocus lenses don't have an autofocus motor. Instead, they rely on an autofocus motor built into the camera body that links to the lens when the lens is mounted on the camera.

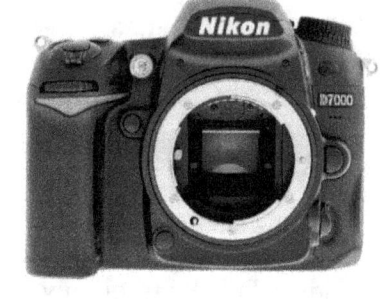

The photo of a Nikon D7000 shows the autofocus motor linkage in the lower-left corner of the lens mount.

Nikon also offers the mirrorless Z mount. Z-mount has many mirrorless lenses designed specifically for it, and those Z lenses cannot be used on their DSLRs. However. The Z-mount can use F-mount DSLR lenses when you add the FTZ adapter. The FTZ adapter does not have a focusing motor built into it, so older AF lenses will be manual focus only.

This table shows the compatibility between these mounts, and adds the now-dead Nikon 1 mirrorless mount.

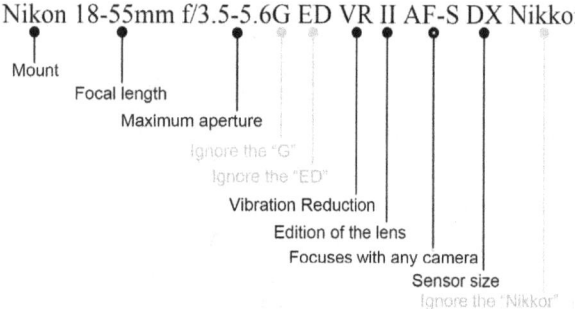

		Lenses			
		F (DX)	F (FX)	Z	1
Cameras	F (DX)	✓	1.6X crop	X	X
	F (FX)	With 55% megapixels	✓	X	X
	Z	With adapter	With 55% megapixels	✓	X
	1	1.8X crop	2.7X crop	X	✓

Here's a typical Nikon lens description diagrammed. Nikon lenses tend to have much more complex descriptions than Canon lenses because Nikon makes more of an effort to maintain lens compatibility with older film cameras:

Nikon 18-55mm f/3.5-5.6G ED VR II AF-S DX Nikkor

- Mount
- Focal length
- Maximum aperture
- Ignore the "G"
- Ignore the "ED"
- Vibration Reduction
- Edition of the lens
- Focuses with any camera
- Sensor size
- Ignore the "Nikkor"

The useful information is:

- "Nikon" and "Nikkor" mean that it's designed for Nikon SLRs.

- "18-55mm" is the effective focal length. Because DX lenses have a 1.5X crop factor, the zoom range is equivalent to 29mm-88mm on a full-frame camera.

- "f/3.5-5.6" is the maximum aperture. Because the description lists a range of apertures, the lens is a variable aperture lens that changes the maximum aperture throughout the zoom range. Therefore, the maximum aperture is f/3.5 when zoomed out to 18mm, and f/5.6 when zoomed in to 55mm.

- "DX" indicates a lens designed for Nikon DX DSLRs with a smaller sensor size. You can use DX lenses on full-frame FX Nikon cameras, but you'll get a smaller picture.

- "II" means that this is the second edition of the lens.

- "AF-S" means the lens has a focusing motor built in and can autofocus on camera bodies that don't have a built-in focusing motor, including the D40, D3000, D3100, D3200, D5000, D5100, and D5200. If you have one of those bodies, be sure you buy AF-S lenses. You can use non-AF-S lenses, but you will need to manually focus.

- "VR" is an acronym that means Vibration Reduction, Nikon's version of Image Stabilization, which reduces camera shake.

Notice that I've asked you to ignore several parts of the description:

- "G" after the aperture indicates that the lens isn't completely compatible with some very old film cameras that lack aperture control on the camera body.

- "ED" means that extra-low dispersion elements are used in the lens—Nikon's now outdated attempt to indicate lens quality.

- "Nikkor" is simply Nikon's lens branding; Nikkor just means it's a Nikon lens, which we already knew.

You might see other acronyms in Nikon lens descriptions, all of which you can ignore:

- "D" after the aperture dates back to 1992, and it indicates compatibility with very outdated camera features.

- "IF" or "IF-ED" indicates that the lens uses internal focusing, which means the focus ring won't spin when you auto-focus.

- "S" indicates Nikon's high-end Z-mount mirrorless lens lineup.

Sony E-Mount Lens Varieties

Sony launched the Alpha E-type bayonet mount in 2010, making it extremely young for a lens mount. It's a mirrorless mount and supports both APS-C and full-frame lenses. It's backwards-compatible with the Sony A mount, though that lineup seems to have been discontinued, and Sony E-mount users don't need to adapt lenses as frequently as Canon and Nikon mirrorless users.

Just as you can adapt Canon EF lenses to the mirrorless Canon RF platform, you can adapt Canon EF lenses to Sony E-mount. Several companies offer adapters, however, our testing has shown that the Sigma MC-11 adapter performs best.

SONY

		Lenses			
		A	E (APS-C)	FE	Canon EF
Cameras	A	✓	X	X	X
	E (APS-C)	With adapter	✓	1.6X crop	With MC-11 adapter
	FE	With adapter	With 55% megapixels	✓	With MC-11 adapter
	Canon EF	X	X	X	✓

Sony uses model numbers to identify their lenses, such as "SEL1635GM", "SEL1635Z", "SEL100F28GM", or "SELP1650":

- SEL is an abbreviation for Sony Alpha E-mount Lens (SAL refers to A-Mount).

- P, if present in the model number, indicates that the lens is a power zoom. Power zooms are useful for smooth zooming during video.

- Next, they simply list the focal lengths. A 10-18mm lens is SEL1018. It's convenient that Sony does this, making it easier to casually identify different lenses without needing to write out the full (and very long) name.

- Sometimes, they list the f/stop, such as "F28" in "SEL100F28GM" or "F18" in "SEL85F18". They do this to distinguish multiple lenses with the same focal length.

- For macro lenses, they will list the f/sotp with an M, such as "M28" in "SEL90M28G".

- Depending on the model, Sony might add a letter or two to the end:
  - **No letter:** Entry-level lenses
  - **LE:** Mid-range super-zoom lenses (like SEL18200LE)
  - **G:** Mid-range lenses (like SEL1224G)
  - **Z:** Zeiss-branded mid-range lenses (like SEL1635Z)
  - **GM:** G-Master high-end lenses (like SEL100400GM)

The lens product names list additional features:

- **E or FE**: Indicates whether the lens is designed for APS-C (E) or full frame (FE). Technically, you can use any E-mount lens on any E-mount body, but you'll get much sharper results when you use E lenses on APS-C bodies (such as the Sony a6x00 cameras) and FE lenses on the full-frame bodies (such as the Sony a7 and a9 cameras).

- **OSS:** A stabilized lens that reduces camera shake. You don't necessarily need lens stabilization on bodies like the a6500, a7R II, or a9, because they have sensor stabilization.

- **PZ**: Power zoom, which is useful for smooth zooming while recording video.

- **ZA**: Indicates a Zeiss-branded lens, which doesn't necessarily convey sharper results. Zeiss lenses often have Zeiss-specific technology brands, such as Planar T*, Sonnar T*, Distagon T*, and Vario-Tessar T*. I haven't noticed that these terms significantly impact the final results.

- **STF**: The lens has a special apodization element that makes particularly smooth bokeh. Currently they only offer the FE 100mm f2.8 STF GM OSS.

- **GM**: G-Master lenses are Sony's top-end lenses. We've found them to be their sharpest offerings.

- **LE:** Mid-range super-zoom lenses (like SEL18200LE)

Sony A-Mount Lens Varieties

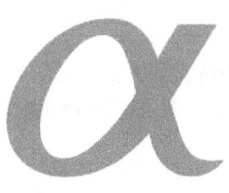

The Sony Alpha's A-type bayonet mount dates back to 1985, when it was introduced by Minolta. In 2006, Sony bought Minolta and designed their new camera bodies to work with the Minolta A-mount system. As a result, new Sony DSLRs work with many Minolta lenses from the last 30 years, and vice-versa. Compatibility isn't always 100%, so check your camera's documentation and search the Internet to determine whether any given combination will give you the features you need.

Like Canon and Nikon, Sony makes lenses optimized for cameras with compact sensors as well as full-frame lenses. Those lenses designed for cameras with compact sensors use the DT designation, which stands for Digital Technology. Like Nikon (and unlike Canon), you can you DT lenses on full-frame bodies, but you should avoid it whenever possible.

In the current Sony SLT lineup, the a99 and a99 II are the only full-frame cameras. Therefore, if you have any other Sony body (the a58, a65, or a77) you should buy DT lenses whenever possible. If you have an a99, be sure to avoid DT lenses.

Here's a typical Sony lens description diagrammed:

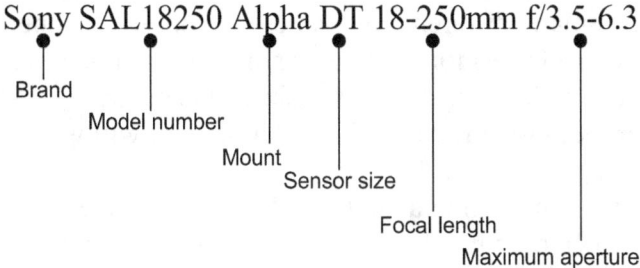

The useful information is:

- "SAL18250" is the model number. SAL is an abbreviation for Sony Alpha A-mount Lens (SEL refers to E-Mount). Next, they simply list the focal lengths. The high-end Sony lenses (G and Zeiss Alpha) have a model number that ends in G or ZA.

- "DT" indicates a lens designed for Sony Alpha DSLRs with a smaller sensor size. You can use DT lenses on full-frame Sony cameras, but you'll get a smaller picture cropped from the center of the frame.

- "18-250mm" is the effective focal length. Because DT lenses have a 1.5X crop factor, the zoom range is equivalent to 27mm-375mm on a full-frame camera.

- "f/3.5-6.3" is the maximum aperture. Because the description lists a range of apertures, the lens is a variable aperture lens that changes the maximum aperture throughout the zoom range. Therefore, the maximum aperture is f/3.5 when zoomed out to 18mm, and f/6.3 when zoomed in to 250mm.

You might see other acronyms in Sony lens descriptions:

- "G" refers to high-end lenses, much like Canon's "L" series.

- "ZA" refer to "Zeiss Alpha," a product of Sony's partnership with Carl Zeiss, a maker of prestigious lenses. ZA lenses are designed and built by Sony, but Zeiss approves the designs to verify that the lenses meet Zeiss' quality standards. These high-end lenses are priced from $1,000 to $2,000.

- "SAM" (Smooth Autofocus Motor) and "SSM" (SuperSonic Motor) refer to specific types of focusing motor. SSM is better than SAM, but don't let that factor alone change your purchasing decision.

- "ED" refers to Extra-low Dispersion. Ignore this.

- "E," "E-Mount," or "NEX" refers to lenses designed for Sony's mirrorless cameras, rather than their DSLRs. You can't use these lenses on an Alpha DSLR, but you can use the Alpha lenses on the NEX cameras with an LA-EA1, LA-EA2, or LA-EA3 adapter.

- "OSS" (Optical SteadyShot) refers to lenses that have image stabilization built in.

Micro four-Thirds Lens Varieties

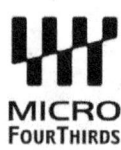 Unlike the other major brands discussed here, Micro Four-Thirds isn't a company—it's a standard that many different companies (including Olympus and Panasonic) create bodies and lenses for. Unlike Canon, Sony, and Nikon, the Micro Four-Thirds brand only has a single sensor size. This makes choosing lenses much easier, because any Micro Four-Thirds lens will work without you having to worry about cropping.

While Panasonic lenses will work on an Olympus camera, and vice-versa, choose lenses and bodies from the same manufacturer when possible. Each manufacturer only seems to optimize their cameras for their own lenses, so some Olympus cameras with sensor stabilization will work together with lens stabilization—but only with specific Olympus lenses. If you use a Panasonic lens on an Olympus body, either the sensor or the lens stabilization must be disabled. The same applies for using stabilized Olympus lenses on stabilized Panasonic bodies.

Additionally, Olympus' heavily marketed weather-sealing might be compromised with Panasonic lenses.

Aperture

Aperture, measured in f/stops, is the most important quality of a lens. Lenses with lower f/stops are heavier and cost more, but they focus faster, blur the background better, and let you handhold the camera in less light. To understand the cost difference, compare Canon's three commonly used 50mm lenses. Each lens is one f/stop faster than the previous, passing twice as much light to the sensor:

- Canon 50mm f/1.8: $100

- Canon 50mm f/1.4: $350

- Canon 50mm f/1.2: $1,500

As you can see, doubling the light roughly quadruples the cost. Size and weight also increase, especially with telephoto lenses. These three lenses are each one f/stop faster than the one previous:

- Canon 400mm f/5.6: $1200, 2.8 lbs.

- Canon 400mm f/4: $5,800, 4.3 lbs.

- Canon 400mm f/2.8: $7,200, 11.8 lbs.

For professionals, the extra cost and weight is worth it. If you're an amateur, I'd recommend starting with an inexpensive lens at the focal length you need and upgrading only when you're frustrated with the maximum aperture—you can usually sell lenses for close to their original cost, so the risk is minimal.

For information about how to use aperture creatively, read Chapter 4 in *Stunning Digital Photography*.

Variable Apertures

Most consumer zoom lenses have a *variable aperture*, which means the maximum aperture when zoomed in is smaller than the maximum aperture when zoomed out. You can recognize variable aperture zoom lenses because their name has two f/stop numbers listed, such as "f/4.0-5.6."

For example, the Canon PowerShot SD950 P&S camera has a zoom lens with a focal length of 7.7-28.5mm and a maximum aperture of f/2.8-f/5.8. That means at its widest angle (7.7mm) the maximum aperture is a respectable f/2.8. However, when zoomed in to 28.5mm, the maximum aperture is f/5.8—requiring more than *four times* more light than f/2.8.

The smaller aperture when zoomed in means your camera will have a harder time focusing and shutter speeds will be much slower. Handholding telephoto lenses requires faster shutter speeds, meaning many of your telephoto pictures will be shaky.

So, should you always avoid lenses with variable apertures? Not necessarily, but you should understand the limitations. Variable aperture lenses are much less expensive to make than constant aperture lenses. For example, the variable aperture Canon 28-135mm f/3.5-f/5.6 lens above is about $430, but the constant aperture Canon 24-105mm f/4.0 lens costs more than twice that.

Focal Length

Telephoto lenses have a narrow field of view, allowing light from a very small area in front of you to the sensor and blocking all other light. Wide-angle lenses focus light from a much broader area.

Wide-angle lenses have short focal lengths, and telephoto lenses have longer focal lengths. When photographers talk about focal lengths, they always measure them in millimeters (mm), and they usually discuss them in full-frame 35 mm equivalents—even if the sensor isn't 35mm-sized. If you have a compact camera, divide the 35mm focal length by 1.6 (for Canon) or 1.5 (for Nikon and other manufacturers).

The table below shows common full-frame 35mm focal lengths, their equivalents on APS-C cameras, and how you might use them. Most lenses are *zoom lenses*, which cover a range of focal lengths. If a lens doesn't zoom, it's called a *prime* or *fixed focal-length* lens.

Full-Frame Focal Length	APS-C Focal Length	Typical Usage
8mm	5mm	Fish-eye views that distort the world around you.
16mm	11mm	Super wide-angle views for photographing nearby large objects, such as buildings in narrow streets.
24mm	16mm	A wide-angle view good for photographing groups of people indoors.
35mm	23mm	A moderately wide-angle view good for landscape photography or photographing a single person indoors.
50mm	33mm	Called the "normal" lens, the field of view is roughly equivalent to how the human eye sees.
85mm	56mm	A good focal length for photographing individuals outdoors, where you might stand farther from the person.
120mm	80mm	A moderate telephoto view good for portrait work, photographing children and pets at play, and photojournalism.
200mm	133mm	A telephoto view for headshots and close-range sports such as basketball.
400mm	266mm	A super telephoto view good for larger animals, such as deer or bear. 400mm is perfect for zoos and sports with larger fields, such as football.
800mm+	586mm	An extreme telephoto view used for birding, long-range sports, and spying on celebrities.

The descriptions in the table give a general idea of common uses for different focal lengths, but you can always move closer to or farther from a subject, allowing a wide-angle lens to be used for wildlife, or a telephoto lens to be used for landscape work. There's nothing to stop you from taking a picture of a bear with a 50mm lens, except, perhaps, the bear—you'd have to be about four feet away to fill the frame.

Tip: Big telephoto lenses are expensive. If you see a cheap telephoto lens with a focal length of more than 400mm (250mm on a compact camera), it's probably a mirror lens. Mirror lenses are cheap for a reason: quality is low, there's no autofocus, and light blurred in the background (known as *bokeh*) takes on some really weird shapes.

To give you a sense for different focal lengths, the following sequence of pictures covers a range of 17mm to 400mm with a full-frame camera (11mm to 266mm on an APS-C camera).

17mm FF/11mm APS-C/8mm MFT

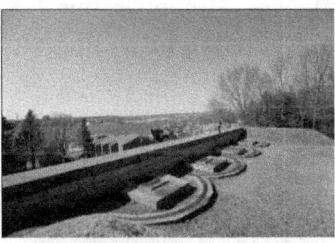

24mm FF/16mm APS-C/12mm MFT

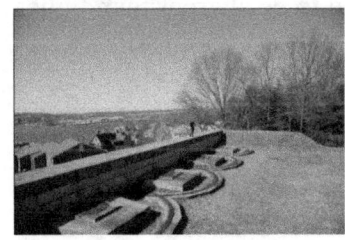

50mm FF/33mm APS-C/25mm MFT

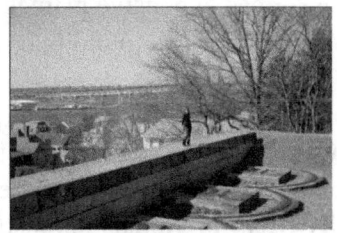

100mm FF/67mm APS-C/50mm MFT

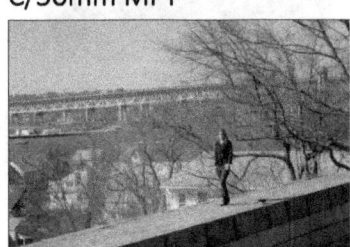

200mm FF/133mm APS-C/100mm MFT

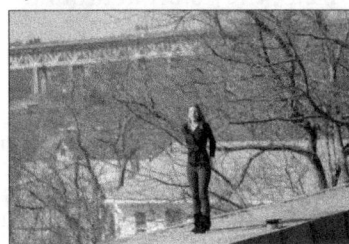

400mm FF/267mm APS-C/200mm MFT

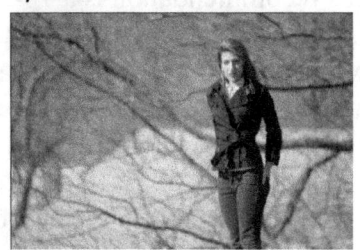

Zooms vs. Primes

Zoom lenses can change their focal length. *Zooming out* feels like moving away from a subject, showing you a more wide-angle perspective and allowing you to see more of the background. *Zooming in* feels like moving towards a subject, showing you a more telephoto perspective and hiding more of the background.

Prime lenses have a single, fixed focal length. If you want to make your subject larger in the frame, you need to either change lenses or move closer to your subject.

The reality is, zoom lenses give you more options. If you use primes and you're doing travel photography, you might pack 14mm, 35mm, and 50mm lenses. What happens when the best framing for a shot requires 24mm? Zooms allow you to pick the perfect focal length.

Also, zooming to change focal length allows you to control the background, something you can't do when you move closer to or farther from your subject. For more information about how you can use zoom lenses to control the background in your pictures, refer to Chapter 2 in *Stunning Digital Photography*.

Zooming also changes the proportions of your subject. This is critical in portraiture, where using a wide-angle lens gives your subject cartoonishly large features, while using a telephoto lens gives your subject smaller, more flattering features. For detailed information, refer to Chapter 6 in *Stunning Digital Photography*.

Of course, you can change your focal length using prime lenses—but you have to change the entire lens. That also means that you need to carry multiple lenses with you. Prime lenses do have several key advantages, though:

- They are smaller and lighter than zooms.

- They tend to be "faster," which means they have a smaller minimum f/stop number, providing faster focusing, faster shutter speeds, and nicer background blur.

- Less expensive primes tend to be sharper than typical kit lenses. However, our testing has found that professional zooms (like a 24-70 f/2.8) tend to be as sharp as a prime lens.

I think everyone's general-purpose lens should be a zoom. However, all camera manufacturers make an excellent and inexpensive 50mm f/1.8 prime lens, often known as the "nifty fifty" or "fantastic plastic." That's the lens I'd grab if I needed to take pictures in a dimly lit environment, such as a bar, without a flash. While my professional portrait lens recommendation is a 70-200 f/2.8, they start at around $760. You can achieve a similar effect using the much less expensive 85mm f/1.8, which starts at less than half the price of the zoom.

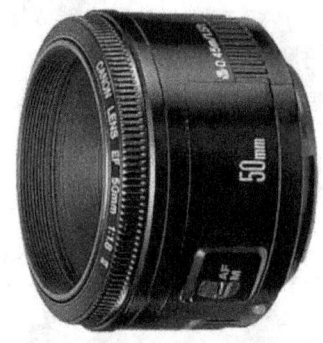

Ultimately, zoom lenses are superior for casual and candid amateur pictures and fast-paced professional work (such as weddings). Prime lenses are better for carefully planned photos. Prime lenses are also the best choice for low-light and action photos, such as concerts, wildlife, and some sports, because the prime lenses tend to let in far more light.

Different Types of Zooming

Zoom lenses change focal length in three different ways:

- **Circular zoom**. Turn a ring on the lens clockwise or counter-clockwise to mechanically adjust the lens' optical elements. This is the most common method of zooming, and in my opinion, it's by far the best.

- **Push-pull zoom**. Some telephoto lenses, such as the Canon 100-400mm L IS, are push-pull zooms. Instead of turning a ring to zoom, you push the lens in and out. This allows you to zoom and manually focus at the same time. It takes a little getting used to, but it works fine.

- **Power zoom**. With power zooming (also known as motorized zooming), the zooming is controlled by a motor in the lens. Power zooming is standard on P&S cameras and a common feature on mirrorless lenses, but it's uncommon on DSLRs. Power zooming is slower than other types of zooming, which makes it less than ideal for photography. However, the smooth motion of power zooming makes it an excellent choice for video. Power zooming can also allow for one-handed operation.

Focusing Speed

While the camera body contains the sensors and logic to focus an image, the lens impacts the focus speed in two ways:

- For most lenses, the lens contains the electric motor that turns the lens elements. The faster the motor is, the faster the camera can focus. Some motors are louder than others, too, and a quiet autofocus system is nice.

- Lenses with small maximum f/stops, such as f/2.8, allow more light to reach the camera's autofocus sensors. Therefore, an f/2.8 lens will focus much faster than an f/5.6 lens, especially in low-light.

In our tests with dozens of lenses in sports scenarios, we never found a difference in focusing speed between different lenses of the same focal length and f/stop. Switching between bodies—such as moving from a D5600 to a D500—makes huge difference. Switching between lenses? Notsomuch.

One exception to that is with some older lenses, and third party lenses, like the YongNou 50mm f/1.8 for Canon. While Tamron and Sigma seem to have focusing speeds comparable to Canon and Nikon, other third-party manufacturers haven't yet mastered it. Other than focus speed, consider whether a lens offers full-time manual focusing. Higher-end lenses allow you to adjust the focus even when the lens is set to auto-focus, which can be convenient for fine-tuning. Less expensive lenses require you to flip a switch to manually focus the lens.

Minimum Focusing Distance

All lenses have a minimum focusing distance. If you're closer than the minimum focusing distance, the camera will hunt for focus for several seconds, and then give up. All you can do is to move farther from the subject and re-focus... or add an extension tube, which will allow you to focus closer.

To get great close-up pictures, however, you'll need a true macro lens. Unfortunately, because true macro lenses are always prime, they aren't good general-purpose lenses. My advice: wait until you get frustrated with your primary zoom because you can't focus close enough, and then add a 100mm or 150mm macro lens to your collection.

Sharpness

Some lenses are razor sharp, while other lenses always produce pictures that are a bit blurry—even when the subject is in focus. As a general rule, the more expensive, heavier, and larger the lens, the sharper it will be. To get a more objective comparison of the sharpness between two lenses you're considering, just search the Internet—the photography community examines the sharpness of all new lenses.

> **Tip:** Prime lenses are usually sharper (and lighter, and cheaper) than zoom lenses. However, many professional zoom lenses are as sharp as the best primes. Either way, I prefer the flexibility of zoom lenses.

Please don't obsess about lens sharpness, though. Most lenses are *just fine*, and your technique is much more important than the lens' optical quality. Read Chapters 4 and 5 of *Stunning Digital Photography* to improve the sharpness of your images without buying a more expensive lens.

Contrast

Consider these two unedited pictures of Chelsea in natural lighting, backlit by a window. The first is severely washed out because the inexpensive lens diffracts the bright light across the entire picture. The second, taken with a DSLR and professional-quality lens, is high contrast and sharp. Not all pictures will show this severe of a difference; backlit pictures are particularly challenging for inexpensive cameras.

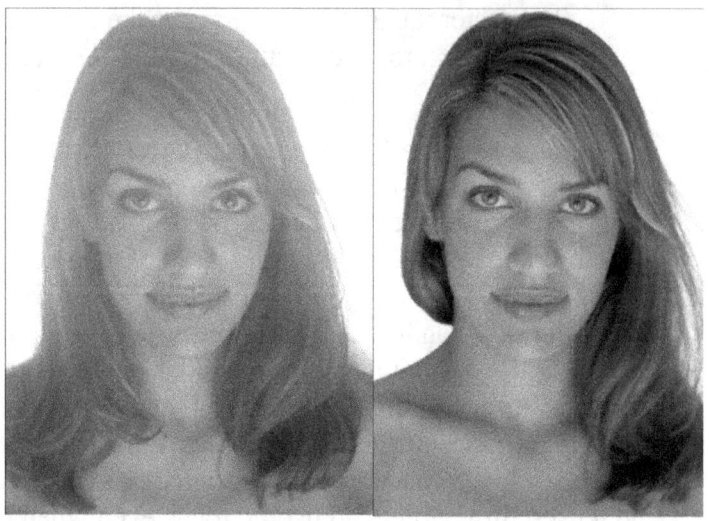

Image Stabilization

Image stabilization helps to prevent shaky shots by counteracting the movement of your hands. Basically, image stabilization lets you handhold your camera when you would otherwise need a tripod. Image stabilization gives you two or three stops more handholding capabilities, meaning you can use a shutter speed four to eight times slower without creating a shaky picture.

I can't recommend image stabilization enough. If your budget allows for it, it's the single most important feature on any lens. Image stabilization will save you countless blurry shots, allow you to use lower ISO settings (and thus reduce the noise in your pictures), and allow you to focus more on composition and lighting than camera settings.

If you're shooting moving subjects, such as animals or sports, image stabilization becomes less important because you will need to use a faster shutter speed to prevent motion blur, and that faster shutter speed will also eliminate camera shake. Image stabilization still helps, but if you plan to shoot flying birds, a 400mm or 500mm lens without image stabilization can still get the job done.

Though they all have similar effects, image stabilization is known by different names depending on the camera manufacturer:

- **Canon and Fuji**: Image Stabilization (IS)

- **Nikon**: Vibration Reduction (VR)

- **Sony**: SteadyShot

- **Sigma**: Optical Stabilization (OS)

- **Tamron**: Vibration Compensation (VC)

Stabilization can be built into the lens, the camera, or both. I always choose image-stabilized camera bodies and lenses when I have the choice—though it's not particularly important (nor generally offered) on wide-angle lenses. Image stabilization is also not required when using a high shutter speed, such as for sports photography or when photographing flying birds, though it does make it easier to look through the viewfinder.

While image stabilization generally improves the sharpness of images by reducing camera shake, the design requires manufacturers to add additional complexity and lens elements. While there are many very sharp stabilized lenses, this fundamentally decreases the sharpness of a lens while it's on a tripod. Therefore, stabilized lenses might test as being less sharp than their unstabilized counterparts. One example of this is the older, unstabilized Nikon 24-70 f/2.8 vs the newer, stabilized Nikon 24-70 f/2.8 VR. The newer, stabilized lens tests as less sharp than the older lens, however, in practice, the VR lens will produce more sharp shots by reducing camera shake.

Sensor stabilization provides a great compromise by reducing camera shake without compromising the lens design. Sensor stabilization can also cancel out rotational camera movement. When a photographer presses the shutter, it twists the camera a little bit, which decreases sharpness. Lens stabilization can't counteract that movement, but sensor stabilization can.

For more information about camera shake, refer to Chapter 4 of *Stunning Digital Photography*. For more information about the causes of blurry pictures, refer to Chapter 5 of *Stunning Digital Photography*.

Mirror lenses

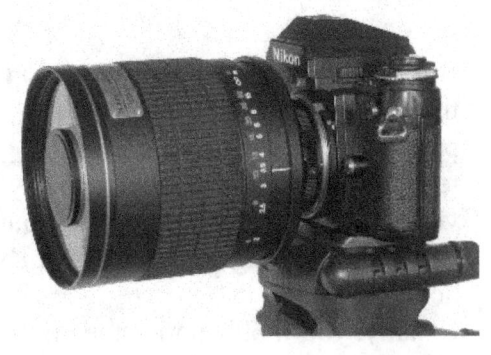

Mirror lenses, such as the Rokinon 500mm f/6.3 (shown here), Rokinon 800mm f/8, and Polaroid 500mm f/6.3, seem to offer incredible value for the money. These lenses each cost about $190 dollars, whereas the Canon 500mm lens costs over $10,000. The mirror lenses are also less than one-third the length and weight, making them easier to carry. Yet, even with a 98% discount, no professionals and very few amateurs use mirror lenses. Why not?

- **Manual focus**. Mirror lenses cannot autofocus, making them useless for moving wildlife. You can still use them for stationary subjects, but focusing can quickly become annoying.

- **High minimum f/stop number**. Mirror lenses have a high minimum f/stop number, which means they don't allow as much light in as other lenses. This requires you to either use a much longer shutter speed (if your camera is on a tripod) or a much higher ISO (which increases noise in the picture). Also keep in mind that the real f/stop number is probably higher than what the lens manufacturer advertises; some mirror lens manufacturers exaggerate this number.

- **Fixed aperture**. Almost all mirror lenses have a fixed aperture, preventing you from choosing a higher f/stop number to increase the depth-of-field.

- **Lack of image stabilization**. Most low-cost telephoto lenses don't have image stabilization. This, combined with the high minimum f/stop number, will require you to either use a tripod or use a very fast shutter speed to prevent camera shake.

- **Donut bokeh**. Out-of-focus parts of a picture taken with a mirror lens have an unusual "donut" look to them, just like the shape of the lens' front element. People make a big deal out of this, but I actually don't think it's too disturbing.

- **Low contrast**. The mirror reflections reduce contrast in the photo, so pictures are a bit washed out. You can improve this with some processing, as described in Chapter 5 of *Stunning Digital Photography*.

The need to manually focus, combined with the shallow depth-of-field of all telephoto lenses and the fast shutter speed required to prevent camera shake, means that you almost always need to use mirror lenses with a tripod. Some people do take hand-held photos of still subjects in bright daylight, but it's challenging.

In a nutshell, you get what you pay for. Usually, you'll get better overall images by cropping photos taken with an inexpensive telephoto zoom (such as a 75-300mm zoom) than by using a mirror lens. While they're inexpensive, it's still a waste if you can't get pictures that you're happy with.

If you do get a mirror lens, don't go too inexpensive. Models costing less than $150 will give you terrible image quality.

Bokeh

Video: Sigma 50mm f/1.4 Art Review
18:59 - *sdp.io/s50review*

Bokeh is the Japanese word for the appearance of out-of-focus parts of your picture. Often, the term is used simply to mean a blurry background, but that's not quite accurate, because you can have a very blurry background with either good or bad bokeh. From my experience talking on the Internet, I've learned that there is no correct pronunciation for bokeh.

Usually, you'll only notice the difference between good bokeh and bad bokeh in specular highlights; those bright lights in out-of-focus areas. The most striking example are out-of-focus city lights, but you'll often see good or bad bokeh by looking at bright leaves with sun reflecting off of them.

The average person looking at a picture will never notice bokeh. I've never had a client, personal or commercial, complain about bad bokeh. Only photographers and videographers notice bokeh. If you're taking pictures to please other photographers or videographers, this might be a consideration—otherwise, I don't recommend spending more money just to change the shape of out-of-focus specular highlights.

For an example of the difference between good and bad bokeh, watch the following comparison video of the Sigma 50mm f/1.4 Art lens (which has amazing bokeh) vs. the Canon 50mm f/1.8 (which has lousy bokeh).

Video Support

For stationary, tripod-mounted videos (like most of those I've created for this book), just about any lens will work for video. However, if you plan to use your DSLR as a family video camera, you should consider how well the lens supports video. Keep in mind these factors:

- **Quiet image stabilization**. Image stabilization is very useful for handheld video, and the image stabilization systems designed for still shots work remarkably well. However, most of them make an awful clicking sound that the on-camera mic will record. If you need image stabilization that's not recorded by the on-camera mic, consider the Canon's STM series of lenses.

- **Smooth zooming and focusing**. Most lenses don't focus or zoom all that smoothly; as you turn the dials, they're a bit jerky. That's fine for still photos, but it looks awful during video. Look for lenses that have particularly smooth video and focus rings. If you plan to use the on-camera microphone, also look for lenses that are quiet while focusing, such as the Canon STM lenses.

- **Motorized zoom**. Some mirrorless camera lenses have motorized zooms, which provide much smoother zooms than are possible with a traditional zoom ring. No DSLR lenses have motorized zooms.

- **Parfocal (constant focus while zooming)**. If you plan to zoom while filming, look for parfocal lenses that maintain their focus during zooming. Most consumer lenses need to be re-focused after zooming, which will cause your video to become blurry if you zoom in on a subject. Currently, this limits you to very few lenses: Canon 17-40mm f/4, Canon 16-35mm f/2.8, Canon 70-200mm f/2.8 (without IS), Nikon 17-35mm f/2.8, Nikon 24-70mm f/2.8 AF-S, Nikon 70-200mm f/2.8 VR Mark I, Panasonic Micro Four-Thirds 7-14mm f/4, and Olympus Four-Thirds 11-22mm f/2.8-3.5. Other lenses are not truly parfocal, but they might be close enough for video use, including the Canon 24-105mm L IS (one of the most popular video lenses).

- **Breathing**. Most photography lenses "breathe," which means the image zooms slightly when you refocus. To test this, put your camera on a tripod and point it at something with detail, like a bookcase. Manually focus near and then far, and watch the edges of the frame to see if objects are moved in and out of the frame while you refocus. It's not a problem for amateur videos, but it's a factor to consider when assessing a lens for serious video production use.

The Canon STM lenses are designed to be used with video, but currently your options are limited to the EF-S 18-55mm f/3.5-5.6 IS STM, the EF-S 18-135mm f/3.5-5.6 IS STM (both for compact DSLRs with the smaller APS-C sensor), the EF-S 55-250 f/4-5.6 IS STM, and the EF 40mm f/2.8 STM (for all Canon DSLRs). These lenses support smooth, silent autofocusing, image stabilization to reduce shakiness in handheld video, and electromagnetic diaphragms to smooth adjustments to the aperture while filming. Note, however, that the 18-135mm is not a parfocal lens, so if you zoom, you'll also need to refocus.

Teleconverters

Teleconverters cost between $125 and $500, and they're a relatively inexpensive way to increase your focal length and get closer to wildlife or sports. However, they're not a good investment for most photographers.

Teleconverters connect between your lens and your camera body, increasing the focal length and minimum f/stop number by 1.4x or 2x. For example, if you were to connect a 1.4x teleconverter to a 70-200mm f/2.8 lens, the lens would function as a 98-280mm f/4 lens. Connect a 2x teleconverter to the same lens, and it would become a 140-400mm f/5.6 lens.

Teleconverters work by capturing the center part of the image as it passes between your lens and your camera and then spreading that center part of the image across the entire sensor. Essentially, it's exactly like cropping to the center half or one-quarter of your picture. However, because teleconverters work by taking half or one-quarter of the light from your sensor and spreading it across the entire sensor, they have some rather nasty side effects (described in more detail in the sections below):

- Teleconverters drastically reduce the light coming in, requiring you to use a slower shutter speed or a higher ISO. With living subjects such as sports or wildlife, using a slower shutter speed isn't an option, so you need to use a higher ISO, which increases the noise in your picture and degrades image quality.

- Teleconverters slow down autofocusing and, in most cases, completely prevent your camera from autofocusing. I never recommend using teleconverters unless your lens has a minimum f/stop number of f/4 or lower.

- Teleconverters reduce the sharpness of an image. Unless you are using a professional-quality lens, the final image won't show any more detail. If you are using a professional-quality lens, you can often get more detail from distant subjects by using a teleconverter.

- Teleconverters narrow your field of view, which makes wildlife photography more challenging. It's hard enough to keep a flying bird in view with a 500mm lens. Add a 1.4x teleconverter, and it becomes about twice as hard to keep the bird in the frame.

However, if you have a professional telephoto lens with a minimum f/stop number of f/4 or lower and you shooting a faraway subject, a teleconverter can produce more detailed images than simply cropping a picture taken without the teleconverter. One good example of this is astrophotography; if you're taking pictures of the moon or other faraway objects, a teleconverter will show more detail than you would get by cropping your pictures.

Using a 1.4x Teleconverter

When using a 1.4x teleconverter, you'll lose the ability to autofocus unless:

- You're using a lens that has a minimum f/stop number of f/4 or lower, OR

- You have a Canon 5D Mark III, Canon 1 series, Nikon D600, Nikon D800, or Nikon D1-D4. With the Canon 5D Mark III and 1 series, you will be limited to the center focusing point. With the D600, you will be limited to the 7 center focusing points. With the D800, you will be limited to 11 autofocus points. With the Nikon

D4, you can use any focusing point, but you should limit yourself to the center 9 to provide more reliable focusing.

Many modern mirrorless cameras, such as the Sony a9, Canon R6, and Canon R5 can autofocus with any combination of lens and teleconverter but autofocus is slower. Additionally, a 1.4x teleconverter blocks half the light from reaching your lens. Even if your camera can autofocus with the teleconverter attached, it will be much slower and might be too slow to keep up with moving subjects.

I only recommend using a 1.4x teleconverter with the following lenses: 70-200 f/2.8, 70-200 f/4, 300mm f/2.8, 300mm f/4, 400mm f/2.8, 400mm f/4, 500mm f/4, 600mm f/4. Notice that those are all telephoto lenses; you shouldn't use teleconverters with wider angle lenses. Instead, simply buy a telephoto lens.

If you have a full-frame camera, you might consider using a camera with a compact (crop/APS-C) sensor instead. The smaller sensor provides a 1.5x or 1.6x increase in focal length and often provides greater detail than a similar full-frame camera. For example, I often use a Canon 7D compact camera instead of a much more expensive Canon 5D Mark III for photographing wildlife, because the 1.6x crop of the 7D captures more detail in faraway subjects.

Using a 2x Teleconverter

When using a 2x teleconverter with a DSLR, you'll lose the ability to autofocus unless:

- You're using a lens that has a minimum f/stop number of f/2.8 or lower, OR

- You're using a lens with a minimum f/stop number of f/4, and you have a Canon 5D Mark III, Canon 1 series, Nikon D600, Nikon D800, or Nikon D1-D4. As described in the previous section, you will be limited to fewer focusing points.

Additionally, a 2x teleconverter blocks three-quarters of the light from reaching your lens. Even if your camera can autofocus with the teleconverter attached, it will be much slower and might be too slow to keep up with moving subjects.

I only recommend using a 2x teleconverter with the following lenses: 70-200 f/2.8, 300mm f/2.8, 400mm f/2.8.

Focus Breathing

With many zooms, the closer you focus, the shorter your focal length is. This is easier to experience than to explain.

With any zoom lens (but especially a telephoto zoom), manually focus to infinity and zoom all the way out to the most telephoto setting. Look through the viewfinder at a door frame with the edges of the door near the left and right edges of the frame. Then, manually focus closer and closer until your lens is at its minimum focusing distance. If your lens exhibits focus breathing, the edges of the door frame will seem towards the center of the frame.

Basically, as you focus closer, you also zoom the lens out to a wider angle, even if you don't adjust the zoom. Because of focus breathing, zooming and focusing are always linked, even on many prime lenses.

This isn't a problem with your lens. In fact, most zooms exhibit focus breathing to some extent. Some primes do, too, though it tends to be more severe with zooms than with primes.

For example, I strongly recommend a 70-200 f/2.8 zoom lens for portraits. To maximize the background blur and compress the facial features, a portrait photographer will zoom all the way to 200mm. However, when the photographer is focused close enough to the model to take a headshot, the maximum zoom on a 70-200mm lens is actually about 150mm.

To put that another way, a 70-200mm lens only zooms to 200mm when it's focused at a distant subject. When focused on something close, the maximum focal length is typically closer to 150mm.

Focus breathing isn't typically a problem for photographers. However, it can be an issue with some professional videography. Imagine a scene where the director wanted to change focus from a subject in the distance to a nearby subject. If you were to change the focus while filming using a lens that exhibited focus breathing, the lens would also seem to be zooming to a wider angle, which could be noticeable in the video and be distracting.

For still photography, however, you can simply adjust your focal length or distance to the subject to get the composition you need. The one time focus breathing becomes a problem for still photographers is when performing focus stacking. If you perform focus stacking with a lens that exhibits focus breathing, you will need to compose the picture with the lens focused at the most distant part of the subject. As you take a series of pictures and focus closer with each picture, the lens (even if it's a prime lens) will seem to zoom out to a wider angle. If you were to compose your picture focused on the nearest part of the subject, the lens would zoom in and crop your picture more than you wanted. For more information about focus stacking, refer to Chapter 12 of *Stunning Digital Photography*.

Chapter 5: Flash Features

If you're interested in Wireless Flash Triggers, visit sdp.io/triggers to download an entire chapter on the topic. Since the original release of this book, wireless triggers built into flashes have become standard, therefore, I placed that chapter online to reduce the page count.

Light is the single most important element in a photo, and flashes are the most portable way to control light. While many photographers swear by natural light, I love my flash. A bounce flash can fill a room with natural-looking light, allowing you to create noise-free images in dark environments. Outdoors, you can use fill flash to create flattering lighting without getting that overexposed "flash" look.

Every portrait and real estate photographer should have at least one flash, and serious photographers will own several that they trigger remotely. However, choosing the right flash can be really complicated. The low-end models will quickly frustrate many photographers, and the high-end models cost more than many camera bodies.

Video: Buying a Flash
8:28 - *sdp.io/BuyingFlash*

In this buying guide, I'll list the current flash models for Nikon, Canon, and Sony flashes, along with my recommendations. If you want to better understand the different flash features available so you can make a more educated buying decision, the rest of this guide gives you an overview of flash features found on different models.

Before you buy a flash, read Chapter 3 of *Stunning Digital Photography* so you have a better understanding of what flashes are capable of. For information about flashes for use with macro photography, refer to Chapter 12 of *Stunning Digital Photography*.

Recommended Features for Different Types of Photography

If you have an unlimited budget, it never hurts to buy the most expensive, fully featured flash. Then, you'll be prepared for every situation. Most of us have a limited budget, however, and every dollar we can save on a flash could be put towards a nice tripod, lens, or other accessory more likely to improve our pictures.

Here are the features that you absolutely need for common types of flash photography:

- **General candid family pictures**: TTL metering, tilt bounce head

- **Weddings and events**: TTL metering, tilt & rotate bounce head, external battery pack support

- Posed portraits with on-camera flash: Tilt & rotate bounce head

- **Sports**: TTL metering, high-speed sync

- **Real estate**: Tilt & rotate bounce head

- **Off-camera flash (for use with multi-light setups)**: Tilt & rotate bounce head, manual controls, optical slave or other wireless control, audible ready notification

These features are described in more detail later in this section.

Brand

Most people shouldn't buy a Nikon, Canon, or Sony flash. They're wildly overpriced and most people will never use most of the sophisticated features.

Instead, I recommend most photographers buy a generic flash that supports TTL metering and has manual controls and a bounce head, such as those by Godox, Neewer or Yongnuo. They typically cost less than a third of the name-brand models.

Here's the catch: the generics aren't as good as the Canon and Nikon models. They simply don't have as much testing and engineering, so they're never quite as reliable, and you might have to manually tweak the power output more often. If you're a professional photographer, especially a wedding photographer, you can't take the very slight risk that one might flake out while the bride walks down the aisle. Some of the generic models lack features that might be important to you, such as TTL, built-in wireless support, and high-speed sync.

However, for most of us (including most professionals), a generic flash is the best value. Because of their low prices, you can buy several and use them in multi-light setups (and also have a backup if one dies), and if you ever decide you need the official Canon or Nikon flash, they'd still make an excellent second flash for off-camera use, so your small investment won't be lost.

The remainder of this section will discuss these features in more detail, and at the end of the section, I'll recommend specific generic and name-brand models with different feature sets.

Output

Some flashes are more powerful than others. A flash with more power allows you to illuminate subjects that are farther away, fill a larger room with bounce flash, and use lower ISOs. More powerful flashes also tend to recycle faster, so they're ready for your second or third shot faster than flashes with lower outputs.

Flash output is measured by the Guide Number (GN), and flash manufacturers always tell you the GN. The GN is the distance that a flash can illuminate a subject. It's really difficult to compare the GNs of flashes from different manufacturers, however, because it varies depending on the position of the zoom head, the size of the sensor, the camera's ISO setting, and the manufacturer's testing procedures.

Generic flash manufacturers, in particular, have been known to exaggerate the GN. For example, the Sigma EF-610 claims to have a higher GN than just about any other flash on the market. Independent tests, however, show that it has only about half the stated output. Canon and Nikon, however, tend to understate their GNs, and testing shows that the name-brand flashes are even more powerful than the manufacturers claim. The tables at the end of this section show real-world GNs where available.

If you use a flash with a lower output, you can simply choose a higher ISO on your camera's settings. Each time you double the ISO, you double the effectiveness of your flash. Increasing the ISO also increases the noise in your picture, so you can think of more powerful flashes as reducing the noise in your images by allowing you to use lower ISOs. For most photographers, the GN isn't nearly as important as other factors, such as cost, having a bounce and zoom head, reliability, and recycle time.

Through-the-Lens (TTL) Metering

The farther you are from your subject, and the darker the subject is, the more flash you need. So, your flash can't simply fire with the same strength all the time.

It's your camera's job to tell the flash how brightly to fire. To determine that, the camera puts out some light from the flash (it usually appears as a flickering) and measures the light as it bounces off the subject and back through your lens. This process is known as through-the-lens (TTL) metering. Canon and Nikon have proprietary systems that they call evaluative-TTL (E-TTL), digital-TTL (d-TTL), and intelligent-TTL (i-TTL).

TTL support is extremely important if you plan to mount the flash to your camera. If your flash doesn't support TTL metering, you'll need to adjust the flash output for every picture you take. If every photo is planned, or if you plan to use the flash off-camera, you'll probably end up manually adjusting the flash output, anyway, so it's not important. However, if you plan to take candid or spontaneous photos, you need a flash that supports TTL.

As long as a flash has TTL support, the difference between TTL, E-TTL, d-TTL, and i-TTL are relatively minor. Yes, the newest TTL technologies from Canon and Nikon can improve your flash exposure in some scenarios. If you're a wedding photographer or a photojournalist and you won't have time to review your exposure and re-shoot if you need to, these improvements might be worth the extra expense. For most of us, however, generic TTL flashes work well enough.

Bounce Head

The single most important feature of an external flash is a bounce head. Bounce heads allow you to point the flash in different directions to bounce the light off the ceiling or walls. They're tremendously useful, and they allow you to add light without having that "flash" look: an overexposed foreground with a dark background.

Not all bounce heads are created equal, however. Some bounce heads only tilt up. If you hold your camera vertically, the flash would be pointed at the wall instead of the ceiling, completely changing your lighting. Therefore, the ability to rotate the flash side-to-side is very important, though it's often reserved for higher-end flashes, so choose a flash with a bounce head that both tilts and rotates.

Zoom Head

Zoom heads focus the flash's light beam to match the focal length of your lens. Without a zoom head, your flash would always attempt to fill the view of a wide-angle lens, even if you were zoomed in on a faraway subject. Therefore, the zoom head uses your flash output more efficiently, allowing you to reach greater distances and reducing the output required. The less output a flash uses, the faster it will recycle, and the more flashes you will get out of your batteries.

High-end flashes have zoom heads and will automatically zoom to match your lens' focal length. They have limits, however. For example, the top-end flashes from Canon and

Nikon both zoom up to 200mm, but lower-end flashes might only zoom to 50mm or 105mm. If you are taking a picture of your friends with a 100mm lens and your flash only zooms to 50mm, light will still reach the subject, but some of the flash will illuminate the scene outside of the frame and will be wasted. Your subjects will probably still be well illuminated, but your flash batteries won't last as long and you will need to wait longer between taking pictures.

High-end flashes have automatic zoom heads. Your camera communicates your current focal length to the flash, and the flash automatically zooms to match the length, taking into account your sensor size. This is very important for candid shots where you don't have time to plan. It also means you won't forget to change the flash zoom. Lower-end flashes might have manual zoom heads, and sometimes they support only two different focal lengths, such as 24mm and 50mm.

Some flashes include a diffuser that flips down in front of the flash head to distribute the light across a wider area. Without the diffuser, most zoom flashes only zoom out to 24mm. Therefore, if you were to take a picture at 18mm, only the center of the picture would be illuminated. Flipping down the diffuser spreads the light a little wider, fully covering an 18mm scene for super-wide angle shots.

Zoom heads are only useful when your flash is pointed directly forward. Therefore, you won't bounce the flash and use the zoom head at the same time. To shoot a variety of different situations, however, it's important to have both a zoom head and a bounce head.

Recycle Times

The recycle time is the amount of time you have to wait after taking a flash picture. If you take a picture before the flash has recharged completely, the flash might fire at less than full power, leaving your picture underexposed.

Recycle times are very important, yet there's no consistent way to compare recycle times between different flashes. In general, flashes with a higher GN have faster recycle times. Adding an external battery pack can also improve recycle times. Therefore, if you need a fast recycle time (which is extremely important for wedding and event photographers), choose a flash with a high GN and support for an external battery pack.

External Battery Pack Support

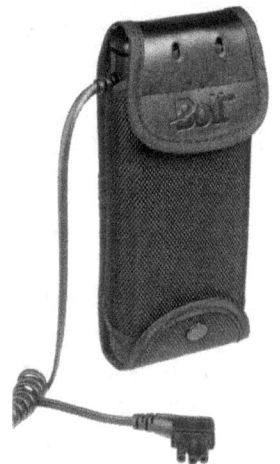

Higher-end flashes have a jack for connecting an external battery pack. The extra batteries extend your battery life, allowing you to take more shots before changing the batteries, and they reduce recycle time, allowing you to take photos faster.

Most people don't need a battery pack for their flash. Therefore, the battery pack jack is a feature found only on higher-end flashes.

Wedding and event photographers, however, should choose a flash that supports connecting an external battery pack because those scenarios often require you to take flash shots as fast as possible, and you might miss a shot in the time it takes you to change your batteries.

Rather than buying a battery pack from your flash manufacturer, look for a generic battery pack designed for your flash. They cost a fraction of the name-brand version and work just as well.

Diffusers and Reflector Cards

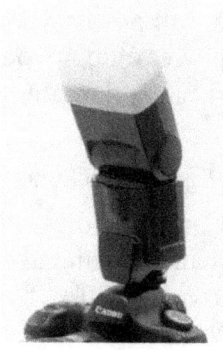

Some flashes have light modifiers built directly into them. For example, they might have a built-in diffuser that you can flip down to cover the flash head and distribute the light across a wider area. Other flashes have a small white card that bounces some of the light forward—useful for creating a catch light when using a bounce flash. Built-in diffusers and reflector cards are convenient because they're always there. If you don't have these features, you can buy attach a third-party diffuser or reflector card to your flash; the only downside is that you have to carry it with you.

PC Sync Jacks and Cords

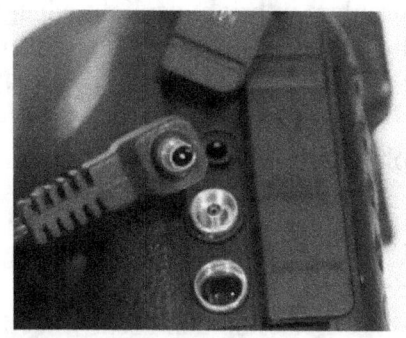

Sync cords (often known as PC cords) are an 80-year old standard for triggering a flash. Basically, you can plug one end of a sync cord into your flash and the other end into a camera or remote trigger, and the flash will fire whenever you take a picture. If you just connect your flash to the hot shoe on top of your camera, you never have to worry about a sync cord. However, sync cords can be useful for triggering an off-camera flash.

Sync cords don't communicate much information; they really just tell the flash when to fire. On the other hand, the flash hot shoe (the bottom part of the flash that mounts to the top of your camera body) can communicate a great deal of information about the distance to the subject, ambient lighting, focal length, and desired amount of flash output. Because the flash hot shoe is so much more powerful, most photographers trigger remote flashes using the hot shoe, either by using an off-camera flash cord or by using a wireless trigger.

Some flashes and camera bodies have a PC/sync jack that you can connect a sync cord to. If your camera or flash doesn't have this feature and you need it, you can get an inexpensive ($10) adapter that uses the hot shoe to connect to the sync cord. But you'll probably never need it, so don't pay more for a flash just to get a PC/sync jack.

Manual Controls

All flashes allow you to control their output using flash exposure compensation built into your camera. Lower-end flashes designed for the casual photographer don't have any extra buttons or controls. Therefore, if you want to adjust your flash output up or down, you need to adjust the flash exposure compensation using your camera's controls.

Higher-end flashes designed for enthusiasts and pros include controls directly on the flash. These aren't really necessary as long as your flash is attached directly to your camera.

However, if you want to use your flash off-camera, those manual controls can be very useful. For example, if you're doing a portrait shoot and you move a flash behind your subject to provide rim lighting, you could use the buttons on the flash to reduce the flash output to get exactly the lighting you need. Flashes that lack manual controls typically fire at full power when they are not connected to a camera or wireless trigger, and the full-power output would quickly wear through your batteries, take much longer to recycle the flash, and might be much more light than you need.

High-Speed Sync

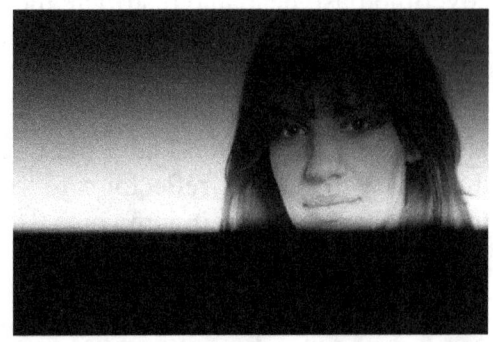

Camera bodies can't use a traditional flash faster than the camera's X-sync speed, which is usually about 1/200th. If you use a faster shutter speed, the flash might not evenly illuminate the entire frame, as shown in the example photo.

Because flash is primarily used for portrait work where high shutter speeds aren't necessary, the need to use a slower shutter speed only becomes a problem in bright daylight when you might want to also use a small f/stop number to blur the background. If that's a scenario you need to be prepared for, you should choose a flash that supports high-speed sync (HSS).

No generic flashes support high-speed sync for Sony or Micro Four-Thirds cameras, so photographers on those systems will need to choose a name-brand flash.

Audible Ready Notification

Some flashes include an audible beep to let you know when they have completely recharged. You don't really need this when you're using an on-camera flash because you can hear the flash recharging, and you'll soon become accustomed to the sounds your flash makes. However, the audible beep can be useful when using a flash off-camera.

Focus Assist Light

Some flashes include active focusing capabilities, which transmit a focus assist light (usually red or infrared) that helps the camera focus on nearby subjects in dark rooms. Some camera bodies have a focus assist light built right into the body, as well.

Focus assist lights are limited in their usefulness. First, they generally only transmit a beam across the center of the image, so you need to be using the center focusing point. Second, the beam only reaches a few feet, so you need to be within about ten feet of your subject (but sometimes closer). Third, many modern cameras are quite capable of focusing in dimly lit environments without an assist light. Nonetheless, focus assist lights can be useful when taking pictures in dimly lit bars and restaurants. Typically, however, I wouldn't pay extra for the feature.

Video Lights

Some low-end flashes include an always-on LED light that you can turn on for use with video. Basically, it's a small flashlight that lights up nearby subjects if you're filming. When built into flashes, these video lights are very small and tend to look terrible. I don't recommend buying a flash with a video light. Instead, buy a dedicated video light, such as the *Neewer CN-160 ($25)*. The larger surface area and brighter output will light a larger area and provide a much more flattering light.

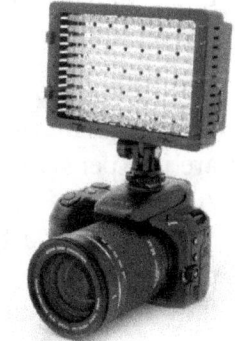

Modeling Lights

A modeling light stays on constantly and is intended to simulate the light that the flash will produce so that you can adjust the light or your model without taking test shots. Because modeling lights are not as bright as the actual flash, you can't use them to assess the brightness or power output of the flash. However, they're usually bright enough to see if your flash position is making an ugly shadow on your model's face. A modeling light is a nice-to-have feature, but it's not particularly important because modeling lights are never all that accurate. Especially outside of the studio, I prefer to simply take a test shot, make adjustments, and then re-shoot.

Light Modifiers

You can buy a variety of diffusers and soft boxes that fit onto your flash head to modify the light. For detailed information about their use and example pictures, refer to Chapter 3 of *Stunning Digital Photography*.

Radio

Photographers often want to move flashes off-camera to achieve different lighting effects. Ideally, the photographer would be able to control the power of each of the remote flashes, so they don't have to physically move between flashes to make adjustments. Flashes with radio remote capabilities provide this.

In the past, we used remote triggers that attached to the bottoms of flashes. Remote triggers were a pain, though. They were one extra gadget that you had to carry, along with extra batteries, and if the trigger failed or the batteries died, your flash stopped working. This buying guide had an entire chapter dedicated to understanding remote triggers, and you can download that chapter at *sdp.io/trigger*.

In recent years, flash manufacturers have begun building radio triggers into flashes, making remote setups simpler, less expensive, and more reliable. If you're interested in working with remote flashes, you should choose a flash with a built-in radio trigger, such as those available from Godox, Cactus, or Phottix.

We've tested most popular radio flashes, and settled on the Godox wireless flash system for our own purposes, both in and out of the studio. Godox 2.4G radio flashes start at about $85 (less expensive than any other manufacturer) but you can choose from a wide variety of higher-end flashes and strobes that are compatible. Godox offers the X1 and XT32 transmitters, which cost about $46. The X1 supports TTL, whereas the XT32 is manual-only. Many photographers prefer the XT32's user interface, however.

Batteries

The vast majority of flashes take 4 AA batteries. That's convenient, because AA batteries are inexpensive and easily replacement. The next section provides recommendations for AA batteries.

Some newer flashes, such as the Neewer NW 870 and Godox V860II come with proprietary Lithium-Ion batteries. I generally prefer the proprietary batteries, because they're more compact than 4 separate AA batteries, and they're easier to quickly swap because you don't have to worry about putting 4 different batteries in the correct direction. Therefore, I recommend choosing a flash with a single battery pack and purchasing an extra battery pack for extended shooting such as a wedding.

Flash AA Battery and Charger Recommendations

If you use flash regularly, batteries can get really expensive, and I know you'd rather spend that money on new camera equipment. For most of us, rechargeable batteries are the right choice. Rechargeables tend to last longer and recycle flashes faster, too.

The _Sanyo eneloop XX AA batteries_ are the best rechargeables you can buy for flash. However, the _standard Sanyo eneloop AA batteries_ are the best value: they're almost as good and the price is much better, so buy three sets. Keep one set in your flash, keep another set in your camera bag so you can swap the batteries out if recycle times get too long, and keep the third set in the charger at all times.

The eneloops give you about 30% faster flash-recycle times than other batteries. If you need to buy batteries and you can't find the eneloops (they're not commonly available in stores), buy standard Duracell batteries. Ignore higher-priced varieties like the Duracell Ultra Power or Energizer Max.

For the casual photographer, any AA charger should be fine, including the _optional charger with the eneloop batteries_. However, most AA chargers have a drawback that can reduce your flash performance: they charge the batteries in pairs. This can result in one of the batteries in the pair not being completely charged. I recommend the _La Crosse BC-700_ ($40) because it charges each battery separately, it's small enough to travel with, and it has features to maximize and monitor your battery life.

Chapter 6: Studio Lighting

Flashes are every photographer's first step into controlling light. We start with on-camera flash and then move the camera off flash to achieve different lighting effects.

Video: Studio Tour
8:54 - *sdp.io/TourStudio*

The next step is always to add multiple light sources, and this is where photographers divide. Some photographers add multiple flashes and control them using the technologies described in the previous chapter. Other photographers keep their single flash for on-camera use, and add larger monolights to create studio-quality lighting, whether in the studio or on location.

The previous chapter described technology designed to allow you to create multi-light setups with standard flashes. Monolights, however, are a much better choice for most portrait, wedding, product, and commercial photographers, whether amateur or professional. In fact, I believe that far too many photographers invest in complex multi-flash setups when they should choose less expensive, more powerful, and often just as portable studio lighting.

This chapter provides quick recommendations for beginner studio lighting. For those of you who want to understand the technology so you can make your own educated buying decisions, I'll go into detail about the most important considerations for choosing studio lighting and give you an overview of the most popular brands of studio lights.

Quick Recommendations

Depending on your budget, here are recommendations for systems for use either in a fixed studio or on location when you can plug them into a wall outline. Most of these are four-light systems. It's common for beginner kits to have only one or two lights, and indeed, I recommend mastering the use of a single light before adding more lights. Ultimately, though, you'll want four or five lights for a decent studio setup. Feel free to choose the budget you're most comfortable with, buy just one or two of the lights and light stands, and add the rest of the system later.

Notice that I no longer need to provide separate recommendations for Canon and Nikon; all monolights will work equally well with either system. You can use adapters to connect monolights to almost any other type of cameras, too, including Sony and Minolta (and just about any camera except your smartphone).

- **$100 (AC powered):** Godox MS200 strobes with an X-Pro trigger. You might want 2 or 3 lights. These inexpensive monolights accept Bowens light modifiers, allowing you to use the same softboxes and beauty dishes that professionals use in their studios. If you want more power, you could upgrade to the slightly more expensive MS300 or DS400. Pick the model that fits your budget; they'll all get the job done.

- **$600 (battery powered):** Godox AD600BM with an X-Pro trigger. These wireless are compatible with the previous suggestion, so you can mix-or-match them. Being battery powered is useful in a studio, too, because it reduces the risk someone will trip over a cord. The batteries last through multiple portrait sessions, but you should remember to plug them back in to charge overnight. You do have the option of adding an AC adapter to these strobes if you decide charging batteries is too tedious. I do not recommend upgrading to the higher-end Pro models; we've found their batteries to be extremely unreliable.

> **Tip:** Fill your sandbags with small gravel, not sand. Sand is messy.

The sections that follow provide more information about each of these components. Take a quick look at the *Wireless Flash System Cost Estimates* section of the previous chapter (which only provides for three lights and does not include light stands or light modifiers) and you'll see that these studio lights provide much more flexibility, control, and power at a far lower cost. For example, a four-light system based on the Canon 600EX-RT flash would cost you $2,285 for the lights alone, not counting batteries, light stands, or light modifiers. Even the least expensive monolight recommendation has about 2-3 times more power than the most expensive speedlight recommendation.

Flashes vs. Monolights

Flashes (also known as speedlights or speedlites) are designed to be connected directly to the flash hot shoe on top of your camera. Most flashes have sophisticated intelligence to allow them to communicate with your camera to automatically control the amount of light, giving you perfect settings entirely automatically.

Because they're optimized to fit on top of your camera, they're lightweight and portable. Unfortunately, they also have several huge drawbacks compared to monolights:

- **Flashes cost more than monolights**. Believe it or not, most studio monolights are less expensive than most flashes.

- **Flashes have far less power than monolights**. Monolights produce 4 to 16 times more light than flashes, allowing you to use a lower ISO on your camera (reducing noise in the image) and allowing you to overpower the sun when working outdoors.

- **Flashes take far longer to recycle than monolights**. Flashes are typically powered by 4 AA batteries, whereas monolights are typically powered by electrical current from your wall. With access to all that extra power, monolights can recycle much faster. It also helps that you won't have to use monolights at full power every time, allowing you to snap multiple pictures without waiting for the flash to recharge.

- **Flashes produce rectangular light; monolights produce circular light**. Flash heads are designed to be mounted to the camera. To minimize the amount of light wasted, the shape of the flash head is the same as the shape of your camera's sensor. That works perfectly when your flash it attached to your camera, but once you move it off-camera, the rectangular shape is less-than-ideal for lighting your subjects. Monolights produce a circular light.

- **Monolights are designed for use with light modifiers**. You can attach softboxes and beauty dishes directly to your monolights, and many of them allow you to attach an umbrella directly to them. With flashes, you need separate, clumsy adapters to connect them to light modifiers.

- **Flashes have small batteries**. Monolights can be powered either by battery or wall current. Batteries for monolights tend to be much larger and more powerful than AA batteries, and when you have access to power and can plug your monolights into the wall, they'll never run out of power.

- **Monolights support more groups than flashes**. Most remote control flash systems allow you to configure only three groups of lights. That's enough to let you control the main, fill, and background light—but what about your hair light, or your kicker light? Many monolight systems are designed for more serious studios and can support independently controlling larger numbers of lights.

If you're using it on-camera, or if you need to rely on auto-exposure, buy flashes. If you're putting your light on a light stand, even if you're travelling with your lights, you should probably choose monolights instead.

This chapter will provide quick recommendations for those looking to buy monolights, along with an overview of the most important monolight brands.

Flash Power vs. Monolight Power

Flashes use Guide Numbers (GNs) to measure their output, and monolights use Watt-Seconds (WS). Unfortunately, there's no easy way to convert between the two units; they're entirely different measurements. GNs estimate the reach of a flash, but that factors in the flash's zoom head. GNs assume that the flash head is zoomed all the way in, focusing the light into a narrow beam. With a monolight, the light focusing is controlled by a light modifier that you attach to it.

To make this more complicated, flash manufacturers exaggerate their GNs, and monolight manufacturers exaggerate their WS outputs. So, the numbers can't be converted or trusted. We do know this from testing out flash outputs: monolights are more powerful than flashes. You'd expect that, given their much greater weight and size—and the fact that they're plugged into the wall rather than running off of four tiny AA batteries.

One estimate from Paul C. Buff, the genius who designed the Alien Bees, White Lightning, and Einstein monolights, is that the low-end monolights (specifically his own Alien Bees B400) have about 2 3 stops more output than the top-end flashes (specifically the Nikon SB-900). As an example, the Einstein e640 costs about 10% less than a Nikon SB-910, but the Einstein produces about 32 times more light.

Continuous vs. Strobe

I see many beginning photographers interested in buying continuous lights rather than strobes. Continuous lighting seems to have many advantages over strobes:

Video: DIY Ring Light
3:50 - *sdp.io/ring*

- They can be less expensive.

- They don't have to flash.

- You don't have to connect them to your camera.

- You can use them for video.

However, I never recommend continuous lighting to photographers. Continuous lighting is for videographers, and photographers should always buy strobes.

I personally do both video and photography, and I have continuous lighting for video and a separate strobe system for photography. I'm a huge cheapskate, so if there were a way that I could use one system for both, believe me, I would.

But I can't. I've tried. Continuous lights have these disadvantages:

- They're not nearly as powerful as strobes. Depending on the conditions, you'll often be forced to use high ISOs, such as ISO 3200. This will introduce a great deal of noise in your pictures. You also won't be able to overpower ambient light, even indoors, making it difficult to get deep shadows and dark backgrounds.

- You don't have many options for light modifiers. First, they generally don't physically connect to beauty dishes and softboxes. Even if they did, you'd find the light modifier blocked the light output even further.

- They waste a great deal of power. Continuous lighting uses electricity continuously, even though you only need it for that split second you're taking a picture.

- They're hot. Even the new cool LED lights give off quite a bit of heat.

Fortunately, you can buy strobe systems that are extremely inexpensive.

Most photographers shouldn't use continuous lighting if you plan to use light modifiers such as softboxes. Continuous lighting is better for video, and it isn't bright enough to get great image quality for portraits. I do enjoy using continuous lighting for controlled, short-range portraits, such as headshots. For an example of how I used under-cabinet LED continuous lights to make a custom ring light, watch the following video.

Monolights vs. Pack-and-Head Systems

Monolights, also known as monoblocks, monoblocs, compacts, or self-contains, plug directly into the wall, and provide individual controls on the back of each light. Pack-and-head systems, also known as flash packs, have a single power source that plugs into the wall. All the lights plug into the power pack,

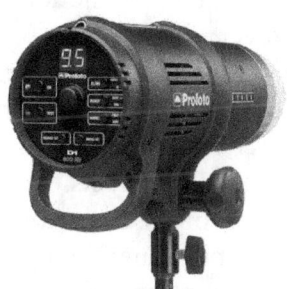 VS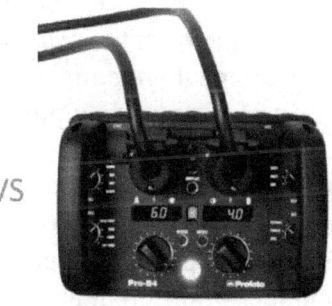

and you control all of the lights from the power pack.

Most lighting companies offer products that use both technologies. For example, most of Paul C. Buff's products are monolights, but they also offer the Zeus pack-and-head system. Most of Profoto's products are pack-and-head, but they also offer the D1 Air monolights. In general, monolights offer these advantages:

- **Control over power output at low costs**. Even budget monolights allow you to control the exact power output of each individual light. Only higher-end pack-and-head systems offer that much control.

- **Better cable organization**. With a pack-and-head system, all heads have to plug into a pack, and then the pack plugs into the wall. This means that you have more cables on your floor. If you're using a single pack, you have to route the cords for all the heads to that single pack, meaning you'll be running cords across your studio floor. With monolights, you simply plug them into the nearest outlet. This also means that you can move your lights as far apart as you need, which is useful when lighting large rooms, such as at a wedding reception.

- **Independence**. If one monolight fails, it won't impact your other monolights. If a pack fails, you won't be able to use any of your heads.

On the other hand, pack-and-head systems offer these advantages:

- **Lighter heads**. Monolights have all the electronics in the head, whereas lights designed for pack-and-head systems have most of the heavy electronics on the head, which is on the floor. With lighter heads, you can use lighter booms to support the heads. For ring flashes, which you often need to handhold, the lower weight can greatly improve the handholdability. For example, the ABR800 ring flash monolight is 2.5 lbs., whereas the Zeus ring flash head (which is almost identical) is only 1.5 lbs.

- **Higher outputs**. While a typical monolight has about 600 Ws of output (compared to 50-75 Ws for a typical flash), typical packs have outputs of 1,200-4,800 Ws (and much higher). And, you can channel all that power through a single head if you want to fill a room—or permanently blind a model.

- **Lower head cost**. Because heads for pack-and-head systems don't need the electronics, individual heads can be substantially less expensive. Again comparing

ring flashes, the ABR800 monolight is $400, whereas the comparable Zeus ring flash head is only $300. In practice, monolights tend to cost less overall once you factor in the pack, however.

I recommend monolights to photographers with a lighting budget under $5,000—and that's most of us, including professional portrait photographers. While there are pack-and-head systems for all budgets, the lower-end systems have very limiting and frustrating controls. Commercial photography studios with budgets over $5,000 should consider pack-and-head systems such as those offered by Profoto. These high-end systems offer bulletproof reliability, power outputs that can permanently damage your retinas, and an incredibly number of different heads for different lighting requirements.

Brand Overviews

Modern flashes and lenses have complex, proprietary communications with the cameras. If a new company is going to make a flash or lens, it has to spend millions reverse-engineering these communications and testing its products with every different camera it hopes to be compatible.

Monolights, however, have traditionally used a much simpler communication mechanism: the PC sync cord. It's just a cord that carries a simple electrical signal that tells the lights when to fire. This simple communication standard has allowed dozens, perhaps hundreds, of different companies to build monolights.

I won't be describing them all in this book. Instead, I've chosen four of the most popular lighting brands: Elinchrom, Paul C. Buff, and Profoto. I'll give you

> Video: Studio Lighting Comparison
> **31:36** - *sdp.io/studiolights*

quick recommendations for beginning studio setups and then provide an overview of the products offered by each of the three brands.

Just as choosing Canon, Nikon, or Sony is a big decision because you'll be stuck with that company's accessories, choosing the brand of your lighting locks you into that system's wireless remote controls and light modifiers. If you need to add a light, you'll certainly want to use one from the same system. Therefore, you should first choose a brand that you want to invest in, and then select lights and light modifiers from that brand's offerings.

In other words, you should evaluate the brand's entire offerings. Even if you're just looking for a single monolight for your home studio, if you plan to add more lights later, and you think you might want more power, you should evaluate your future costs and upgrade options within that brand.

The following table samples the different makes and models for each brand to give you a sense for what a 4-light studio setup with wireless control over the light output will cost you.

Make	Model	Output	4 lights	4 wireless lights	Replacement bulb cost	Computer controlled?
Godox	AD200	200 Ws	$1,340	$1,340	$40	
Godox	AD600BM	600 Ws	$2,260	$2,260	$100	
Elinchrom	D-Lite	200 Ws	$1,360	$1,480	$58	X
Paul C. Buff	Alien Bees	320 Ws	$1,120	$1,660	$35	
Paul C. Buff	Einstein	640 Ws	$2,000	$2,340	$35	
Elinchrom	BRX	500 Ws	$2,700	$2,820	$110	X
Elinchrom	RX	600 Ws	$4,000	$4,120	$97	X

| Profoto | Air | 500 Ws | $4,800 | $5,100 | $156 | X |

Since the introduction of the Godox strobes, we personally have switched both in-studio and on-location to the same set of Godox AD600BM lights (with an extra AD200). They present the best value because they have built-in battery power. In recent years, more and more portrait work is done outside the studio, and we can easily pick up Godox and take it anywhere, without having to learn or maintain a second system.

Be sure to factor in maintenance costs. To help you estimate this, I've included the cost of a replacement bulb for each of the systems. Repair costs tend to be proportionate, with Paul C. Buff equipment being the least expensive to repair.

The sections that follow provide a more detailed overview of each lighting brand.

Godox and Flashpoint

We've tested all the greatest studio light models, ranging from $50 strobes to those costing over $10,000. For our own purposes, both in and out of the studio, we settled on Godox AD600BM strobes with the Bowens mount. These same strobes are sold in the US by Adorama using the Flashpoint name. You can mix-and-match Godox and Flashpoint devices. In the US, I recommend choosing Flashpoint models so that you can use Adorama support if needed. Avoid the AD600Pro; the batteries are extremely unreliable and cost $180 to replace. We are now on our third battery, so I've spent more than $500 on batteries alone for a single strobe.

Godox is a really new name in the studio lighting world, and they definitely haven't earned the trust of experienced commercial photographers. However, we've used them in a wide variety of real-world conditions and they've never failed us. They're proven to be far more powerful and versatile than any other model.

Transmitter

Godox transmitters work with most popular camera models and are compatible with their entire range of wireless flashes and strobes. Currently, transmitters are natively compatible with Canon (X1C), Nikon (X1N), and Sony (X1T-S). Godox has announced upcoming support for Fujifilm and Panasonic/Olympus, so compatible transmitters might be available by the time you read this. The transmitters are not the most intuitive to use, but they're not terrible, and within 5 minutes you'll be adjusting the power output of your strobes without 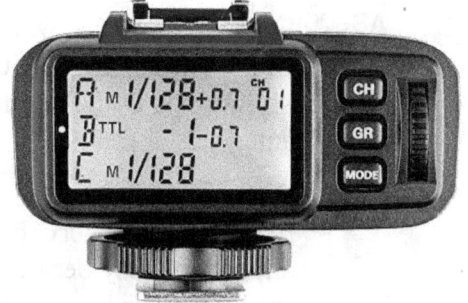 interrupting the flow of your shoot. They don't support app control, but we prefer to control strobes directly from the camera, anyway.

The following strobes have built-in receivers that work with any of the Godox X1 transmitters. In other words, any of these strobes will work regardless of the type of camera that you have.

AD200

The AD200 ($300) is a portable strobe in the shape of a conventional on-camera flash. While most conventional flashes use AA batteries, the AD200 has a proprietary but very powerful flash that lasts longer than any shoot we've ever done. Of course, you can buy extra batteries if you need them.

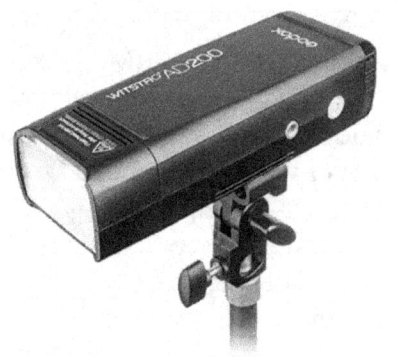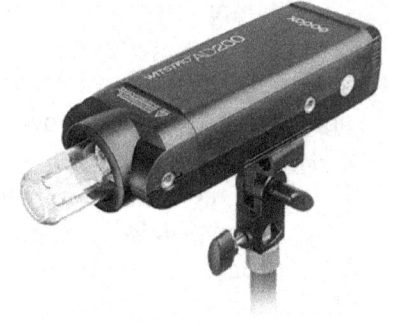

The AD200 is outrageously powerful with incredibly fast recycle times.

The AD200 supports both a square flash head like a traditional on-camera flash, so you'll need to use flash modifiers designed for on-camera flashes. It also supports a bare-bulb flash that transmits light more efficiently, but is unfocused. Use the bare bulb option with light modifiers like umbrellas and soft boxes.

AD360 II

If want a more traditional on-camera TTL flash with a bounce head that can also be used wirelessly with the X1 transmitters, the AD360 II is a powerful, professional option. The round flash head is bulky, but creates more pleasing catch lights than a traditional rectangular on-camera flash. Combined with the PB960 external battery pack, you'll have plenty of power to show even the longest weddings.

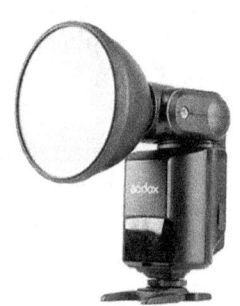

AD600BM

We chose the AD600BM ($550) with the Bowens mount for most of our studio lights. Previously, we had used the Paul C. Buff Einstein strobes, but the AD600BM has several big advantages:

- **Battery power**. Each strobe includes its own rechargeable battery that seems to last forever and has almost no delay between flashes. In the past, location shoots required us to either use conventional flashes (which are clumsy to connect to flash modifiers like large soft boxes) or to use our Einsteins with a separate battery pack for each strobe (which requires carrying extra stuff). On location, the battery-powered AD600BM is easy to setup and carry. In the studio, there are no cords to trip over, and recharging the batteries is simple.

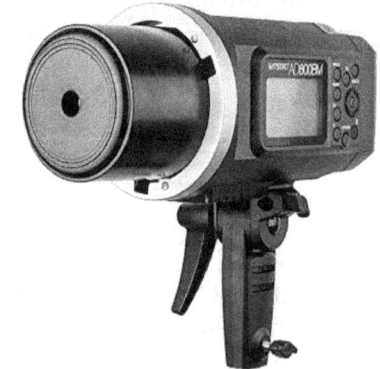

- **Fast recycle time**. Other battery-powered strobes had long recycle times, requiring us to slow down our shooting pace. The AD600BM strobes keep up even with really quick shooting.

- **Short flash duration.** Like the Einstein E640s, the AD600BMs have a very short flash duration that's useful for freezing motion, creating sharper pictures. The flash fires faster than your camera's shutter, even without using high-speed sync mode. The AD600BMs have proven faster than more expensive battery-powered Profoto strobes.

- **High-speed sync**. When shooting outdoors, you often need to use a fast shutter speed like $1/2000^{th}$ of a second. Just a few years ago, only small on-camera flashes supported high-speed sync.

- **Bowens mount.** They support the industry-standard Bowens mount, giving you access to hundreds of flash modifiers, ranging from entry-level to high-end.

Godox provides powerful accessories that extend the functionality of this strobe, such as the AD-H600B ($70) which allows an assistant to carry the weight of the strobe and battery on a belt clip, while holding the bulb on a boom. You can also by adapters (Godox AD-AC, $100) to run the strobes from wall power, though we prefer to be cordless even in the studio.

AD600B

The AD600B is exactly like the AD600BM, but it supports TTL metering and costs a little more. We don't typically use TTL metering with off-camera strobes, but you might when shooting events and weddings. For our own purposes, we bought one AD600B for those situations, and saved money by using AD600BM for every other situation.

Paul C. Buff (Alien Bees)

The Paul C. Buff brand manufactures three lines of monolights and a less popular power pack system. All the Paul C. Buff lights use the common Balcar mount for light modifiers. The Balcar mount is popular primarily because of the popularity of the Alien Bees lights. You can use the CyberSync system to control any of the Paul C. Buff lights, which is one of the greatest benefits. PocketWizard also offers adapters to provide remote control over the Alien Bees, White Lightning, and Einstein lights.

The Alien Bees monolights are my recommendation for beginner and intermediate photographers thanks to their low cost, performance, reliability, and availability of light modifiers and remote controls. The Einstein monolight is my recommendation for advanced photographers.

Make	Model	New	Used	Wireless	Output (Ws)	Model Tracking	Beep Ready	Model Ready
Alien Bees	B400	$225	$190	$90		X		X
	B800	$280	$215	$90		X		X
	B1600	$360	$320	$90		X		X
	ABR800	$300	$250	$90		X		X
White Lightning	X800	$390	$250	$90		X		X
	X1600	$440	$300	$90		X		X

	X3200	$550	$350	$90		X		X
Einstein	e640	$500	$430	$30		X	X	X

Buff only sells its lights through the Paul C. Buff websites, directly to consumers, so you can't buy them at Amazon, Adorama, or B&H. While a nuisance, this is also the secret to its success, as it allows it to sell its lights at lower costs. Outside the US, the lights might be substantially more expensive.

Wireless Control

You can control the output of all modern Paul C. Buff monolights using the CyberSync system. You'll need a Cyber Commander transmitter attached to your camera and one receiver attached to each light. The on-camera Cyber Commander transmitter ($180) is far more sophisticated than the remote control systems offered by other brands:

- It has a large, full-color display.

- You can control up to 16 lights independently (or combine them into groups).

- You can increase or decrease the output of all lights at once, allowing you to maintain the same exposure when you change your camera's ISO or aperture.

- You can control all lights simultaneously, allowing you to keep lighting consistent while adjusting your ISO or aperture.

- It has an extremely useful flash meter built in to calibrate your light output. When you meter a light, the Cyber Commander will show you the actual light output on your subject relative to other lights, factoring the distance and impact of any light modifiers. With this, you can instantly balance your main and fill light for instant, even lighting.

- It shows you the maximum and minimum output of different lights, so you can mix-and-match different Paul C. Buff lights.

- With a sync cord, you can fire other brands of flashes and monolights—though you can't remotely control their output.

- The Cyber Commander rotates up and down, allowing you to see it when your camera is at eye level or at waist level.

The capabilities combine to allow you to do some amazing things in the studio. It really will drastically reduce the time you spend setting up your lighting and solving lighting problems, and that will lead to you getting more, and better, photos.

Unfortunately, it also has significant disadvantages compared to other brands:

- The monolights don't have built-in receivers, so you have to buy a separate receiver for each light. These are $90 for each Alien Bees or White Lighting monolight, and $30 for each Einstein monolight.

- The setup is both complex and non-intuitive. You definitely have to read the manual, and getting your lights set up the first time will take you at least a couple of hours.

- The screen can be hard to read in full daylight—a problem you won't run into on systems that don't have LCD displays.

Alien Bees

These budget-oriented monolights are my recommendation for beginner and intermediate photographers. They're powerful, durable, lightweight, and inexpensive. They sell used for about 85% of their new value, protecting your investment.

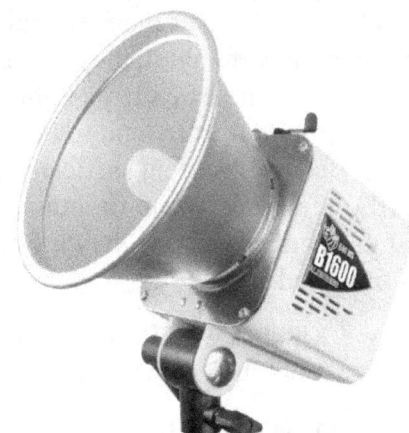

You can buy the Alien Bees in different, and crazy, colors. You can also just get them in standard black. I find it useful to have them in different colors, however, because it makes it easy to direct assistants to the "yellow light" or the "green light".

Buff offers heads with three different outputs, all with 6 f/stop power variability:

- **B400 ($225 new, $190 used)**. Power output from 5 Ws to 160 Ws. 2.5 lbs.

- **B800 ($280 new, $215 used)**. Power output from 10 Ws to 320 Ws. 2.9 lbs.

- **B1600 ($360 new, $320 used)**. Power output from 20 Ws to 640 Ws. 3.7 lbs.

A B800 at half power performs exactly like a B400 at full power, with identical flash duration and recycle times. A B1600 at half power performs exactly like a B800 at half power.

Generally, the higher-end lights are better than the lower-end lights, just because you have the option of using more power when you might need it. The B400 is an excellent choice for a hair light because its light weight makes it easy to put on a boom. The power of the B1600 is really only needed if you're using a very large light modifier (such as the 86-inch PLM umbrella), if you need to use it as a bounce flash to light a very large room, or if you need to use it outdoors to overpower the sun.

White Lightning

A step up from the Alien Bees monolights, the White Lightning lights are a bit more expensive for the same level of output. They offer a sturdier build and manual control over the modeling light, neither of which is a particularly important feature. While they're advertised as indestructible (and indeed they are extremely durable), they're also quite a bit heavier than the Alien Bees lights, making them more difficult to carry and to put on booms, and the Alien Bees lights seem sufficiently durable.

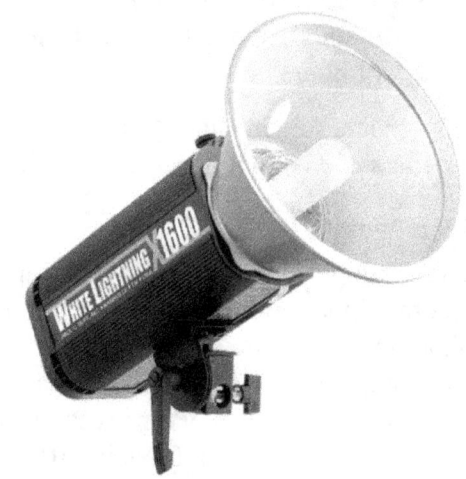

Buff offers three White Lightning models. I don't list used prices because they are relatively difficult to find used:

- **X800 ($390)**. Power output from 10 Ws to 330 Ws, making it equivalent to a B800 (but much heavier at 4.1 lbs., and more expensive). The B800 is a better value.

- **X1600 ($440)**. Power output from 5 Ws to 660 Ws. The maximum power output is similar to the B1600, but you can also push a button on the back of the unit to cut the power by 75%, allowing you to instantly decrease the total output. That can be quite useful when using light modifiers very close to your subject for soft lighting. However, the Einstein e640 is a better value than this unit, especially when you consider that they cost the same once you add the remote control unit. 4.9 lbs.

- **X3200 ($550)**. No Paul C. Buff light offers more power output than this light, which provides 10 Ws to 1320 Ws, allowing you to easily fill the largest of rooms or overpower the sun with even a very large light modifier. Weighing in at 7.1 lbs., you'll need a sturdy light stand for it, and you probably won't enjoy carrying it on location.

Einstein

The top-end monolight from Paul C. Buff, the Einstein e640 is the only member of the lineup, and it's my recommendation for most advanced studio photographers. It offers both a shorter flash duration and better white balance than the White Lightning and Alien Bees lights. It also offers a nice digital display on the back of the unit. The e640 offers power output from 2.5 Ws to 640 Ws, matching the maximum output of the B1600 and X1600, but offering a useful lower output, as well. It weighs 4.4 lbs.

The $500 price tag is softened by the fact that remote control transceivers are less expensive. If you use the CyberSync system, adding remote control to an Einstein only costs you can extra $30. Adding remote control to an AlienBees or White Lightning costs you an extra $90. Therefore, an Einstein with remote control costs you $530, and a White Lightning X1600 ($440) with remote control also costs you $530. However, the Einstein is a much more capable monolight, making it a much better value with remote control.

The color LCD display on the back on the Einstein looks cool and does provide you with a great deal of information and flexibility. However, in practice, I only ever push the big up and down buttons, and I actually prefer using the analog sliders on the back of the Alien Bees and White Lightning units. The LCD screen is hard to read in full sunlight, too.

Vagabond Mini

Though it's not a light, the _Vagabond Mini ($240)_ can be the most important tool you have for location lighting. Attach it to your light pole, plug your monolight into it, and you can use your studio monolights anywhere. It will power almost any monolight, including Paul C. Buff, Elinchrom, and Profoto lights—the brand doesn't matter, because the Vagabond simply provides power in a portable package (though it provides less power than your wall outlet, so your lights will take longer to recycle). In fact, I often see photographers with very expensive Profoto monolights who strap the Vagabond Mini to their light stands rather than buying one of the $1,800 Profoto BatPac battery packs.

They are good for about 400 to 500 flashes with a 640 Ws light at full power. With the aid of extension cords, you can connect up to four monolights to a single Vagabond Mini, though that will also increase your recycle time (which will already be much longer than normal).

The Vagabond Mini is only 3.5 pounds, so you can easily throw it into your travel bag and use it to charge your laptop, phone, or tablet. It even has a USB charger port built into it. It recharges in about 3 hours.

Balcar Mount

Used by all Paul C. Bluff lights, the Balcar mount is neither the easiest to use nor the most durable, but it's a reasonable compromise. Four brackets grip the circular ring of your light modifier. You'll need to move a lever or squeeze a pair of levers on your light in order to slide the speedring over the brackets, and then springs push the brackets back out and hold the light modifier in place.

It works well, but it requires you to have one hand on the light and one on the modifier. This can be a clumsy process with heavier modifiers, and it's definitely more graceful with two people.

Alien Bees has become popular enough that just about any type of light modifier is available for the Balcar mount, both from Paul C. Buff and from less expensive third parties.

Elinchrom

Elinchrom offers a wide variety of both monolights and pack-and-head systems. Elinchrom uses the extremely popular Bowens S-Type mount of lighting modifiers, meaning you'll always be able to find any type of light modifier at the price point you need.

Make	Model	New	Used	Wireless	Output (Ws)
D-Lite	RX ONE	$252		Free	10-100
	RX 200ws	$378		Free	20-200
	RX 400ws	$439		Free	40-400
BRX	Style 250	$525		Free	25-250
	Style 500	$675		Free	50-500
BXRi	Style 250	$640		Free	25-250
	Style 500	$750		Free	50-500
RX	Digital Style 300RX	$885		$139	9-300
	Digital Style 600RX	$1,000		$139	18-600
	Digital Style 1200RX	$1,400		$139	37-1200

All the Elinchrom monolights that I'm describing here except for the RX models have wireless capabilities built-in, so keep that in mind when comparing their pricing to competitors that don't built-in wireless. For example, adding wireless capabilities to Alien Bees lights costs $90 per unit, or $30 per unit for the Einstein e640 light. That's included in the price of the Elinchrom lights.

Like the Paul C. Buff lights, Elinchrom lights all have model light tracking, which automatically adjusts the modeling light according to the monolight output, and they'll use the modeling light to indicate when the flash has completely recycled. Unlike the Alien Bees and White Lightning lights, the Elinchrom lights also offer an audible beep when the monolight has recycled.

Skyport RX Wireless Control

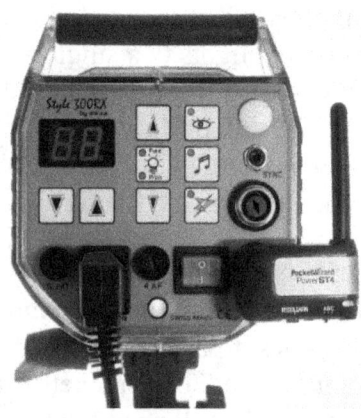

Elinchrom offers a fantastic variety of both wired and

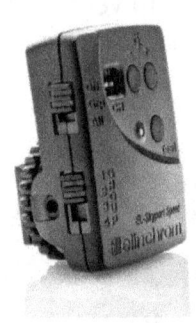

wireless remote systems, and you can use the PocketWizard PowerST4 receiver ($130) to control the Elinchrom RX monolights, if you've already invested in PocketWizards. If you choose Elinchrom, my recommendation is to invest in their Skyport wireless system. Many of their monolights have Skyport receivers built-in.

You will need to buy a transmitter (shown here) and attach it to your camera's flash hot shoe. You can use the transmitter to adjust the output of the lights in four different groups. It's simple and works wonderfully, despite the primitive interface. Be sure to buy the EL-Skyport Transmitter SPEED model ($120), and not the Eco model, because only the SPEED model provides control over light output.

You can use a second transmitter to allow an assistant to make adjustments to the lights off-camera while still triggering the lights using the on-camera transmitter.

In comparison to the Paul C. Buff Cyber Commander ($180), with its control of up to 16 lights, and a full color LCD display, the EL-Skyport seems extremely primitive. However, all I ever really need is to adjust output up or down, and the EL-Skyport does that perfectly well. Four groups of lights is sufficient for most studios, too.

If you use the top-end RX monolights ($900-1,400), you also have the option of using the PocketWizard system with one PowerST4 adapter per monolight ($130). I prefer the straightforward control of the PocketWizard AC3, but it's limited to three zones.

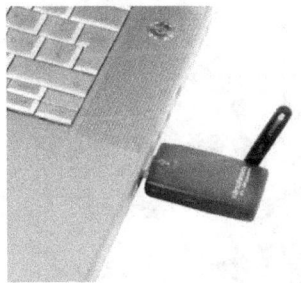 You can also control you lights from a computer running Windows or Mac OS. That sounds very cool, but I consider it to be more trouble than it's worth.

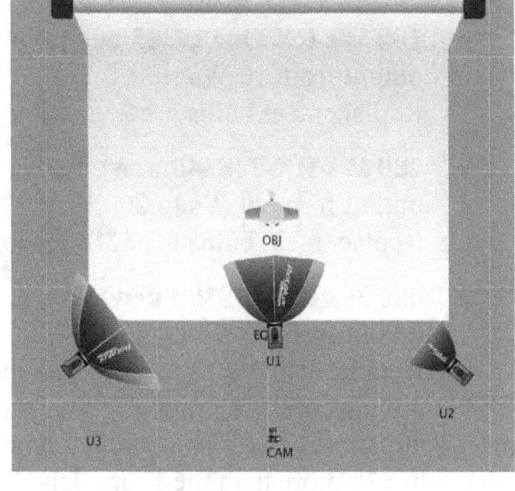

Setting up the Skyport system can be difficult. You might need to update the firmware on each of your lights. The documentation is poor and confusing.

However, once you get it running, the software provides full configuration over your lights from a computer running Windows or Mac OS. As shown in this screenshot, you can create diagrams that show the layout of your lights, helping you recreate a specific lighting setup. Like the transmitter, the software supports up to 4 different groups of lights, which is sufficient for most studios.

Skyport RX Wi-Fi Control

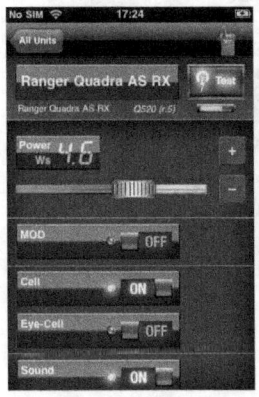
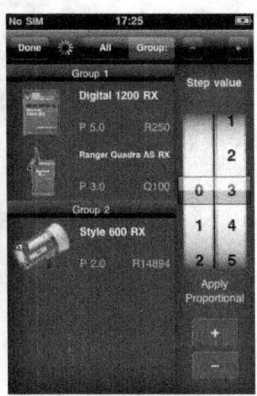

Elinchrom also offers an app that allows you *to control the lights from an iPhone, iPod, or iPad*.

However, none of the monolights have the required Wi-Fi support built in, so you need to add the SkyPort Wi-Fi adapter, shown here, for $200 per light. You might also need the Skyport USB RX Speed adapter ($30) to update the firmware on your lights to support the SkyPort Wi-Fi adapter.

This seems terribly cool, but I don't recommend bothering with it. It's unreliable, clumsy, and because of the cost of the adapters, very expensive. Setup is hard, and the app won't always find all your lights. It's also difficult to tell which light is which. In short, don't bother.

The sections that follow describe Elinchrom's monolights, which they refer to as Compacts.

D-Lite

The base-level Elinchrom monolights, D-Lites are offered at several different power ratings:

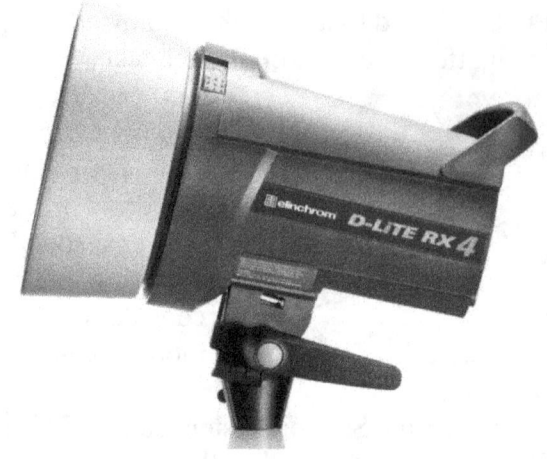

- **100 Ws RX One ($225 new)**. Power output from 10 Ws to 100 Ws. Replacement bulbs are $58. 2 lbs.

- **200 Ws RX 2 ($340 new)**. Power output from 20 Ws to 200 Ws. Replacement bulbs are $58. 2.9 lbs.

- **400 Ws RX 4 ($395 new)**. Power output from 40 Ws to 400 Ws. Replacement bulbs are $58. 3.3 lbs.

Elinchrom also offers IT models, but I recommend choosing the RX models because they offer compatibility with Elinchrom's amazing RX wireless system, and the RX models cost only $25 more than the IT models.

BRX

The BRX monolights are heavier duty than the D-Lite units, and therefore are designed for all-day use in portrait and product photography studios. Functionally, they are very similar to the D-List models.
You should only spend the extra money on the BRX models if you plan to use them constantly and you're afraid you might wear out the D-Lites.

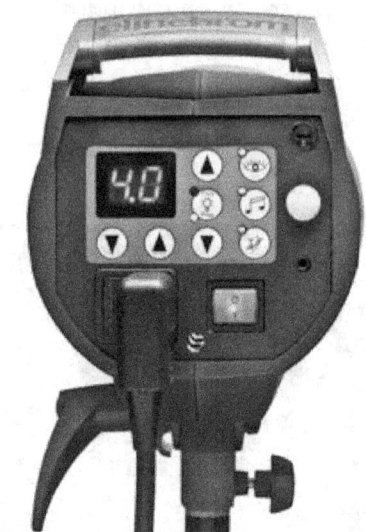

- **Style 250 ($525 new)**. Power output from 25 Ws to 250 Ws. 4 lbs. Replacement bulbs are $110. Minimum flash duration of about 1/2000th.

- **Style 500 ($675 new)**. Power output from 50 Ws to 500 Ws. 5.5 lbs. Minimum flash duration of about 1/500th. Replacement bulbs are $110.

BXRi

The BXRi monolights are the predecessor to the BRX lights, and I list them here only for completeness and to help reduce confusion if you're browsing Elinchrom's offerings. You might find good bargains on used models, but I don't recommend buying these new, because the BRX monolights are a better value.

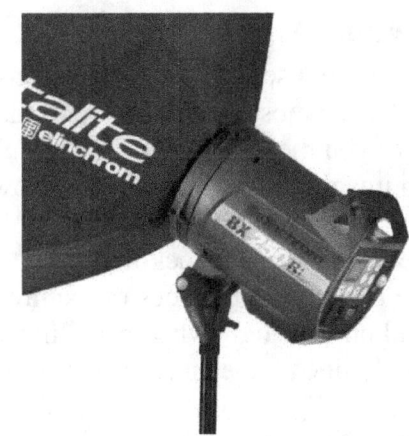

- **Style 250 ($640 new)**. Power output from 16 Ws to 250 Ws. 4.1 lbs. Replacement bulbs are $90.

- **Style 500 ($750 new)**. Power output from 31 Ws to 500 Ws. 4.5 lbs. Replacement bulbs are $90.

RX

The higher-end RX models are digital, allowing you to precisely specify the output and control it in $1/10^{th}$ increments, much like the Paul C. Buff Einstein e640. Unlike the Einstein unit, the RX monolights can be digitally controlled from your computer, which is only convenient if you already use a computer as part of your studio workflow. Elinchrom offers three RX models:

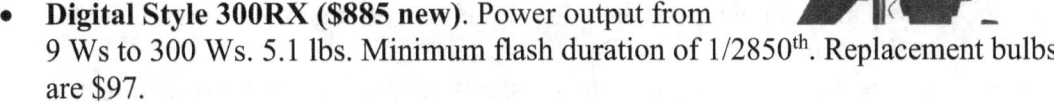

- **Digital Style 300RX ($885 new)**. Power output from 9 Ws to 300 Ws. 5.1 lbs. Minimum flash duration of $1/2850^{th}$. Replacement bulbs are $97.

- **Digital Style 600RX ($1,000 new)**. Power output from 18 Ws to 600 Ws. 5.7 lbs. Minimum flash duration of $1/2050^{th}$. Replacement bulbs are $97.

- **Digital Style 1200RX ($1,400 new)**. Power output from 37 Ws to 1200 Ws. 7.5 lbs. Minimum flash duration of $1/1450^{th}$. Replacement bulbs are $155.

Before buying the RX system, compare it to the Paul C. Buff Einstein e640. The e640 is very similar in output and features to the 600RX model, but offers reduced output down to 2.5 Ws and costs about half the price. The Einstein unit cannot be controlled from a computer, but the Cyber Commander controls will be sufficient for most studio needs.

Bowens S-Type Mount

Elinchrom uses the popular Bowens S-Type mount.
By far the most common mount, the Bowens mount
gives you the most options. Any light modifier you
can imagine is available for the Bowens mount, at
any price range, from high-end professional gear to
cheap third-party knockoffs.

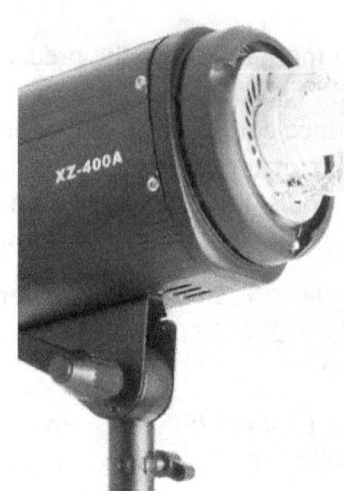

The Bowens mount doesn't require you to have a
hand on your light when installing a light modifier;
simply line up the three pins, push them in, and then
give it a slight twist. It locks into place. When you
remove it, you have to push a button on the light to
release the modifier.

Besides Elinchrom, many other lighting
manufacturers have adopted the Bowens S-Type mount. Therefore, if you decide to switch
to another system later, you might be able to continue to use your existing light modifiers.

Profoto

Profoto makes professional lighting equipment, and it's priced that way. It has a long
history with studio photographers, and as a result, many commercial and fashion studios
use Profoto equipment.

Profoto lights use Profoto's own speedring for attaching light modifiers such as soft boxes.
However, its system is popular, and there are more than enough name-brand and third-
party light modifiers available for the Profoto system.

If you have a large, fixed studio, multiple assistants, a constant stream of commercial
work, medium-format digital cameras, and $10,000+ budget for lighting, Profoto's system
is unparalleled. With practice, it will streamline your workflow like no other system and
have you shipping a constant stream of images to your art directors.

However, the system isn't the best use of budget for amateur and portrait photographers.
Instead, I recommend starting with the Paul C. Buff Alien Bees or Einstein lights, which
offer most of the capabilities of the Profoto system at about one-third the cost.

The sections that follow describe the key components of the Profoto system.

Air System

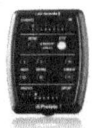

The Profoto Air Remote ($300) attaches to your camera's flash hot shoe and can
remotely trigger and control the output of up to six groups of lights. Though the
unit is overpriced compared to the competition, it's reliable and simple to use. If
you are invested in the Profoto system, it's a necessary accessory.

Profoto Studio Air and Capture One Plugin

Profoto also offers free software for Windows or Mac OS that allows you to control your lights from a computer. Unfortunately, this "free" software requires the purchase of a $500 USB dongle. Remember, the Elinchrom's USB adapter costs only $30.

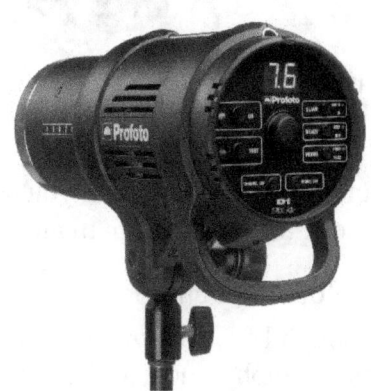

For most, the Profoto Air Remote is easier to use. However, the computer software provides control over more groups of lights.

If you use the Capture One software, Profoto offers a plugin for controlling its lights. Capture One is most often used with medium format digital cameras, such as those made by Mamiya and Hasselblad. Capture One allows you to remotely control and trigger your camera, and when you add the Profoto plugin, your camera settings can be linked to the light output. If you already use Capture One and Profoto equipment, the plugin is a great addition to your workflow. However, most photographers won't need it.

Air Monolights

Profoto offers five monolights, three of which support their wireless Air system:

- **D1 Air 250 ($1,100)**. Power output from 3.9 Ws to 250 Ws. Minimum flash duration of 1/1400th. Replacement bulbs are $156. 4.9 lbs.

- **D1 Air 500 ($1,200)**. Power output from 7.8 Ws to 500 Ws. Minimum flash duration of 1/2600th. Replacement bulbs are $156. 5.4 lbs.

- **D1 Air 1000 ($1,750)**. Power output from 15.6 Ws to 1,000 Ws. Minimum flash duration of 1/1800th. Replacement bulbs are $156. 5.4 lbs.

All the Profoto equipment is designed to be used in professional environments and is known to be reliable and sturdy.

Profoto Off-Camera Flash

Profoto's B1 500 AirTTL is a battery-powered 500Ws monolight that supports TTL, allowing your camera's autoexposure system to work correctly. Or, if you look at it a different way, it's a monster flash that's designed to be mounted on a light stand or tripod and can support a beauty dish, softbox, or other light modifier.

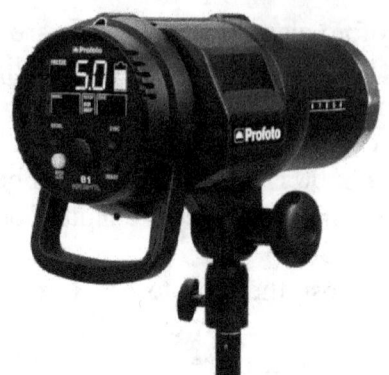

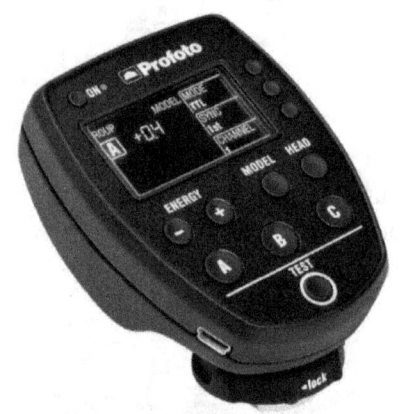

It can work as a standard cordless monolight, syncing in all the standard ways. However, its power is the ability to communicate exposure to your camera's TTL system, allowing you to immediately adapt to changing lighting conditions. In order to use TTL, you'll need to attach the Air Remote TTL-C (for Canon cameras) or the TTL-N (for Nikon cameras, when it's finally released) to your camera's flash hot shoe.

The B1 is quite possibly the world's greatest speedlight, but it's not for most photographers. Each head costs about $2,000, and the transmitter will run you another $400.

However, there's really no competition in this segment. If you need real monolight output and light modifiers, and you need to setup in seconds (without taking the time to manually tweak the output from manual lights), the B1 is your best and only choice.

Profoto Pack-and-Head Kits

Though I typically recommend monolights over pack-and-head kits to the up-and-coming photography, most of Profoto's lights are part of their pack-and-head systems. The sections that follow give a brief overview of the different systems.

You should note that Profoto calls their packs "generators." In the US, the term "generator" is typically used to describe an engine that turns gasoline into electricity. Profoto's generators, just like every other brand's pack systems, are mostly just capacitors that store electricity and then send it to the heads to be converted into light.

Profoto offers a flexible assortment of heads, including a ring flash ($1,700) and many specialty heads. While each system has heads designed specifically for that system, you can use lower-end heads on higher-end generators/packs. For example, you can use the Acute heads with your D4 generator, but you can't use the ProHead plus (designed for the D4) with the Acute system; it would overpower it.

Profoto D4 System

Profoto's mid-level system offers increasing power, flexibility, and cost:

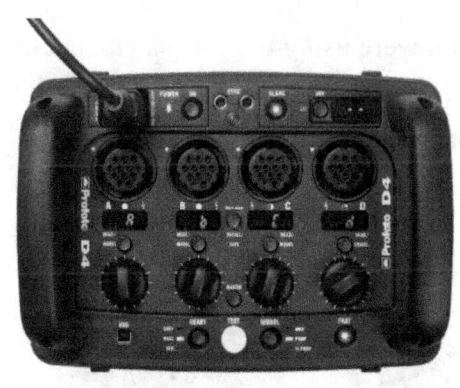

- D4 Air 1200Ws ($7,350)

- D4 Air 2400Ws ($8,400)

- D4 Air 4800Ws ($11,123)

There are several good reasons to upgrade to the D4 system from the Acute system:

- Greater output for large studio spaces or for shooting in full sunlight and longer distances.

- Individual control over the output of each head. You have monolight-like flexibility.

- Full remote control using the Profoto Air system.

- Higher frames per second.

- Increased color stability.

None of the D4 heads offer battery power.

Profoto Pro System

Profoto's high-end system is the ultimate in pack-and-head lighting. You can choose from four heads:

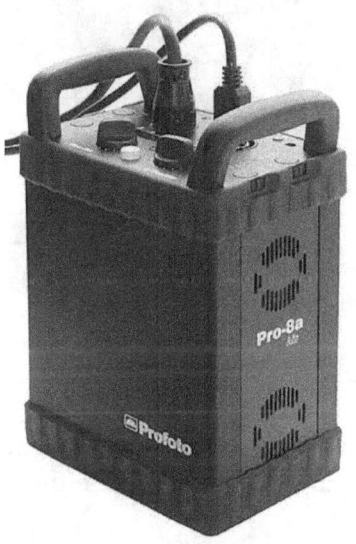

- **Pro-8a 1200 ($11,300)**. Shoots up to 12 frames per second with a 9-stop power range. Provides individual control over two lights.

- **Pro-8a 2400 ($13,000)**. Shoots up to 12 frames per second with a 10-stop power range. Provides individual control over two lights.

- **Pro-B3 1200 AirS ($5,300)**. Offers an 8-stop power range and up to 1,200 Ws of output. Up to 300 flashes at maximum power.

- **Pro-B4 1000 Air ($7,900)**. Offers an 11-stop power range and up to 1,000 Ws of output. Up to 220 flashes at maximum power.

Each pack supports only two heads, so you'll need twice as many packs as you would if you were using the D4 system, but each head can receive up to half the pack's total output. All the heads can be controlled by the Profoto Air system.

Profoto Specialized Monolights

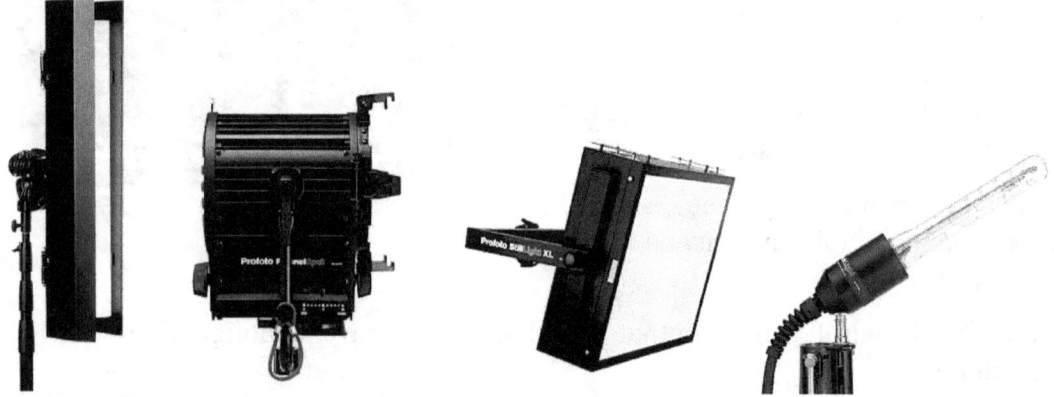

Profoto offers a *wide variety of specialized heads* for their pack-and-head systems, including narrow striplights, spotlights, and large light sources. This wide variety of heads is part of what makes Profoto the choice for professionals. Most photographers will never need a tiny stick light ($2,400) that they can wedge into a small place to light something from the inside out; but when you need it, you need it, and Profoto has the tool for you.

Profoto Mount

Profoto's mounts are generally considered the most solid. They'll hold onto even the heaviest light modifiers for years without showing any signs of age. The mount itself is quite simple—it's simply a cylinder, and you slide your light modifier over it (and, depending on the modifier, you might need to tighten it down). The mount also has an important functional advantage: you can slide the light modifier closer to or farther from the head, while still keeping it secure against the light. It works much like a flashlight with a zoom head. This allows you to zoom many light modifiers, changing the quality of the light by changing the position of the light head inside the light modifier. In the previous picture of the Profoto monolight, you can see a series of numbers on the side of the light, indicating different positions you can slide the light modifier to change the zoom effect. Unfortunately, it's not as popular as the Balcar or Bowens mounts used by Paul C. Buff and Elinchrom lights. Therefore, the mount type will limit your options for third-party light modifiers. However, if you have the budget for Profoto lights, you probably also have the budget for Profoto modifiers.

Chapter 7: Portrait Studio Equipment

Studio lights, by themselves, are almost useless. To get beautiful light, you need light modifiers, such as umbrellas, softboxes, and beauty dishes. This chapter explains the difference between these light modifiers and discusses other gear you might need to get great photos in your studio.

Light Modifiers

Without a light modifier, strobes produce a very harsh, uncontrolled light that spills throughout the room. You will almost always attach a light modifier to your strobes. They come in several different varieties:

Strobe Reflectors

Typically included with a strobe, the reflector is a dish that simply bounces the light straight forward. Some accessories, such as grids, attach to reflectors.

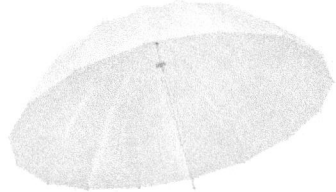

Shoot-Through Umbrellas

Cheap and portable, umbrellas are the first light modifier that you should master. Attach a shoot-through umbrella to your light stand and point the light directly into it. The umbrella softens and diffuses the light as well as casting it around the room.

Bounce Umbrellas

Attach these umbrellas to your light stand, point the light into them, and then point the light directly away from the area you want to highlight. Bounce umbrellas increase the size of the light source, but generally do not diffuse it. Huge focusing bounce umbrellas, known as parabolics, are popular in professional fashion studios but cost thousands of dollars.

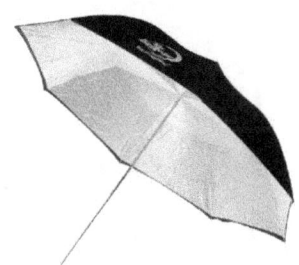

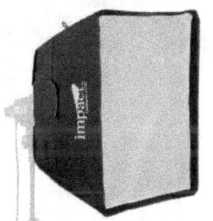

Softboxes

Soft boxes do a much better job of diffusing and softening light than umbrellas. The bigger the soft box, the softer the light will be. It's much more work to collapse a soft box than an umbrella, so soft boxes usually stay in the studio. Soft boxes are usually square, but you can also get octagonal softboxes (known as octoboxes) to create a differently shaped catch light.

Octoboxes

Octoboxes are simply eight-sided softboxes. They work exactly the same, but provide a round catchlight rather than a square catchlight.

Strip Lights

Strip lights are tall, narrow soft boxes. They're useful for evenly lighting a subject's entire body. They're also nice for creating a straight reflection in a polished surface.

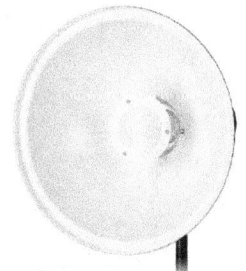

Beauty Dishes

Large reflectors with a diffuser in the center, beauty dishes have become the preferred main light for most types of portrait photography because they create a very even light with soft edges and a round catch light. Beauty dishes are easier to work with in the studio than softboxes, but they're not collapsible. You can put a diffuser (known as a sock) over a beauty dish to soften the light, or add a grid to limit the amount of light that spills.

Snoots

Snoots are cones or tubes that create a small tunnel of light. Snoots are most often used as hair lights, though they can be used any time you need to create a small spot of light.

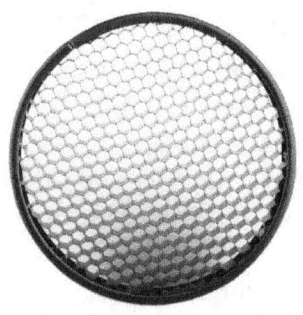

Grids

Grids, also known as honeycombs, narrow a beam of light, reducing the amount of light that spills. If you notice that a light is spilling onto a black backdrop, add a grid. When shooting high key, use grids to light the backdrop without spilling light onto your subject. Grids are measured in degrees, with smaller degrees producing a narrower beam of light. The figure shows the light cast by a strobe without and with a 30-degree grid.

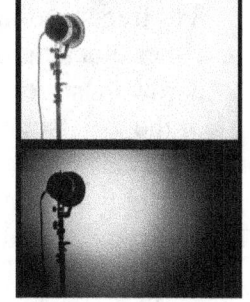

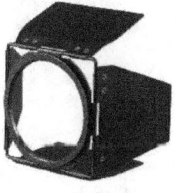

Barn Doors

Like grids and snoots, barn doors are designed to prevent light from spilling into unwanted areas. Barn doors can be individually adjusted to shape light the way you need it. In a pinch, you can tape a piece of cardboard to your reflector to act as a barn door—just make sure it doesn't get too hot.

Scrims

A scrim is a large sheet of fabric supported by a rigid frame (often made from PVC) that is used to diffuse light. Photographers sometimes use scrims indoors in rooms with direct light, but usually scrims are used outdoors and held between the model and the sun. By casting the model in partial shade and diffusing the direct sunlight, you both soften and reduce the light. Not only does this make the lighting more pleasant on the model, but it's easier to overpower the sunlight with your own flashes. You can buy scrims, but they're also easy to make.

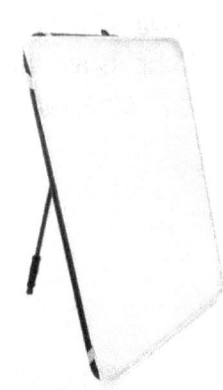

Gobo

A *gobo* (short for "go between") stands between the light and the subject, altering the light in some way. Technically, a scrim is a gobo. Most of the time when photographers use the term gobo, they mean a big piece of cardboard or foam core that blocks light and prevents it from spilling onto something that shouldn't be lit. For example, you might put a gobo between a kicker light and the backdrop to prevent the kicker from illuminating the backdrop.

V-Flats

V-flats are made from two large pieces of rigid foam-core board taped together so they can stand freely. 4x8 feet is a great size: taped together, they'll form an 8x8 free-standing and lightweight wall. V-flats are cheap yet tremendously useful for either blocking or reflecting light. You can buy the foam core you need at craft or hardware stores, though it can be difficult to find large enough sizes. If you cover one side of the v-flats with black, you can use that side to absorb light and prevent it from reflecting back.

Figure 6-40 shows several of these light modifiers. Of these different light modifiers, everyone should start with a big soft box, a snoot, and a reflector. If you enjoy portrait work, you should definitely get a beauty dish with a grid and a sock. As you use your studio lights, you'll no doubt discover that you need the other modifiers at some point to get the lighting you imagine.

Backdrops

Backdrops provide a perfect, non-distracting for portraits, product photography, and other commercial work. Regardless of the size of your studio or the quality of your gear, a simple backdrop creates instant, professional results.

Backdrops are available in a variety of different sizes and are made from many different materials. The sections that follow provide more information about backdrops.

> **Tip:** If you have a dedicated studio, paint the walls and floor a bright white color and use that as the default backdrop. While you're at it, paint the ceiling white, too: it'll soften the light in the entire room. If you're really committed, build a cyclorama wall,

which provides a seamless connection between wall and floor. For complete details, watch "How to Build a Cyclorama Wall" at http://vimeo.com/16778474.

Paper Backdrops

Paper backdrops are the least expensive, costing $20-$40 for a roll. You can buy big rolls of paper—usually white or black—and either suspend them from a stand system (costing around $70) or just tape them to a wall. The rolls are usually 6 to 9 feet wide and quite heavy, however, so be prepared to store and transport the roll. The paper will probably get creased and marked during the photo shoot, and re-rolling paper after a shoot is 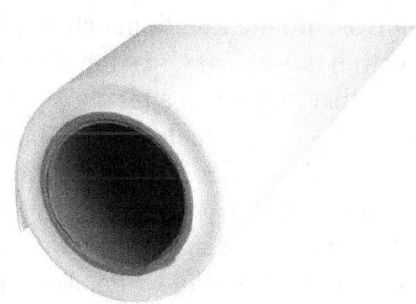 nearly impossible, but it's cheap enough that you can just tear off what you use and throw it away afterwards.

Paper backdrops provide a solid, smooth background that you just can't achieve with cloth. For that reason, paper is perfect when you need a solid white background for web images. Once you have a picture with a solid white background, it's easy to change the backdrop to anything in Photoshop: transparency, a different color, a texture, or even another photo. To prevent a white backdrop from appearing grey or showing the model's shadow, point one or two strobes at the backdrop. Ideally, the entire backdrop will be slightly overexposed, so make sure the background is blinking in your camera's display after you take the picture.

Vinyl Backdrops

Vinyl backdrops are a worthwhile upgrade to paper backdrops. They're several times more expensive than paper, costing around $150 for a traditional 8x20 foot backdrop. While paper is disposable, vinyl backdrops are more permanent. Rather than throwing them away when they get dirty, you can clean them.

Vinyl backdrops are less likely than paper to wrinkle, which allows you to produce cleaner pictures. Any wrinkles you do get will disappear after a few days of hanging.

Cloth Backdrops

Cloth backdrops, including muslin and canvas, are the traditional choice for portrait work. They're so traditional, in fact, that they seem a bit cheesy and dated nowadays. Nonetheless, cloth backgrounds are still the choice for many school and business photos.

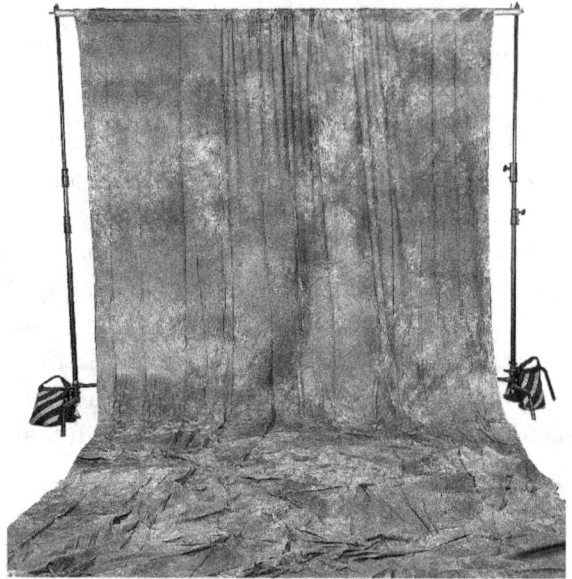

Cloth backdrops always provide a texture, even if they're solid white. To minimize the texture, leave at least four feet between your model and the backdrop. To prevent shadows and to create a pleasing halo effect behind the model, point a strobe at the backdrop behind the model's head.

Cloth backdrops are more expensive than paper, running $50 to $200 depending on the size. You'll also need to spend about the same amount on a stand system to suspend the backdrop—they're typically too heavy to tape to the wall.

Cloth backdrops are far more portable than paper because you can fold them up into a small space and toss them into your bag. However, you might also have to bring an iron to steam the creases out of the backdrop.

Collapsible, Portable Backdrops

If you don't have room to store a paper backdrop, but you want a flat background, you can buy portable backdrops that fold to less than one-eighth their full size. These backdrops are also useful if you travel to a client's home or office and you want to be able to provide a clean background.

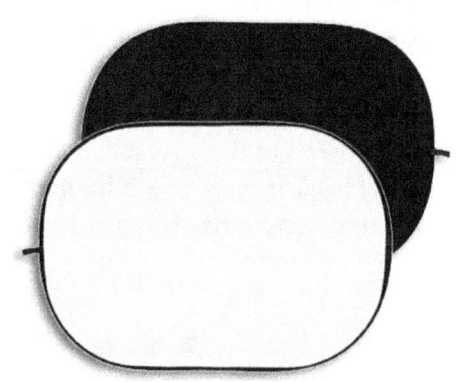

While useful, these expandable backdrops are less than ideal. Though they have a frame designed to stretch the background fabric, the fabric will still be wrinkled.

That doesn't have to ruin the picture, but it is a challenge to make the wrinkles not appear in the pictures. If you're shooting a white background, shine a light directly on it to completely overexpose it. If you're shooting a black background, make sure no light falls on it so that the backdrop disappears.

Backdrop Support Systems

Of course, you'll need a way to support your backdrops. Most backdrops come on a roll that allows you to place a pole inside and support it from either end.

Backdrop supports can be portable and temporary (shown first), or fixed and permanent (shown second). If you have a permanent studio space, attaching backdrop supports to your wall or ceiling makes backdrops much easier to extend and collapse. Permanent supports also decrease the chance that a child or clumsy photographer will knock over one of the poles, causing the entire system to collapse and possibly ruining your backdrops.

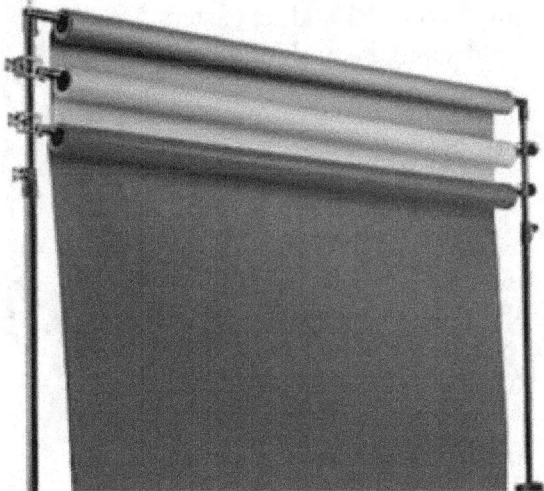 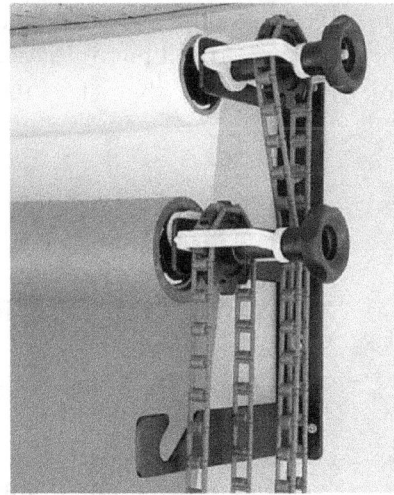

If you do use a temporary backdrop support system, be sure to place sandbags over the base of the supports to make them more stable. You'll probably also need a clamp to stop the paper from unrolling.

Light Stands

Lights, reflectors, and even your computer can be attached to light stands. Light stands are like tripods with a narrower base. You can spend anywhere from $20 to $7,000 on a light stand, depending on the weight, features, and strength.

Your first lighting kit will come with cheap light stands. As you add more lights, you can buy better quality light stands. Better quality light stands are cushioned, which makes it easier to adjust the height without jarring the strobe and possibly breaking the bulb. Monolights and strobes are ready to be mounted directly to light stands. Flashes, however, are designed to be mounted to a camera. To mount a flash to a light stand, you will need to use a *flash light stand adapter* (about $20-$30). Choose one that provides a place to mount an umbrella.

Folding Light Stands

Traditional light stands have three legs that fold for storage or travel, like a tripod with short legs and a very long center column. These light stands are sufficient for most studios. If you frequently move them, you might consider adding casters (wheels) to allow you to roll them. Like with all light stands, you should have a sand bag on them to reduce the risk of the stand falling.

C-Stands

C-stands have fixed, rather than folding, legs. However, they can be disassembled for travel. They also tend to be sturdier and heavier than folding light stands.

C-stands are better than folding light stands. Unfortunately, they're also much more expensive. An 8-foot C-stand will cost $120 to $250, whereas a similar folding light stand might cost $15-$25.

C-stands are a good investment for studios that receive heavy, constant use. Folding light stands can be fidgety, and a C-stand pays for itself if it saves a single broken light.

Microphone Stands

At the other end of the spectrum, microphone stands can be used to support lights or cameras (though you might need an adapter, depending on the threading). Microphone stands can be found slightly cheaper than light stands (as low as $10 each), but the primary benefit is a reduced footprint. Microphone stands have a heavy base that's smaller than a folding light stand, providing stability while allowing you to more easily navigate around your lights in a very small studio.

Booms

Once you get some practice with standard light stands, you'll discover something frustrating: if you want a light directly in front of your subject, the light stand will be in the way of your shot. Booms work around this by hanging the light out on a long arm. I use a _CowboyStudio boom_, shown in Figure 6-38, that costs about $70. While it's inexpensive, it's only suitable for lightweight flash heads and light modifiers; heavier modifiers like large _beauty dishes_ will cause the boom to twist. Booms are particularly unsteady, so use _sandbags_ to both balance the boom and keep it from falling over.

You will need to put sandbags over the base of your light stands (especially booms). Sandbags are surprisingly expensive, but if you don't use them, it's only a matter of time before you knock a light stand over. When that happens, you'll definitely break a bulb (which is expensive to replace). You might also knock over other light stands, damage your backdrop, or hit a person—possibly burning them.

If you use your studio regularly and you have flat, hard floors, put some wheels (known as casters) on them. Casters allow you to more easily move your lights, especially if you have them properly sandbagged.

Apple Boxes

For a traditional portrait, the lens should be slightly above the model's eye level. This is going to be a problem unless every model is about four inches shorter than you.

We studio photographers use apple boxes to change either the subject height or the model height. If the model is more than about six inches shorter than me and I don't want to kneel or crouch, I'll have the model stand on an apple box. If the model is my height or taller, I'll stand on the apple box.

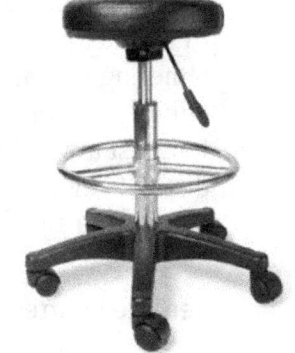

Apple boxes are also critical for equalizing height in family portraits. Have shorter subjects stand on boxes so the height differences are minimized.

You'll need apple boxes in at least three sizes so that you can adjust heights differently. Naturally, you can stack multiple apple boxes as needed. In a pinch, you can use large books, cinder blocks, or bricks.

Posing Stools

Stools used to be a standard element in a photography studio. Nowadays, however, most models simply stand (for an individual portrait) or kneel (for group photos where you need to equalize height). Stools are still useful for children, however, because having a child sit on a stool helps to keep him or her in a single place. Stools are also useful as an alternative to apple boxes for adjusting the height of your subject.

Posing stools should rotate, have adjustable height, and have a foot rest and wheels. They shouldn't have a back (which might appear in pictures).

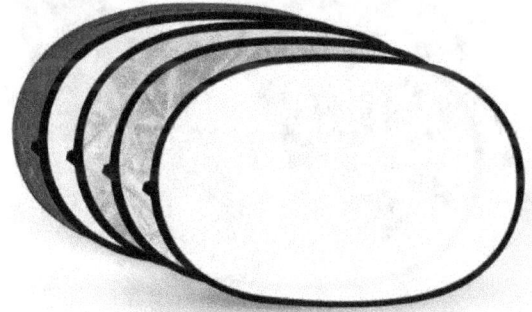

Reflectors

Reflectors bounce light. In a studio environment, photographers often use a reflector as an inexpensive fill light, to reduce shadow depth.

Reflectors are commonly white, silver, or gold. White reflectors add a diffused fill light. Silver reflectors add more light, but the shadows won't be as even. Gold reflectors add a touch of gold color to the light, which might make some skin tones more appealing. I typically use a white reflector first, and switch to a silver reflector only if the white reflector doesn't add enough light. I'll only use a gold reflector if my model has an extremely pale complexion.

Fans and Wind Machines

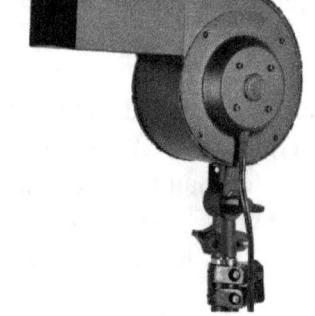

Fans blow the air around, creating movement in the model's hair. A little bit of wind almost always improves photos of subjects with longer hair. It also helps to keep the model cool under the lights. Any standard pedestal fan is good enough for hair.

You can use more powerful wind machines or even an electric leaf blower to add movement to dress fabric.

Tethering

Some photographers keep a computer in their studio and tether their camera directly to it. Tethering allows you to instantly review your photos on a larger screen. When working with Adobe Lightroom, you can even automatically apply some processing to the pictures to better assess the final image.

Tethering is also useful in larger studios with a separate art director. The art director can examine the photos as they are shot, providing instant feedback to the photographer (without having to look over the photographer's shoulder at the back of the camera).

Most cameras can be tethered using Wi-Fi, possibly with a special SD card such as an Eye-Fi card. However, wireless tethering is too slow to be useful in most environments. For tethering to really be useful, you need a camera with wired Ethernet, such as a Canon 1DX, Nikon D4S, or most medium-format digital cameras.

I almost always prefer not to tether when shooting in the studio. Being able to see pictures distracts models and can make them self-conscious. Instead, I prefer to select the best photos and show them to the model on the back of the camera.

Remote Shutter Release

You can use a wireless remote shutter release to trigger your camera while you move around your studio. Obviously, this would only be useful if you had a still subject and had your camera on your tripod. However, that's a common scenario in traditional portrait photography.

Being able to move away from your camera can speed up portraits when working without an assistant. You can adjust a light or your model's hair and instantly take a sample photo without walking back to your camera.

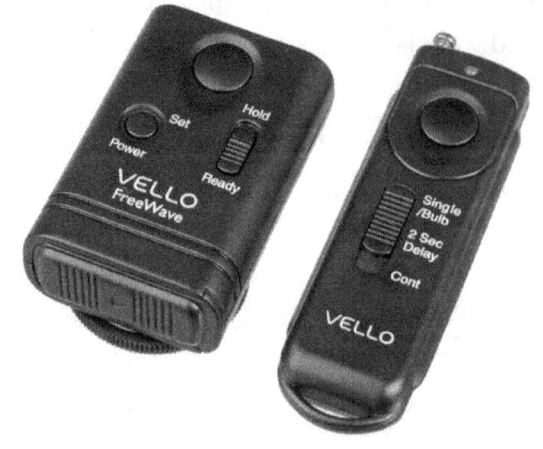

You should only purchase a radio frequency (RF) remote shutter release. Infrared remote shutter releases are too unreliable to be useful.

Chapter 8: Tripods and Monopods

Tripods steady your camera, allowing you to walk away from the camera to pose models, adjust lights, or take a self-portrait. They're also required for long exposures that would induce camera shake, such as pictures taken at night or during the day with a neutral density (ND) filter.

Video: Tripods
9:50 - *sdp.io/Tripods*

This video provides a quick overview of the tripods Chelsea and I use, along with the features that differentiate them.

The sections that follow provide my quick recommendations for those who want to buy quickly and then offer detailed descriptions of different tripod features to help you better make your own buying choices.

Quick Recommendations

If you don't want to learn about tripods and just want something to stick your camera on, here are some quick recommendations for those of you primarily shooting still photos:

- **Cheap first tripod:** _**Dolica GX600B200 ($47)**_. This ball head tripod extends to five feet, making it suitable for photographers up to about 6 feet tall. It has flexible legs that allow you to get low to the ground, and it has a hook that you can hang your bag from to make it more stable. It's small and light enough to travel with. If you're over six feet tall, buy the 65-inch version.

- **Mini tripod:** _**Pedco UltraPod II ($17)**_. There are dozens of mini tripods available, but this model has Velcro that allows you to strap it to poles or branches, giving you a higher viewpoint. It's not sturdy enough for a heavy lens, however.

- **Smartphone tripod mount:** _**i.Trek Super Mount F ($19)**_. These versatile mounts work with any smartphone, making them more versatile than the phone-specific models. I carry one of these with me even when I have my DSLR, because it's a great backup.

- **Professional travel photo tripod:** _**Manfrotto BeFree**_ **($170-290).** Most people won't ever need a professional tripod, but if you're finding your cheap first tripod a bit too cheap, this tripod is light, sturdy, and repairable. Manfrotto sent us one to test, but we bought a second copy because Chelsea & I always fought over it. We prefer the carbon fiber model because it's a tiny bit lighter and we tend to hike for many miles with it, but the aluminum model is a better value.

- **Professional travel video tripod:** _**Manfrotto BeFree Live ($215).**_ The BeFree Live is the video version of the BeFree tripod, including a leveling pan-tilt head.

- **Professional studio video tripod:** _**Manfrotto MVH502A 546BK-1 ($580)**_. This tripod completely lacks a center column, but still sets up quickly. The ball mount allows you to quickly level the head, just like a leveling center column. The MVH502A fluid head provides silky smooth pans and tilts.

- **Luxury tripod: Anything Gitzo.** These $900+ tripods are outrageously expensive but they feel absolutely wonderful. If you have the budget and enjoy a high-quality product, I recommend them.

Types of Tripods

Because photographers shoot different subjects in different conditions, there are many different types of tripods. The sections that follow describe the most common types.

All-in-One

The least expensive and most portable tripods are all-in-one tripods that have the legs and heads permanently connected. These tripods are the right choice for most people's first tripods. Once you've spent some time with your all-in-one tripod, you might discover that it's missing some feature you wish it had, such as a different type of head or more height. However, the money you spent on your first all-in-one tripod won't be a waste, especially if you buy a decent quality all-in-one tripod.

Travel

Travel tripods compromise functionality for portability. Typically, they're quite small, which makes them easy to store in a bag. I look for a travel tripod that fits vertically in my carry-on bag without my having to angle it sideways, where it would consume too much space in my bag.

Inexpensive travel tripods are made from plastic, while the higher-end travel tripods are made from carbon fiber. Either way, they're likely to be light enough that you'll want to anchor the tripod for long exposures, such as those taken at night.

Studio

Studio tripods are strong, sturdy, and heavy. They're definitely the best type of tripod to use, but they tend to be too cumbersome to travel with.

Everyone needs at least one travel tripod, and your travel tripod will probably be good enough for most uses around the house and in studio. Therefore, I recommend starting with a travel tripod, and if it isn't study enough, purchase a separate studio tripod for use when you don't mind the size and weight.

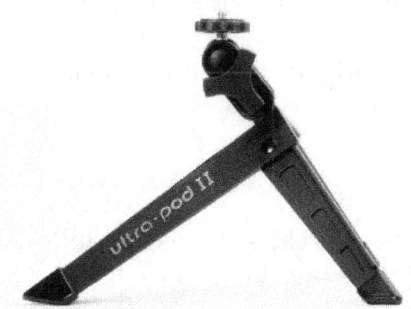

Pocket

Pocket tripods are optimized only for portability. Typically, they stand only 4-8 inches tall and are used for taking quick self-portraits. Because they're so low, they're not designed for standing on the ground—you'll need to find a table you can rest them on. They also don't extend vertically, so how high or low your camera is held will be determined entirely by where you rest the tripod.

The problem with this is that you typically rest them on a table or counter, which is below eye-level. For self-portraits, this results in an uncomfortably low perspective that shows the underside of everyone's chin, making people seem heavier than they are.

Pocket tripods are fairly useless except that you can literally keep them in your pocket or bag, so they're great for carrying with you everywhere. There's an old saying, "The best camera is the one you have with you." The same applies for tripods.

Note that most pocket tripods aren't strong enough to support a full-sized DSLR and lens. They're best used for small mirrorless cameras, point-and-shoot cameras, or smartphones.

Monopods

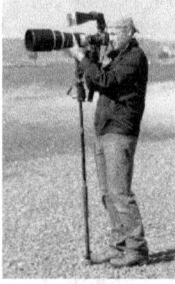

Monopods are tripods with a single leg. They support the weight of your camera and virtually eliminate any camera shake; however, you can't walk away from your camera. They're much lighter and easier to carry than a tripod, making them ideal for wildlife photographers who tire when holding a large telephoto lens, as shown in the following picture. Many photographers will also use a monopod in the studio, because monopods eliminate camera shake while allowing the photographers to move around the studio more freely.

I personally almost never use a monopod. However, my wife, Chelsea, is much smaller than I am and can get tired when using telephoto lenses. Therefore, she prefers to use a monopod in the studio. As small as she is, Chelsea still hates using a monopod for wildlife, because it limits your ability to follow flying birds.

Specialized

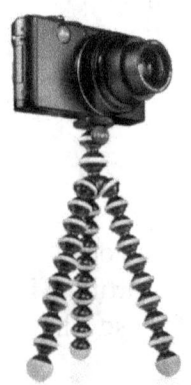

You can also buy several types of specialized tripods. The two types I commonly use are:

- **Suction cups**. You can use suction cup tripods to mount your camera to a car or glass—these are extremely useful for video inside or outside of a moving car. That sounds scary, but I've spent hours driving with cameras stuck to the inside and outside of my car, at highway speeds, and never even had a close call. _Fat Gecko_ makes a series of excellent suction cup tripods. Use them at your own risk, however!

- **Flexible legs**. _GorillaPod_ makes a series of tripods with legs that are so flexible you can bend them into any shape you want, or even wrap them around a pole. While cool (and popularized by vlogger Casey Neistat), I don't find myself using them very often.

These are just two examples—you can purchase specialized tripods that allow you to stabilize your camera in just about any situation.

Legs/Sticks

Tripods have three legs, or as videographers call them, "sticks." The sections that follow discuss the qualities different legs have.

Height

Ideally, a tripod will extend high enough that you'll be able to look through your camera's viewfinder without crouching down. That doesn't mean that the legs need to extend to your standing height. Typically, you can buy legs about a foot shorter than you are and be quite comfortable.

The height of the legs isn't the height of your viewfinder. On top of the legs, you'll add a head, which is typically 4-6 inches high. Additionally, your camera's viewfinder is usually 3-4 inches above the head.

For ideal comfort, I recommend getting legs that extend to about 8-12 inches shorter than you. If you get a taller tripod, it'll be unnecessarily large and heavy. However, I have at times extended a tripod far above my own head. For example, I'll extend my tallest tripod very high when shooting over a crowd. That's a really useful technique when shooting firework shows.

Segments

Most tripod legs have either three or four segments. Legs with four segments allow the tripod to fold smaller and extend longer, but the legs themselves are flimsier. Legs with three segments don't extend as far, but the legs are sturdier.

Therefore, travel tripods typically have four segments, whereas studio tripods typically have three.

Materials

Tripod legs can be made of several different materials, each with their own advantages:

- **Plastic**. Plastic is lightweight and inexpensive. For this reason, it's very common on tripods priced under $30. Unfortunately, plastic tends to be rather flexible, which can cause your camera to wobble a bit, especially in the wind. Nonetheless, a plastic tripod is just fine for taking pictures in daylight or with flash. If you just plan to take some self-portraits, a plastic tripod is perfect. However, if you plan to do long exposures (such as those involved in night photography), a plastic tripod will result in many photos ruined by camera shake.

- **Metal**. Mid-range tripods and studio tripods are often made out of some form of metal, often aluminum. Metal is much more rigid and heavier than plastic, making it ideal for long exposures. When I don't have to carry the tripod somewhere, I always prefer a metal tripod.

- **Carbon fiber**. The most expensive tripods are made from carbon fiber because it's more rigid than metal and as lightweight as plastic. For that reason, carbon fiber tripods are ideal for travel and hiking. However, their light weight means they can shake in the wind. If you do buy a carbon fiber tripod, look for one with a hook that allows you to hang a bag from the center column. The bag will increase the weight of the tripod, reducing the shakiness.

Clasp Types

Legs must lock into place after you extend them. Clasps are either flip or twist, as shown in the next illustration. I find the flip clasps to be much quicker and more reliable; therefore, I always look for legs with flip clasps.

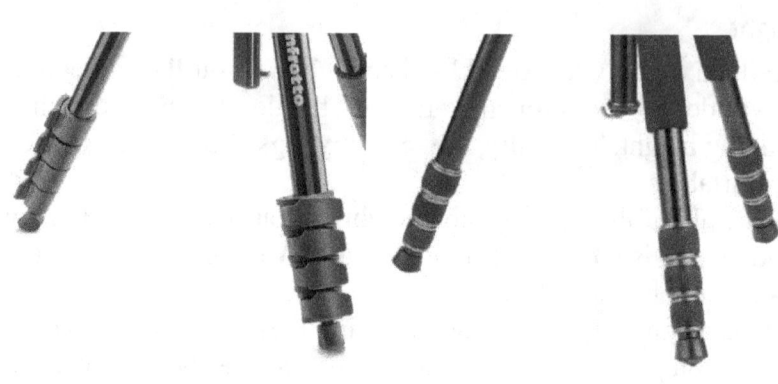

Here are some of the problems with the twist clasps:

- You're never confident that you've tightened them enough to hold them in place. If you've accidentally under-tightened them, the tripod might collapse, causing your camera to fall. Or, if you've stowed the tripod, and under-tightened leg might extend, catching against something as you walk. If you over-tighten a leg, you might crack it.

- You don't have a visual indicator of whether a leg segment is tight or loose. Therefore, the only way to know if a leg is locked or free is to physically twist it with your hand.

With that said, we love the Gitzo twist lock tripods. Loosening the twist locks just a bit allows them to more quickly extend or retract the leg.

One disadvantage to clasp-lock tripods is that the screws need to be tightened every couple of years. I definitely suggest tightening the screws before an important outing to ensure the leg doesn't slip. Often this tightening requires a proprietary tool included with the tripod, so be sure not to lose it.

Center Columns

Most tripods have an extendable center column. Typically, you'll extend the legs as a group to about the right height and then adjust the individual legs to level your tripod on uneven ground. Then, you'll extend the center column so that your camera is at the perfect height.

Center columns have different features, too:

- **Removable**. Some tripods, especially travel tripods, have a removable center column that allows you to extend the center column sideways or even completely upside-down. These are very useful for macro/close-up work, but they're rarely useful for other types of work. Though you won't need the feature often, when you need it, you really need it.

- **Leveling**. Center columns used for video allow you to quickly level the head after you've extended the legs. If you're using a ball head, this isn't useful at all. However, if you're using a pan/tilt head for video, a leveling center column is critical. While you could adjust the length of the individual tripod legs to perfectly level the head, this is difficult and time-consuming. With a leveling center column, a perfectly level head is only a couple of twists away.

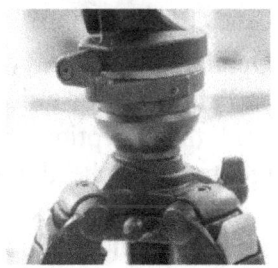

- **Hook**. Center columns with a hook on them allow you to hang a bag from them, adding weight and stability to your tripod. This is a great, but unfortunately rare, feature to have.

Center columns are usually included with the tripod legs because there's no standard way to attach a center column to legs. Therefore, you should consider the capabilities of the center column when choosing legs.

Heads

The tripod head attaches to the top of your legs and provides a connection to your camera. The head gives you the ability to pan, tilt, or twist your camera in different directions without moving the legs. The sections that follow describe different characteristics of tripod heads.

Size

Tripod heads of any type are available in a wide variety of sizes and weights. Smaller heads are obviously better for travel; however, they're also much less sturdy. If your head is too small for your camera weight, it might not hold your camera in place, especially if you have it tilted at an angle or if you're using a large lens.

A bigger head is always preferred when you're taking a photo. They're easier to use and they won't drift as gravity pulls on them. However, they cost more, weigh more, and take up more space in your bag.

For studio tripods, get the biggest head you might ever need. For travel tripods, choose a head just large enough to support the camera and lens you intend on using. Tripod heads typically list a weight that they're rated for, so weigh your camera, lens, and flash, and choose a head that can support that weight (plus a pound or two).

Quick Release Plate

Quick release plates (QRPs) attach directly to your camera, and allow you to quickly attach and detach your camera. With the exception of pocket tripods, I would never recommend a head without a QRP. Fortunately, all tripods with separate heads include a QRP, and only the least expensive all-in-one tripods lack a QRP.

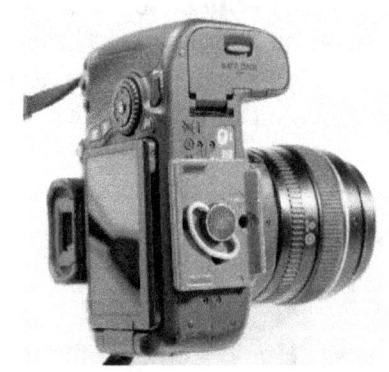

If you purchase multiple heads, choose heads with the same QRPs so that you can easily move between them. It also helps to purchase extra QRPs and keep them in your bag, because it's very easy to lose or forget a QRP, and then you'll be unable to connect your camera to your tripod.

Ball Heads

Ball heads allow you to twist and turn your camera in any direction, including turning it 180 degrees to take a vertical picture. Ball heads are the standard choice for still photography.

There are dozens of variations on the ball head. Most have a knob that loosens the ball head to allow you to adjust the camera angle and then re-tighten it so the camera stays in one position.

Joystick Heads

Joystick heads are a type of ball head designed for one-handed operation. With a typical ball head, you must turn the knob with one hand and adjust your camera with the other hand. With a joystick head, you just squeeze the switch and adjust the head with a single hand.

I generally dislike joystick heads because they tend to be unnecessarily large. However, if you're doing night photography (where you might have a flashlight in one hand) they can make the process much easier.

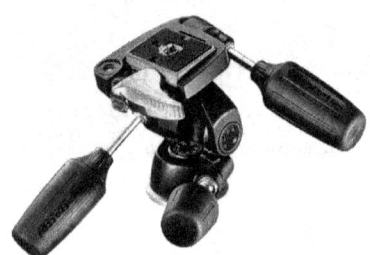

Pan and Tilt Heads

Pan and tilt heads separate horizontal movement (panning) from vertical movement (tilting). This allows you to level your tripod, For video, look for a fluid pan and tilt head. Fluid heads allow you to pan and tilt much more smoothly while recording. If you plan to take only stationary shots, you don't need a fluid head.

Hybrid Heads

Hybrid heads attempt to provide both ball head flexibility and pan-and-tilt capabilities. There's a switch that switches the head between ball head mode (for still photos) and pan-and-tilt mode (for video).

I've tried several different hybrid heads and haven't liked any of them. They're a jack-of-all-trades and master-of-none, and are frustrating whichever mode they're in. They also tend to be more expensive than good examples of either ball heads or pan-and-tilt heads. Therefore, I recommend avoiding hybrid heads and purchasing separate heads for different uses. Swap your heads out as needed; it's rarely necessary to quickly switch from a ball head to a pan-and-tilt head.

Gimbal Heads

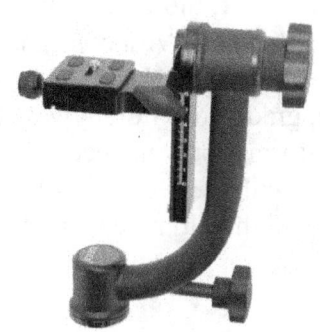

Gimbal heads are designed for use with large telephoto lenses, such as a 300mm f/2.8, 400mm f/2.8, 500mm f/4, or 600mm f/4. Gimbal heads balance these large, heavy lenses better by attaching either at the side or from above.

The advantage of gimbal heads is that your tripod stays better balanced as you tilt your camera and lens up or down. With a traditional ball head, tilting a heavy lens would cause the entire tripod to be off-balance, potentially knocking it over and causing all your expensive gear to crash to the ground.

Many wildlife and sports photographers use gimbal heads. Even though I often use a very heavy 500mm f/4 lens, I don't like gimbal heads for most wildlife work, because I prefer to handhold my lens so that I can easily turn to the side or lift the entire camera up. However, gimbal heads are a necessity if you're going to be staying in one place and you know approximately where your subjects are going to be. This makes them perfect for professional sports work, where you will be assigned a position along the sidelines and you will need to stay in that spot for hours at a time. In that situation, handholding a heavy lens would be impossible. They're also useful for wildlife photographers who are camouflaged and waiting for an animal to appear at a specific location.

Miscellaneous Features

Some tripods offer these features, which vary in usefulness:

- **A level**. In theory, a level allows you to guarantee a straight horizon, which is critical for video and quite important for photography. However, a level that's attached to the legs is usually quite useless, since the head itself might not be level with the legs. Don't bother with tripod legs that have a level—instead, look for a head that has the level built into it, or just use the level that's built into most new digital cameras.

- **Carrying straps**. If you plan to travel with your tripod, a strap that allows you to carry it is extremely useful.

- **Hooks (or attachments for carabineers)**. I love tripods that have hooks or places to attach a carabineer because they allow me to more easily attach the tripod to a bag when carrying it.

Chapter 9: Computer Gear

The more serious you get about photography, the more time you spend editing your photos and the more time you spend waiting for your computer to perform different photo editing tasks. If you're buying a new computer, it makes sense to choose one that will be particularly quick when editing photos.

However, most new computers can do an excellent job with the two most popular applications, Adobe Lightroom and Photoshop. If you're happy with your PC's performance, there's no need to upgrade. However, if things seem slow, or you plan to upgrade your PC anyway, the sections that follow will give you the general background information you need to make educated choices.

As with the rest of this book, my goal is not to simply list all the highest-end equipment to setup the ultimate workstation. Instead, my goal is to get you the most editing power for whatever your budget is.

Quick Answers

The sections that follow provide more detailed information for technical readers, but these tips tell you most of what you need to know:

- **Get at least 8GB of RAM**. More is better, but you probably won't benefit from more than 32GB.

- **Get the fastest processor you can.** Processing power is very important for photo editing.

- **Get an SSD (Solid State Drive) and a PC that supports SATA 3.** SSDs greatly decrease the time it takes to open and save your photos. They're expensive, though, so you might consider getting a high capacity non-SSD drive, such as a 3TB or 4TB drive, and moving your pictures to it after you're done editing them. Lightroom will store your previews on your SSD drive, allowing you to quickly view those images. m.2 NVMe drives, such as the Samsung 960 EVO Series, offer much better disk performance, but we've found minimal impact on Lightroom and Photoshop performance because they're limited by the processor.

- **Get multiple monitors.** Two monitors makes using Lightroom much easier because you can preview the full-size version of your picture on the second monitor. If you have a laptop or other mobile PC, you can connect it to an external monitor or TV and use that as your second monitor.

- **Get big, cheap monitors.** Calibrated monitors are for professional designers who have their own professional printing equipment. Most photographers don't need to calibrate their monitors, and professional monitors are nice, but unjustifiably expensive.

- **Don't waste money on an expensive graphics card.** Any basic graphics card will do. Expensive graphics cards are for gamers and video editors, not for photo editing. Lightroom and Photoshop do offer GPU acceleration, but the current benefits of that are minimal in our testing.

- **Get a desktop.** Unless you need to edit photos while traveling, a desktop always provides better performance than a mobile PC (for similar budgets). Desktops are also much easier to upgrade and repair.

- **Back up your pictures.** Hard drives fail and computers get stolen. If you don't back up your computer, you will lose everything. Online services such as Carbonite.com and Mozy.com will back up your entire PC across the Internet (for a fee). You can also buy an extra drive and back your pictures up to it using the backup software included with your operating system.

Mac or PC?

Either Mac OS or Windows will work fine, and Lightroom and Photoshop function exactly the same on either operating system. Macs are lovely, but they tend to be more expensive for similar performance. Windows 7 and Windows 8 have solved the biggest complaints people used to have about Windows—that it was unreliable and prone to security problems.

If you're an iPhone user, you might appreciate how well Macs integrate into the Apple infrastructure. You can Airdop photos between your computer to your phone in seconds, which is useful for quickly sending an editing picture to your phone for sharing on Instagram.

If you're looking for a laptop, the guidelines discussed in this chapter apply to laptops, too. One particularly important consideration is the resolution of the screen. The higher the resolution, the sharper your pictures will be while you're editing them. Also look for laptops with SSD drives and an i7 processor.

Avoid ChromeBooks. While they're great for what they are, they don't run Lightroom or Photoshop.

Here are some specific recommendations at different sizes:

- **Small 13" ultrabooks.** Consider the _Samsung ATIV 9 Book Plus_ ($1,400) with an amazing 3200x1800 screen. The MacBooks with the Retina displays are amazing for photo editing on the go, too.

- **Standard 15" laptops.** For this size, the 15" Macbooks with Retina displays are the best choice.

Choosing Computer Components

Designing a computer for a particular task involves determining which of the computer components are limiting (also known as bottlenecking) the performance of that task. For image editing, three components equally limit your overall performance:

- **Disk speed**. When you edit a picture, the computer has to read it from the hard disk and copy it into RAM. When you are sorting through hundreds of pictures after a photo shoot, the disk speed is going to be the biggest factor slowing the process down. Therefore, a fast disk (and a fast disk subsystem, which connects the disk to your other computer components) is critical.

- **Memory (RAM)**. After your computer copies the picture from the disk to memory, it will need to work with the memory each time you view or edit the picture. Faster memory will significantly improve the performance while working with a single picture, and you need sufficient memory (4GB or more).

- **Processor**. Processing RAW files, and applying edits to a picture, requires a great deal of processing. Any processor will eventually get through your edits, but the faster your processor, the quicker the edits will take place. It's worth it to spend more to get a faster processor.

The sections that follow describe these components in more detail and provide information about several less-critical components.

Memory (RAM)

RAM (Random Access Memory) is the temporary storage that computers use to store information while an application is actively working on it. RAM gets cleared every time you restart your computer. Everything that survives a reboot, including your pictures and videos, is stored on your hard drive.

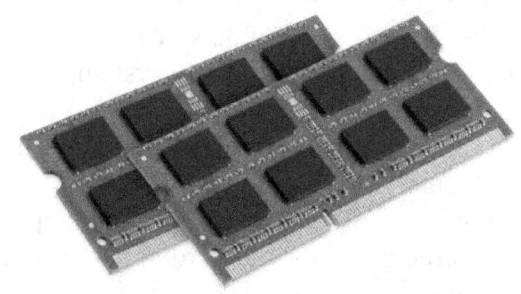

Capacity

For photo editing, 8GB of RAM is typically sufficient. Monitoring my own PCs during extensive photo editing sessions and working in Photoshop with multiple RAW files and dozens of layers, have shown me that Photoshop and Lightroom rarely use more memory anyway.

If you plan to edit video, choose a computer with at least 8GB of RAM.

While those recommendations are sufficient, more RAM is always better. Your operating system (whether Windows, Mac OS, or something else) will use the additional RAM to help improve your computer's overall performance. However, you will see minimal performance gains upgrading from 8GB to 16GB of RAM.

RAM typically costs less than $10 per GB when purchased after-market. When you buy a new PC with extra RAM, the PC manufacturer can drastically overcharge you—often charging over $40 per GB. Therefore, I typically recommend purchasing a new PC with the minimum amount of RAM and then upgrading your memory after you receive your PC. Even if you're not a computer nerd, you're quite capable of upgrading the memory in any desktop computer. Mobile computers are often very easy, too, though some are not designed to be upgraded.

Speed

The speed of your memory is very important, though the speed of your memory is determined by your motherboard. Always purchase the fastest memory supported by your motherboard. When purchasing a new computer, or a separate motherboard if you're building your own computer, it's worth spending a bit more to get a PC that supports faster memory; that upgrade will give you very real performance improvements when editing pictures.

Processor

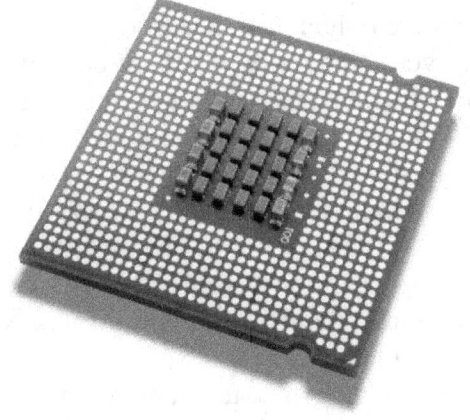

If you spend time editing individual pictures (rather than simply sorting through them), processor speed will be very important. Simple edits like crops and adjusting brightness don't require much processor time, but complex effects like healing spots, changing the perspective of a picture, changing the background, or using warp and puppet effects require a great deal of processor time.

Therefore, if you only perform light editing on your pictures in Adobe Lightroom, a faster processor might not make a huge difference. However, if you use healing or more intensive Photoshop features, investing in a higher-end processor will be worthwhile.

If you take RAW pictures instead of JPG—and you *should* take RAW pictures—your computer will need to do a great deal of processing when you first open your pictures. Therefore, RAW files justify a higher-end processor.

Measuring processor performance is complex, because there are several different factors:

- **Bus speed**. This is the speed at which the processor can exchange information with the computer's memory.

- **Processor speed**. This is the speed (typically measured in GHz) at which the processor performs operations.

- **Cores and hyper-threading**. Modern processors have multiple cores, which is like having multiple different processors. Essentially, this increases the processors' speed, but only for some activities.

To make it easy, choose i7 processors over i5 processors, and get the fastest i7 your budget allows.

Disks

Get two disks: one small SSD and one high-capacity drive. Install Windows or Mac OS and your apps on the SSD drive, and use it to store your newest pictures. When you start to fill up your SSD drive (which only needs to be about 512 GB), move your pictures to your high-capacity drive (which might be 6 terabytes or more). Because your high-capacity drive is only used for long-term storage, it doesn't need to be fast.

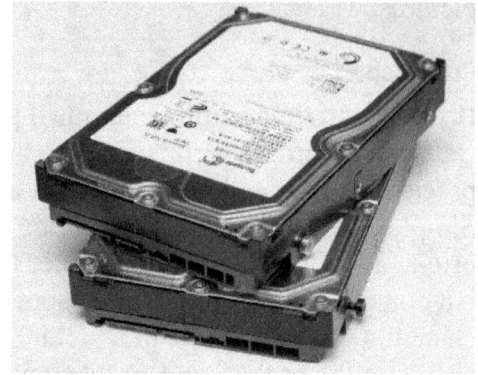

It's very important that your computer supports SATA3. SATA2 is not fast enough to take full advantage of the newest SSDs. Even better, choose a computer with native support for m.2 NVMe drives, which outperform SATA3.

If you're adding an external drive, look for a drive that supports USB 3 (for Windows PCs) or Thunderbolt (for Macs).

Monitors

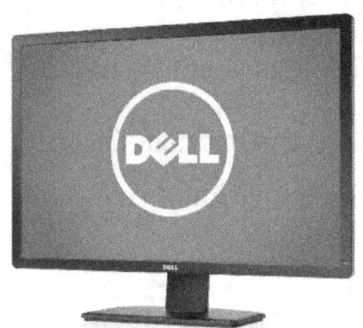

Get two monitors. Having dual monitors really helps with applications such as Lightroom because you can view thumbnails of your library on one monitor and view images full-screen on your second monitor. For photo editing, I'd rather have two smaller monitors than one bigger monitor. For me, more than two monitors doesn't increase my efficiency. When I've used three or more monitors in the past, I never fully utilize them—there are simply too many different screens for me to use efficiently.

The higher resolution the monitors are, the better. Higher resolution screens will allow you to see more detail in your pictures without zooming in, which can greatly speed up the editing process.

Here are specific recommendations at different price points. If you decide to use two monitors, it makes it much easier if they're identical:

- _Dell SE2416HX_ ($130). My producer uses this two of these 22-inch monitors (1920x1080) to edit video, and the glossy screen is bright and beautiful.

- _Dell U2715H_ ($470). This 27-inch 4K monitor quadruples the number of pixels, allowing you to see your images sharply without zooming way as far in. It also allows you to natively watch the 4K video you've recorded.

- _Dell UP3216Q_ ($1,200). If you'd like to freak out nerds who visit your house, get two of these 32-inch 4K monitors. I use an older version, and combined they're just amazing for image and video editing. The 27-inch models are almost as large at one-third the price, however.

Beware of buying leading-edge OLED or HDR monitors; people aren't likely to be viewing your images on these types of monitors, so they might lead you to edit your images with more or less contrast and saturation.

Don't spend money on calibrating your monitors or on buying a professional monitors like the _NEC Multisync_ series. Calibrated monitors aren't for photographers; they're for designers and those in the printing industry. Even if you print your own photos at home, calibration will only make a difference if you have a professional-level printer and you calibrate it to your display.

If you primarily share your images digitally, the people viewing your pictures won't have calibrated monitors. They'll be using smartphones, tablets, and other devices. They'll be viewing your pictures in different lighting conditions.

Instead of calibrating your monitor, carefully set the white balance of your pictures by choosing a custom setting based on a white object in the photo itself. That's the only way to ensure your pictures have the proper color. Also, be sure to check the histogram to ensure your image is bright enough. For detailed information, refer to Chapter 4 of _Stunning Digital Photography_.

Memory Card Readers

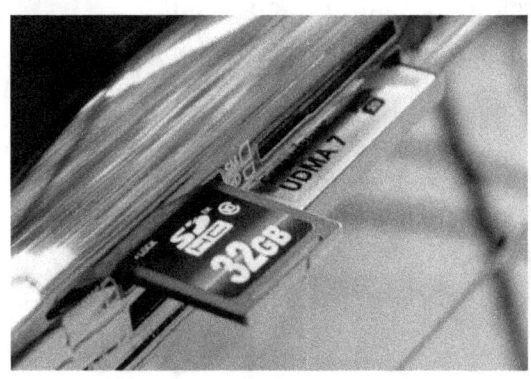

Most photographers copy their pictures from their camera to their computer by using a memory card reader. The memory card reader itself isn't particularly important, but you should choose a USB-C memory card reader so you can copy pictures as fast as possible because I know you'll always be eager to see the pictures you just shot. Also be sure that a new PC supports USB-C.

Many PCs and some monitors have memory card readers built in, and that's definitely a convenience. However, the built-in memory card readers might not be as fast as an external USB-C memory card reader.

Video Cards/Graphics Cards

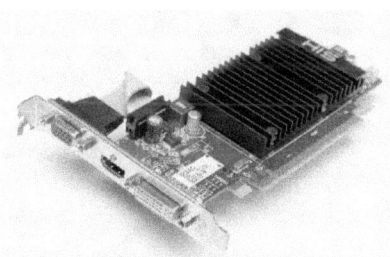

As I mentioned earlier, video card performance isn't a significant factor in photo editing performance. Some Photoshop effects are capable of taking advantage of your video card's graphics processing unit (GPU), but they're not effects that most photo editors use on a regular basis.

However, you should be sure to choose a video card that can connect to your monitors. Specifically, choose a video card that supports connecting to multiple, high-resolution monitors simultaneously. Look for multiple DVI or USB-C ports, and possibly an HDMI port if you plan to connect your computer to a TV.

Drawing Tablets

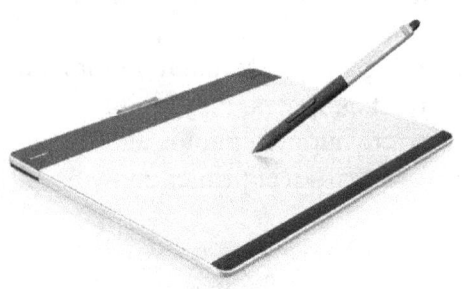

Some people like using tablets to edit their photos. If you like drawing with a pencil, you'll probably like using a drawing tablet. If you're terrible at drawing, like I am, you're probably better off using a mouse.

Wacom makes the best tablets for photo editing. They have several tablets available, at different quality levels and at different sizes. Bigger sizes are nicer to use, but don't choose a size that's too big for your desk.

In order from least to most expensive, and from least to most powerful, Wacom's current lineup is:

- ***Bamboo* ($80-$100)**. Designed more for fun than work, these are good enough for most photo editing tasks.

- ***Intuos* ($100-$500)**. The mid-range Intuos tablets are more sensitive and better support varying pressures, which can be useful when drawing with brushes in Photoshop. The Intuos5 and Pro models offer more controls on the tablets themselves for tasks such as changing brush size. This can improve your workflow and decrease the time you spend editing photos.

- ***Cintiq* ($1,000-$2,500)**. This high-end line combines monitors with drawing tablets, allowing you to draw directly on the screen. It's not as great as it sounds, however, because your hand covers part of the screen as you're drawing. Many people prefer using a separate monitor and tablet.

Chapter 10: Underwater Photography

Whether you want to grab snapshots of their kids in the pool or create ethereal underwater portraits, you'll need gear to protect your camera. This chapter provides

Video: Underwater Gear
13:19 - *sdp.io/underwater*

an overview of the different options available for all different budgets.

Your first choice is to either buy a new camera specifically for underwater use (for as low as $10) or to buy an underwater housing for a camera you already own. Either way, your gear should:

- Allow you to grip the camera while swimming.

- Allow you to see the viewfinder and/or the LCD screen.

- Replicate the buttons on your camera so you can focus, zoom, and change settings underwater.

- Keep your camera dry.

- Float when you drop them.

Disposable Film Cameras

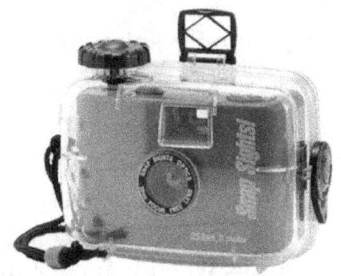

The cheapest way to grab underwater photos is to use a disposable film camera ($7-$20, plus developing costs). It will seem primitive; there's no LCD display, and you'll need to wind the film manually between each shot. The camera won't even have a focusing system. After you take the photos, you'll have to find a store that still develops film, or send your film to an online service (such as thedarkroom.com).

It's inconvenient, but it's not a bad idea for a vacation. You don't have to worry about ruining an expensive camera, and while the quality won't be great, underwater photos rarely have great quality.

Waterproof Sports Cameras

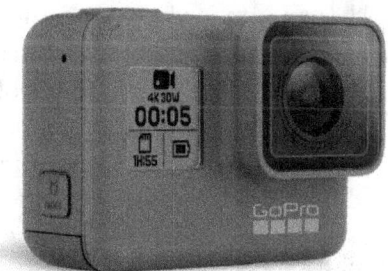

Sports cameras have become very popular in recent years. These compact and durable cameras can go almost anywhere, including underwater. While you wouldn't want to make a portrait with them, they're an excellent choice for casual underwater photography.

While there are many different models of sports cameras available, the GoPro Hero is my top recommendation.

Priced from $200 to $400, all the GoPro models include underwater housings. They're not perfect, however. Some of the challenges are:

- **No viewfinder or LCD display**. Seriously, GoPro cameras do not include an LCD display or a viewfinder. If you want to see what you're taking a picture of, you need to buy the GoPro LCD Touch BacPac for $80. Underwater, the display is very difficult to see. Therefore, I don't recommend it for underwater photography. Instead, I recommend simply pointing the camera in the general direction of your subject, pressing the shutter, and taking as many pictures as you can. It sounds ridiculous, but it works better than using the LCD.

- **An extremely wide-angle lens**. The GoPro lens is essentially a fisheye lens, capturing an incredibly wide-angle view. This is both good and bad. It's very hard to compose a picture carefully underwater, so the fisheye lens ensures you capture everything around you, and you can crop the picture later. The downside of the lens is that you need to be very, very close to your subjects to see much detail. Ideally, you would be about 18 inches (half a meter) away. Most marine life won't let you get that close.

- **Fixed focus**. GoPro cameras have fixed focus. Combined with a super wide-angle lens and a small sensor, they have near-infinite depth-of-field. Everything from about 18 inches away to infinity will be in focus. If you get closer to a subject, it will be blurry. It's not generally a problem with underwater photography, however.

Housing for Point-and-Shoot Cameras

You must find underwater housing made for your specific camera model—that's the only way the buttons can be reproduced. Every important button and dial on your camera will have a copy on the outside of the underwater housing, with springs and levers that push your camera's buttons when you push the corresponding button on the camera housing.

The murkiness of water won't allow you to take advantage of your DSLR's high image quality, anyway, so consider starting your underwater photography with an underwater housing for a point-and-shoot camera, as shown in the following figure. Purchase an underwater housing made by your camera manufacturer. Canon, Nikon, Sony, and Fuji all make underwater housings for their popular point-and-shoot cameras. Check eBay and see if you can find a good price on a used housing. If you have a Nikon, *Fantasea Line* might make a housing for your camera.

Point-and-shoot housings typically add a diffusing screen in front of the camera's flash, as shown in the previous picture. This will help just a bit, but they're not bright enough to light up subjects in deep water. Also, because the flash is placed so closely to the lens, you'll get *backscatter*, which causes particles in the water to reflect the flash back to the camera.

If you go any deeper than thirty or forty feet (or less if it is not sunny or the water is not clear), you're going to need *external underwater strobes*, which are flashes that attach to the bottom of the housing using the tripod connector and are triggered when your camera's built-in flash is fired. You might also need to add weights to your case. While having the case float is helpful when you're snorkeling, when you're trying to dive, a floating case will be fighting you the whole way down.

Dedicated Interchangeable Lens Cameras

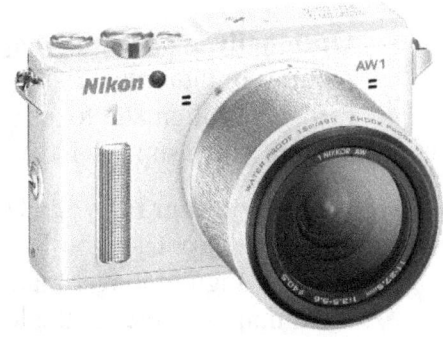

Nikon makes two interchangeable lens cameras: the film Nikonos (starting at $150 used) and the digital AW1 ($750 new). Because they're built specifically for underwater use, they're less bulky and easier to use underwater. You can use either camera above water, too, though I'd only recommend them for the most casual above-water photographers.

Housing for DSLRs

Camera makers don't make underwater housings for DSLRs. Instead, check out *Aquatica*, *Sea and Sea*, AquaTech, and my favorite (for the lower prices), *Ikelite*. You will need:

- **A housing made for your camera body**. This is the most expensive piece of equipment. When comparing housings from different manufacturers, check that all important buttons are replicated and that you can see the viewfinder and LCD clearly. Housings can be acrylic (see-through plastic, as shown in the following figure) or aluminum. Acrylic lets you see inside the housing, which is nice for making sure the o-ring is in place and that the case isn't leaking. Aluminum can be more durable, though acrylic cases are strong enough to handle the inevitable bashing against the rocks. Housings should connect to your camera's flash shoe and provide wiring to fire external strobes when required. Look for a housing that supports through-the-lens (TTL) metering for the external strobes.

- ***A super-wide angle lens (around 16-24mm)***. A zoom lens is useful, and high-end housings provide attachments to adjust the zoom. Wide-angle lenses are the right choice for most outings, especially when you're just beginning. Wide-angle lenses allow you to take scenic views underwater, but more importantly, they allow you to photograph fish just a few inches in front of your camera. You'll need to get that close to take clear pictures underwater.

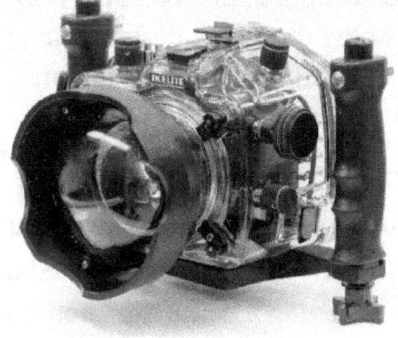

- **A *diopter***. If the lens can't focus closer than 12", you will need a +4 diopter—a filter that screws on the front of the lens and allows you to focus closer. The diopter is required because light passes differently through water. You know how things look closer underwater? Cameras have the same issue. Focusing will work normally underwater with the diopter, but on land, you'll only be able to focus very close to the front of the lens. For that reason, and the fact that the underwater housing is cumbersome, it's a good idea to bring a second camera to take pictures above water.

- **Optionally, a _telephoto macro lens (around 100–150mm)_**. If you want to take pictures of individual fish, you'll need a telephoto macro lens. It can be very difficult to align and focus a telephoto lens underwater, though, so it's a good idea to start with a super-wide angle lens.

- **A _lens port_ that fits your lens (see the following figure)**. Housings ship with a big hole in the front, so you'll need to buy a lens port to cover your lens. The lens port must be slightly longer than your lens (with any diopters attached). Check the maximum diameter of the lens port and verify that your lens will fit through it. For wide-angle lenses, use a _dome port_—a port with a rounded end—along with a diopter filter. For macro lenses, use a _flat lens port_.

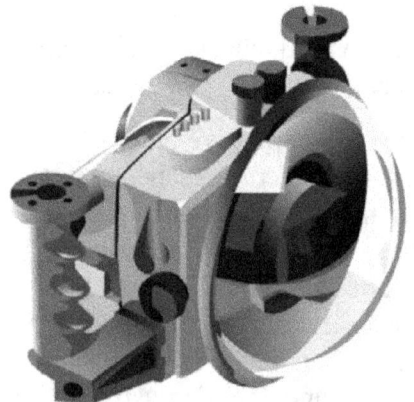 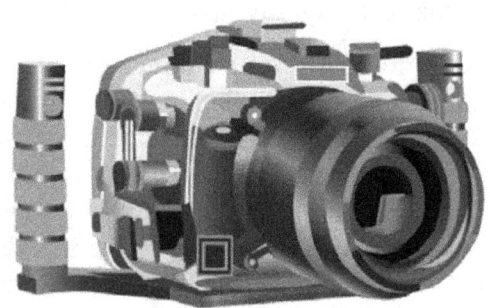

Tech details: Dome ports act as an additional optical element by shaping the water in front of your lens into an arc. This helps to reduce refraction, distortion, and aberrations caused by water. It also requires the lens to focus very closely, as if you were taking a picture of an image projected on the inside of the dome itself. That's why you need a macro lens or a diopter when using a dome port. Flat ports don't correct the problems that occur when light must pass through air, water, and then back through air again inside your housing. The distortion and aberration are unacceptable when using a wide-angle lens, but are mostly hidden when using a telephoto lens. That's why flat ports are limited to telephoto macro lenses.

- _**Underwater strobes**_. Your existing flash won't work; you'll need underwater strobes if you're going any deeper than 40 or 50 feet—preferably, two strobes, attached to the left and right sides of the housing. If you want to take video, be sure to choose strobes that act as hot lights—meaning they can give off light continuously. Be sure you have batteries or chargers for the strobes.

- **Extra o-rings and lube**. The watertight seal is created by an o-ring (a flexible piece of rubber that runs around the edge of the two halves of the housing). It's a good idea to keep an extra in case yours gets damaged.

- **Weights**. On land, your underwater housing might seem incredibly heavy. It's mostly air, though, and underwater, it might be too buoyant to easily dive with. You might, or might not, need to add lead weight to your underwater housing. The only way to find out is to try your gear out, but it wouldn't be a bad idea to bring extra weights with you so you can add them as required.

As you can imagine, all this gear is expensive. If you're using strobes, you'll spend more than $3,000, and you can easily spend $5,000. It's also heavy—you won't want to carry the gear around on land.

Chapter 11: Canon EOS/EF/EF-S DSLR

Canon EOS/EF is the largest and most popular camera system ever made (followed closely by the Nikon F-mount). When you enter the Canon world, you gain access to thousands of bodies, lenses, and flashes.

Canon has, in my opinion, the greatest lens selection of any camera manufacturer. However, their

Video: Canon Wildlife & Sports Comparison
40:53 - *sdp.io/CSports*

camera bodies have fallen behind other manufacturers, especially Nikon.

Most people choose a camera system when they find a camera body they like, and then they choose lenses that work with that body. For some photographers, however, it's smarter to choose a lens you need for your work, and then choose the best body to work with that lens. In the Canon world, there are a couple of lenses that the closest competitor, Nikon, simply doesn't offer a perfect alternative for at the same price point:

- **Canon 50mm f/1.8**. This cheaply made $100 lens isn't technically great at anything, but it's incredibly inexpensive and autofocuses with all Canon bodies. Photographers with budgets under $600 are routinely thrilled with the great background blur and low-light capabilities of this lens. The Nikon version (the 50mm f/1.8G) is better, but it costs more than twice as much. The Nikon 50mm f/1.8D doesn't autofocus with entry-level bodies.

- **Canon 70-200 f/2.8 L IS II**. This is the best lens in the world for professional portraiture. The Nikon variety has severe focusing problems that make it behave like a 175mm lens for close headshots, and an experienced photographer will definitely appreciate the extra length when working at close range with the Canon.

- **Canon 400mm f/5.6**. For wildlife, this lens is relatively inexpensive (around $1,000 used) and lighter and sharper than similarly priced zoom lenses. Getting similar results on a Nikon requires using the 300mm f/4 and a 1.4X teleconverter, which is a significantly more expensive setup.

But before buying your first Canon camera, take a serious look at the Nikon lineup. Nikon camera bodies at a similar price point generally produce technically better pictures, and Nikon has a much newer and fuller lineup of full-frame cameras for professionals. While the casual

Video: Canon Video Comparison
26:55 - *sdp.io/CVideo*

Video: Canon DSLR Comparison
53:12 - *sdp.io/CanonLineup*

photographer will never notice the difference, I now recommend Nikon to all serious photographers who don't need a specific Canon lens.

In other words, if you're shooting landscapes and travel, you should seriously consider Nikon instead of Canon. If you're the type who shoots raw, recovers shadows, and is annoyed by noise in photos, you probably want a Nikon.

While Canon has not officially discontinued the EF and EF-S systems, they're definitely putting the vast majority of their research and development into their newer RF system. Generally, the EF and EF-S DSLRs offer a better value and thus might be a better choice for people who want to spend less than $1,000 or for working professionals who need to make their camera gear profitable. For enthusiasts with a budget of at least $1,500, I would recommend the Canon RF system instead.

Rebel SL1/100D ($220 used)

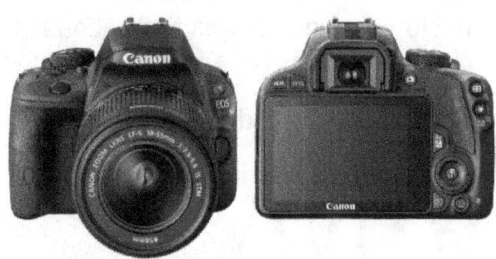

The *Canon SL1* is impossible to compare to the rest of the Canon lineup because it's a very specialized body, optimized for size. It's tiny, offering mirrorless portability with the flexibility of a DSLR.

If you want the smallest Canon DSLR you can buy, the SL1 is the right choice for you. If you want the size benefits of mirrorless cameras with the Canon lens selection, buy the SL1. If carrying a few extra ounces and inches isn't a big deal, you should choose one of the other cameras.

Rebel SL2/200D ($550 used)

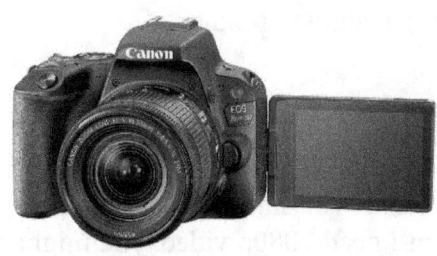

The SL2 is Canon's lightweight option for sale in electronics stores to first-time photographers who haven't done much research. Since you're reading this book, it's probably not the right camera for you. However, if you have an SL1 and love it, but want something newer and don't feel like learning a new camera system, the SL2's articulating screen and Wi-Fi make it a worthwhile upgrade.

Compared to other Canon SLRs, it's only advantage is small size. It's only a little bit smaller than its siblings like the T7i, however.

If you're looking for a small camera, choose an Olympus E-M10 Mark II, which is smaller, cuter, and overall more fun to use. The electronic viewfinder is vastly superior to the SL2's optical viewfinder, especially for beginning photographers who would benefit from the electronic viewfinder's more useful depth-of-field and exposure preview. The Olympus also has sensor stabilization, allowing low-light images with small pancake lenses.

If you have a higher budget, also consider the Fujifilm X-T20.

Rebel T3/1100D ($200 used)

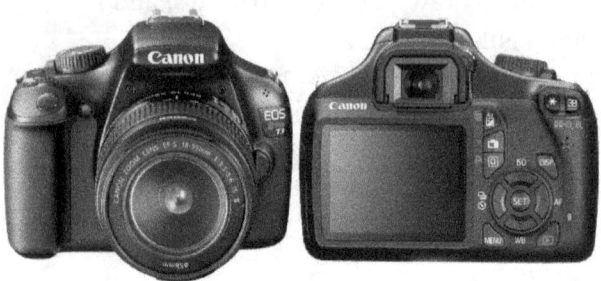

Canon's entry-level camera, the T3, is my standard recommendation for everyone's first camera. Even if you have a budget of thousands of dollars, I'd rather you spend more of your budget on lenses, flashes, tripods, and software.

The T3 gets the job done, but it does have a couple of weaknesses:

- Limited buttons require you to look at the screen to make common adjustments, such as adding exposure compensation. This can slow you down by a few seconds, but isn't a problem for most photographers.

- Video is limited to 720p, whereas all other Canon cameras record 1080p.

- The autofocusing system makes it frustrating to photograph moving subjects.

Rebel T5/1200D ($250 used)

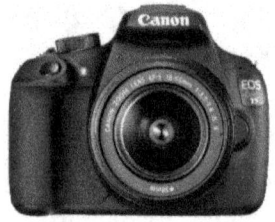 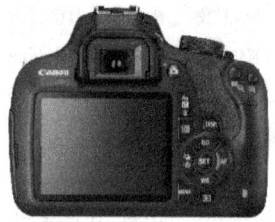

The previous generation of Canon's entry-level camera is hardly changed from the T3. There was no T4 model because tetraphobia (the practice of avoiding the number 4) is common in parts of Asia. The T5 has a couple of minor improvements over the T3:

- 1080p video (instead of 720p in the T3)

- A sharper LCD screen with 460,000 dots, instead of just 230,000

Image quality and focusing speed are essentially unchanged. If you find a good deal on a T5, it's a great camera. If you're shopping around and don't need 1080p video, you might be happier buying a used T3. If you do plan to do some video, a used T3i has 1080p video and adds an articulating screen.

Rebel T6/1300D ($450 new, $300 used)

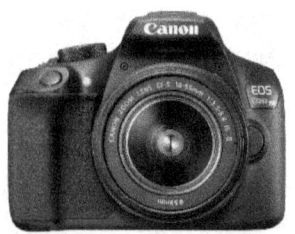 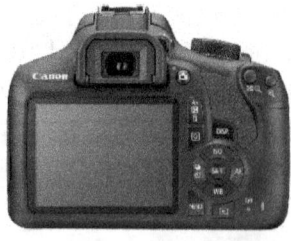

The T6/1300D improves slightly on the T5, adding Wi-Fi and NFC. Those capabilities allow you to more easily copy pictures from your camera to your smartphone, so you can share them on your social networks without using a computer. However, Wi-Fi isn't nearly as convenient to use as you might hope, and as a result, many people don't use it long term.

If Wi-Fi is important to you, get the T6. Otherwise, choose a used copy of the T5, and spend the $200 you save on lenses.

Rebel T3i/600D, T4i/650D, and T5i/700D ($500-700 new, $450-600 used)

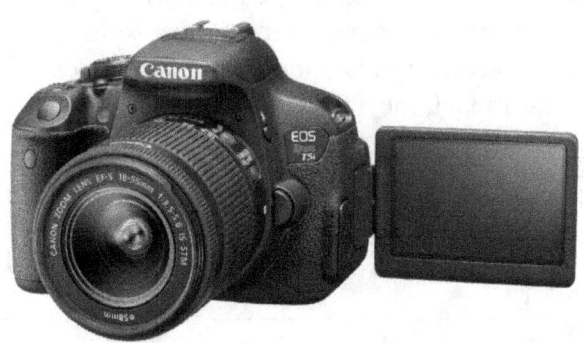

One step up from entry-level, these cameras offer identical image quality and focusing to the T3. However, they do add some nice features:

- An articulating screen, which is useful for self-portraits, video, and shots at high or low angles.

- 1080p video recording

You might want to upgrade if you find it annoying to adjust exposure compensation, if you struggle with focusing on moving subjects, or if you want better image quality.

The only significant difference between the *T3i* and the *T4i/T5i* is that the later models add a touch screen. The image quality and focusing are essentially unchanged.

Before buying one of these Canon cameras, consider the Nikon D3200. While the new Nikons tend to have fewer features, they have substantially better image quality. For example, the image quality of the Nikon D3200 ($400) is about 25% better than the T5i/700D ($650), which is enough to make a visible difference in your photos. However, the D3200 lacks many of the T5i's important features, such as an articulating touchscreen.

Rebel T6i/750D ($700 new, $500 used)

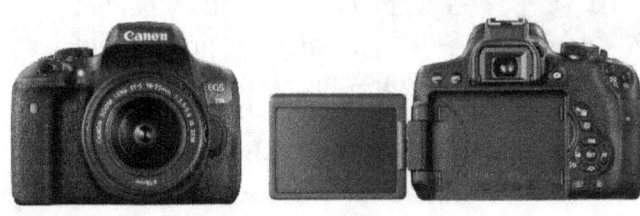

The T6i is one of our top recommendations for casual photographers. With this generation, Canon finally upgraded the sensor, so the image quality is both sharper and cleaner than the previous generation T5 and T5i. With a full 24 megapixels, you can now achieve similar image quality to the competing Nikon D5500, Sony a6000, and Pentax K-S2.

Since you're reading this guide, you're probably fairly serious about photography, and should take a look at the T6s/760D. The T6s adds a secondary dial that makes

Video: T6i & T6s Review **23:14** - *sdp.io/T6iReview*	

adjusting the exposure compensation easier. It also adds a top LCD display to make checking your settings more convenient. The T6s does a better job of tracking action when recording video, too.

Rebel T6s/760D ($800 new, $550 used)

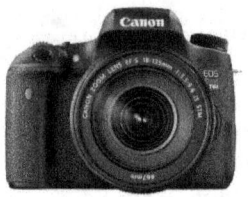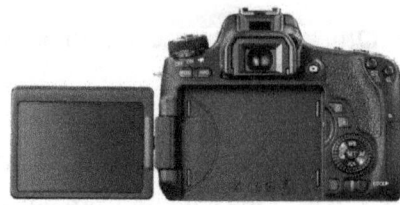

Like it's slightly less expensive sibling, the T6i, the T6s includes Canon's new 24-megapixel APS-C sensor that provides greatly improved image quality compared to the previous generation of cameras. The T6s is our recommendation for most serious photographer's first cameras; the top LCD screen, secondary dial, and articulating screen should last most new photographers several years. In fact, the T6s/760D should last until you decide to upgrade to a full-frame camera.

You should consider upgrading to a 70D for improved video focus tracking capabilities. The 70D also offers direct control over the focusing points, something that requires an extra button push on the T6i and T6s. If you're

> Video: T6i & T6s Review
> **23:14** - *sdp.io/T6iReview*
>

serious about sports photography, the higher frame rate and bigger buffer of the 7D Mark II will make a huge difference. If that's out of your budget, consider buying a used copy of the original 7D, which is much faster than the T6i or T6s, but can't quite match the image quality.

T7i/800D ($750 new, $675 used)

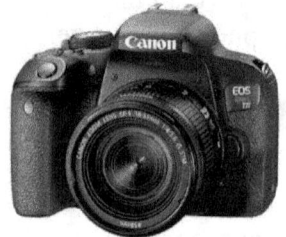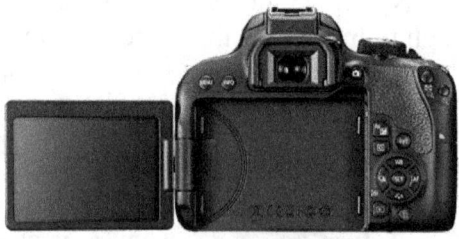

There are SO many mid-range Canon cameras that it gets really difficult to choose between them. Generally, they get better as they get pricier, and that's true of the T7i. It's a good all-around camera, but if you want to spend a bit more, you can get a bit more.

Compared to the older T6i, the T7i adds dual-pixel autofocus. Previously, dual-pixel autofocus was the main distinguishing factor between the T6i and T6s, so it makes sense that we didn't see a T7s model.

Compared to the older T6s, the T7i adds Bluetooth, which I haven't found to be very useful. In other words, if you're choosing between the T7i and T6s, it won't make much of a difference.

Serious photographers should consider upgrading to a 77D for an extra $200. While the cameras are almost identical, the inclusion of the extra control dial at the back facilitates quickly adjusting exposure compensation and manual settings. If you're a casual photographer, save the extra $200 for an extra lens.

60D ($350 used)

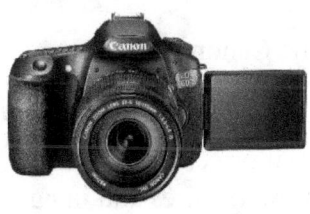

The 60D is no longer being manufactured, but it's a great deal used. The 60D is Canon's mid-range crop camera. Compared to the lower-end cameras, it offers these benefits:

- Better image quality, equivalent to the 7D and 70D.

- A thumb-dial on the back that allows you to quickly adjust your exposure.

70D ($800 new, $500 used)

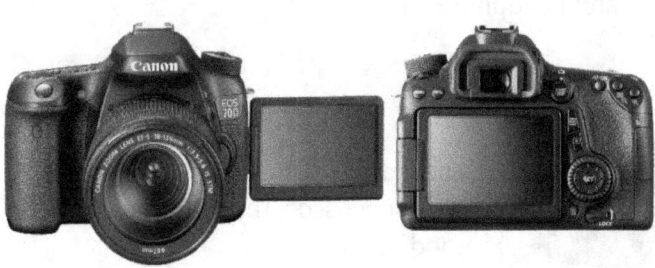

The _70D_ has almost identical strengths to the Canon 7D, though the frames per second is a bit slower (7 fps instead of 8 fps) and the buffer fills up a bit sooner than does the 7D. Because the 70D is technically considered a lower-end camera than the 7D (despite being overall more powerful and expensive), the 70D is smaller and lighter than the 7D, which most people consider to be an advantage.

Here are the reasons to choose the 70D:

- Autofocus while recording video. This is the only Canon camera that can track moving subjects in video. If you want a video camera for general use, the 70D is the right choice for you.

- Compared to the lower-end cameras (excluding the 7D), it offers amazing autofocus and high frames per second.

- Compared to the 7D, it offers an articulating touch screen and Wi-Fi, making it a better camera for general use.

Currently, the 70D is Canon's top-end body with a crop sensor. Before buying 70D, you should seriously consider either a used 5D Mark II or a 6D. For

> Video: 70D Comparison
> **34:46** - _sdp.io/70DReview_

the

just

a few hundred dollars more, those bodies offer significantly better image quality and background blur. However, they are bigger and heavier, have weaker autofocus systems, and lack the articulating touchscreen.

Before buying a 70D, consider a Nikon D7000, which can be bought used for about $600. The D7000 offers about 15% better overall image quality, which is a substantial gain. However, the D7000 isn't as good for action shots, because the buffer fills quickly, and it lacks the 70D's Wi-Fi and video focusing capabilities.

77D/9000D ($750 new, $650 used)

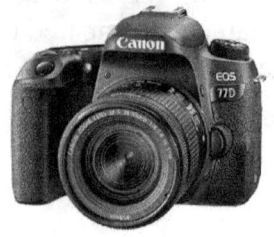 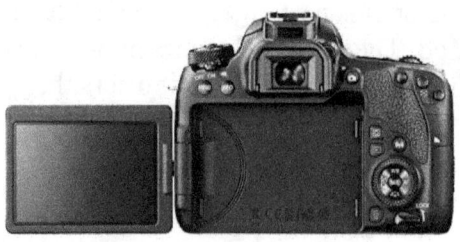

The Canon 77D is an excellent all-around consumer camera at a great price. It offers most of the features and image quality of the more expensive 80D at a much lower price and smaller size. If you're a serious photographer, you'll be happier with the bulkier 80D, with its faster maximum shutter, better controls, faster sync speed, and headphone jack.

80D ($750 used)

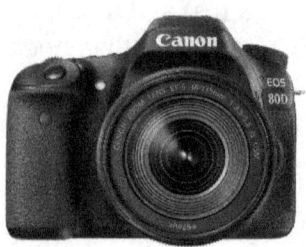 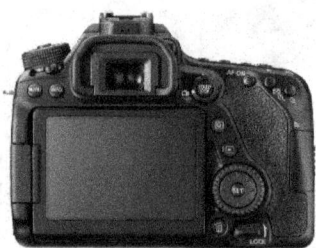

The Canon 80D is almost identical to its predecessor, the 70D. That's both good and bad.
Like the 70D, the 80D is Canon's best all-around camera. Features like an articulating touch screen, solid controls, fantastic focusing for both stills and video, and native compatibility with Canon's amazing lens lineup make this our top recommendation for people who want one camera that can do a bit of everything.

Unfortunately, the 80D doesn't excel at anything in particular. It's a good video camera, but the Panasonic GH4 is better. It's a good sports camera, but the 7D Mark II is significantly better. It's a good landscapes camera, but for $200 more, you could get the full-frame 6D and get much better results. If you're willing to buy used, the full-frame 6D is even less expensive than the 80D.

While those cameras all have specific advantages over the 80D, the 80D is still the best all-around camera. The GH4's small sensor means images are much noisier, and the GH4 is frustrating to use for sports and almost impossible to use for wildlife photography. The 6D and 7D Mark II lack the articulating touch screen that's incredibly useful for day-to-day photography.

Compared to the 77D, the 80D has a bigger battery, a slightly faster shutter (at 1/8000), 7 FPS (compared to 6 on the 77D), and a headphone jack. The 77D is enough for most photographers, but the 80D is the choice for more serious professionals.

90D ($1,200 new)

The 2019 Canon 90D is not just the replacement for the 80D; it's probably Canon's replacement for the ancient 7D Mark II. It's currently the most powerful APS-C sports and wildlife camera… and it will probably hold that title for all eternity, as I doubt Canon will ever upgrade it, and it doesn't seem like Nikon or Pentax will ever beat it. The 90D has 32 megapixels, which is a significant advantage of the 20 megapixel 7D Mark II that had been our favorite wildlife camera for many years. For wildlife photographers who use full-frame lenses but end up cropping, the APS-C sensor provides pixel density equivalent to an 83 megapixel full-frame sensor. That means it will out-resolve every camera on the market.

It offers 10 frames per second still shooting with autofocus when using the optical viewfinder, matching the 7D Mark II and Nikon D500. This can jump to 11 FPS if you don't use autofocus, which is easy to activate if you use back-button focus, as described at sdp.io/bbf. If you switch to live view on the rear screen, it drops to 7 FPS and the autofocus slows down, too.

Speaking of live view, it works better than any other DSLR. Basically, it works like a mirrorless camera. Eye detect works with the rear screen and you have the option of completely silent shooting. Like all DSLRs, the user interface and autofocus system change completely when switching between the optical and electronic viewfinders.

The 45-point autofocus system works well for tracking subjects. It doesn't quite match the 7D Mark II's autofocus system, but it's good enough for most users. The 90D can use many of the center autofocus points with f/8 lenses, so you can attach a 1.4X teleconverter to an f/5.6 lens and still autofocus. I wouldn't recommend that, however, since the extremely high pixel density is already extracting about as much detail as you'll get out of any lens.

The 90D offers full-width 4k/24, 4k/25 or 4k/30 video, a flip screen, and Canon's excellent dual-pixel autofocus. That makes it an excellent vlogging camera. However, the 4k video doesn't process every pixel in the sensor, and thus it's less-sharp than the 4k video from cameras like the Sony a6100.

The 90D supports UHS-II cards, which reduces buffering by allowing faster write times. However, it only has a single card slot. If you require dual card slots for redundancy, the 7D Mark II is still your only choice.

7D ($350 used)

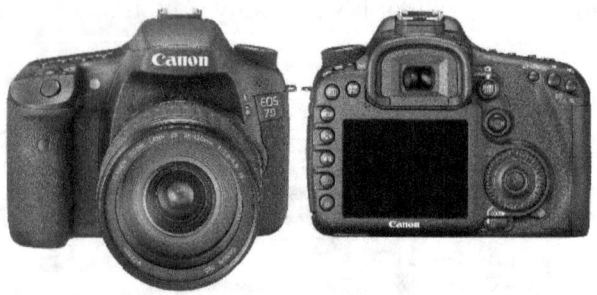

Though technically the _7D_ is still available new, it has been (mostly) replaced by the much newer 70D. If you want a 7D, you should definitely buy a used body because of the steep discount. Most people interested in spending $1,300 on a new body should buy the 70D instead.

A used 7D is an amazing value, however, and they often sell for as low as $550. Its amazing autofocus system has 9 cross-type autofocus points that do an amazing job at tracking moving subjects. In fact, the autofocus system is only exceeded by the much more expensive 5D Mark III and 1D X.

Because the 1.6X crop factor brings you closer to distant sports and wildlife, a used 7D is my recommendation for most outdoor sports and wildlife photographers, regardless of their budget. When I'm shooting distant subjects in good light, I choose the 7D even over a 5D Mark III or 1D X. The 7D's 18 megapixels are all crammed into the center of the frame, providing far more detail than any of Canon's full-frame cameras can provide. Though image quality is no better than that of the 60D, compared to the lower-end cameras, the 7D offers these benefits:

- Autofocus capable of tracking moving subjects using any focus point.

- Higher frames per second, for capturing action.

- Improved durability and weather sealing.

- The 7D is the right camera for well-lit action. If you plan to do portraiture or landscape, you should choose to either save your cash for lenses and buy a used 60D or make the jump to full-frame image quality with a used 5D Mark II or a new 6D.

- Note that the 7D lacks the articulating screen of the lower-end cameras. This improves durability, but I do often miss the articulating screen. The 70D offers most of the capabilities of the 7D with an articulating touchscreen.

- If this is your first camera, you should also consider the Canon 70D and the 7D. The 70D has fantastic autofocus capabilities during video, and the 7D can be had used for $500-$750 (and it doesn't have the D7000's focusing issue). All these cameras have excellent autofocus systems, but the D7000 does have more focusing points (though that will probably never impact your photography) and much better image quality (which will impact your photography).

- Before buying a 7D, consider a Nikon D7000, which can be had for about the same price used. The D7000 offers about 17% better overall image quality, which is a substantial gain. However, the D7000 isn't as good for action shots, because the buffer fills quickly.

7D Mark II ($1,400 new, $1,000 used)

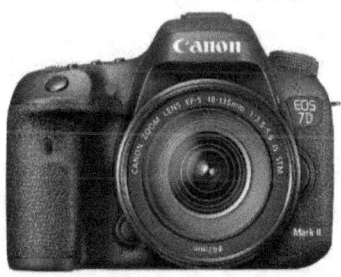 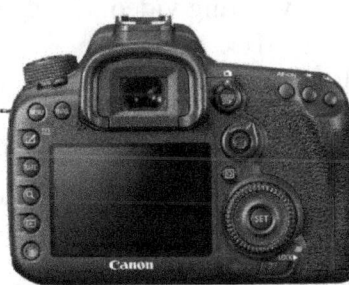

The 7D Mark II is Canon's ultimate sports and wildlife camera, and for many, it's the greatest action camera in the world.

Standing at the top of Canon's APS-C/EF-S lineup, the 7D Mark II has a remarkable autofocus system matched only by the $6,800 1DX. The autofocus points almost fill the entire frame, and each point is an extremely capable cross-type sensor. In short, the 7D Mark II is better than almost any camera at tracking moving subjects.

For most sports and wildlife photographers, the 7D Mark II is a better choice than the higher-end 5D Mark III, primarily because of the smaller sensor. In situations where

Video: 7D Mark II Preview
34:20 - *sdp.io/7D2Preview*

you can't get close enough to your subject and you have to crop by 1.6X anyway, the 5D Mark III would be reduced to only 14 megapixels, while the 7D Mark II still has the full 20 megapixels. Of course, if you can get close enough to fill the frame on your lens with a 5D Mark III, the image quality will certainly be better. However, with wildlife, animals rarely get close enough to allow you to fill the frame, even with massive telephoto lenses, so the extra pixel density does result in sharper pictures.

If you plan to shoot sports and wildlife but the 7D Mark II is out of your price range, the original 7D is still an excellent alternative. It doesn't take pictures quite as fast, the buffer is smaller, and the autofocus system is inferior, but a used 7D costs about 1/3 the price of a new 7D Mark II, and they take similar images.

The following figures compare the autofocus points of the original 7D to the new 7D Mark II. As you can see, the 7D Mark II spreads the autofocus points further around the frame. For action where you don't have the opportunity to use the focus-recompose technique, this provides for more flexible compositions.

For video, the 70D is a better choice, because the 70D has an articulating touchscreen. This allows you to see live view easier from different angles. You can also take advantage of the video autofocus capabilities better by simply touching the screen on the 70D. With the

7D, you need to use the joystick to select an autofocus point in order to refocus, and this is going to shake the camera unacceptably during video. If video is your primary purpose, you might also consider the Panasonic GH4.

The articulating touch screen and lower price also make the 70D a better choice for general photography, other than sports and wildlife. For landscape and portrait photography, I recommend one of the Canon full-frame cameras, such as the Canon 6D, which is about the same price as the 7D Mark II. For landscapes, you might also look into the Nikon DSLR lineup, because their sensors create sharper images with less noise and greater dynamic range.

5D Mark II ($650 used)

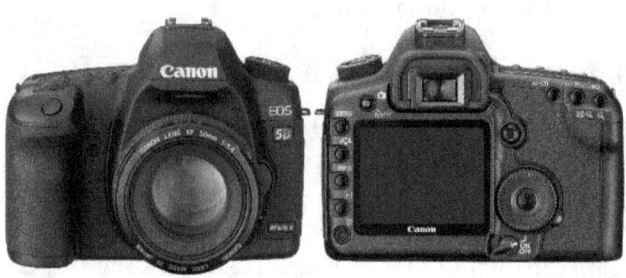

A used _5D Mark II_ is the best value for Canon shooters interested primarily in still image quality. While the autofocus system is less than ideal, the 5D Mark II offers the same image quality as the much more expensive 6D and 5D Mark III. Used, it's about the same price as a Canon 70D, but the image quality *far* exceeds that of the 70D or any crop camera.

For that reason, and because a used 5D Mark II is about the same price as a 70D, I recommend the 5D Mark II to photographers primarily interested in portraits or landscapes.

The 5D Mark II's autofocus system is similar to that of the 6D, though the 6D's center autofocus point does better in very low-light conditions. The 7D and the 70D have far

better autofocus systems for moving subjects, making them better choices for sports and wildlife.

Video: 5D Mark II vs. 5D Mark III *11:46: sdp.io/5D2v5D3*	

6D ($800 used)

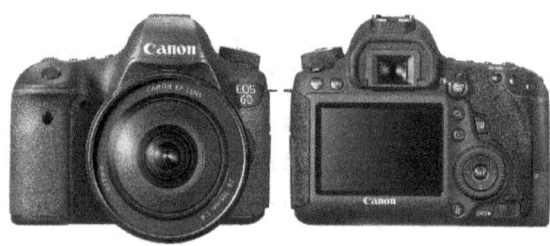

The _6D_ is Canon's entry-level full-frame camera. It provides vastly better image quality than any of the crop cameras. Here are the reasons you should upgrade to a 6D from the previous cameras:

- The full-frame sensor provides 60% better background blur than a crop camera can, improving portraits.

- The full-frame sensor provides MUCH better image quality, especially in low-light, making it ideal for landscapes and indoor photography. For this reason, I recommend it for indoor sports, despite having a weaker focusing system than the 7D or 70D.

- It has a GPS, which no other Canon camera has. It also has Wi-Fi, which the 70D also has.

Despite the fact that it's 50% more expensive than the 70D, however, the 6D lacks several key features of the 70D:

- All of the 70D's focus points are cross-type, making it a better choice for sports and wildlife. With the 6D, you will need to use the center autofocus point for moving subjects, limiting your composition choices.

- The 70D does an excellent job of focusing on moving subjects while recording video. The 6D (and all other Canon cameras) do an awful job.

- The 70D has an articulating screen, which reduces durability but is helpful for shooting self-portraits or at strange angles.

- The 70D has a touch-screen, which the 6D does not.

Also, all lower-end cameras feature a pop-up flash. This is hardly a benefit, however, because pop-up flashes are awful. Additionally, the low-light image quality of the 6D and all Canon full-frame cameras provides better results in situations where you might want a pop-up flash.

The 6D replaced the 5D Mark II, and used prices are within 20% of each other. The 6D is a better camera than the 5D Mark II in almost every way, but these are the key benefits. Use them to determine whether it's worth the extra cash to upgrade over a 5D Mark II:

- Wi-Fi allows you to browse files from your phone or tablet. This is useful for proofing photos in a studio or for posting a picture to Facebook from your smartphone.

- GPS can record your location, which is very useful when traveling.

- The live view display works far better than the 5D Mark II's in low-light environments, making it much easier to compose and focus night photos.

- The mode dial has a locking button, preventing you from accidentally changing modes—a problem that always has plagued the 5D Mark II.

- The center autofocus point works well in much lower light conditions than the 5D Mark II can handle.

Before buying the 6D, consider buying a used Nikon D600 ($1,300) or a new Nikon D610 ($2,000). The Nikon bodies offer about 12% better overall image quality, which is a significant advantage. However, they lack the 6D's built-in GPS and Wi-Fi.

6D Mark II ($1,400 new, $1,000 used)

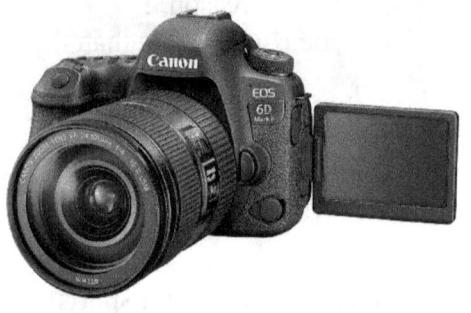 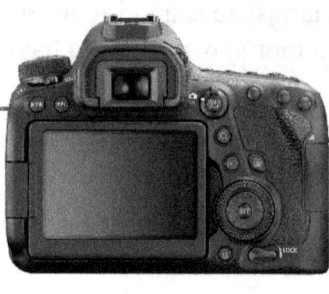

The 6D Mark II is Canon's long-awaited entry-level full-frame camera. Compared to the original 6D, the 6D Mark II has 30% more pixels (26 megapixels, up from 20 megapixels). More importantly, it adds an articulating screen to make video, landscape, and just about every type of photography easier. It also adds dual-pixel autofocus when using live view, so focusing is MUCH faster.

Serious photographers should consider upgrading to the 5D Mark IV for its superior focusing system, thumb stick, dual memory cards, and headphone jack. However, no matter how serious you are, you'll miss the 6D Mark II's articulating screen and GPS. If this is your first full-frame camera, you might consider choosing a mirrorless system instead. Fuji mirrorless cameras, such as the XT-20 and X-T2, have smaller APS-C sensors, but the Fuji f/2.8 zooms produce results with similar noise and depth-of-field to the f/4 zooms most 6D photographers choose. We find the Fuji cameras to be more powerful and enjoyable, the electronic viewfinders reduce exposure problems, and the Fuji's have vastly superior video capabilities.

5D Mark III ($1,250 used)

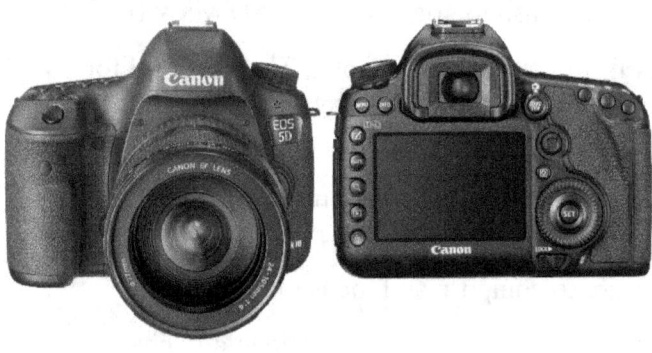

The primary difference between the _5D Mark III_ and the 5D Mark II or 6D is the superior autofocus system, which covers a large part of the frame and tracks moving subjects with amazing accuracy, surpassing even the 7D and 70D.

For that reason, the 5D Mark III is the right choice for sports and wedding photographers, whether serious amateur or professional. Those tasks require tracking moving subjects, often in low-light conditions, and the 5D Mark III will give you more in-focus pictures than any lower-end camera.

However, the 5D Mark III has almost the exact same image quality as the 6D and 5D Mark II, which are far less expensive. Therefore, if

Video: 5D Mark III Long-term Review **12:55** - _sdp.io/5D3Long_	

you're not tracking fast-moving subjects, choose one of those two bodies, and save the rest of your cash for lenses and flashes.

The 5D Mark III isn't the best choice for most wildlife photographers. Instead, I recommend purchasing a used 7D, and saving the rest of your budget for lenses. The 5D Mark III is excellent for wildlife photography when you can get very close, so if you plan to be spending hours camouflaged in a blind, the 5D Mark III might be a better choice for you. The 5D Mark III supports autofocus up to f/8, allowing you to attach a teleconverter to a lens such as the 400mm f/5.6 and still autofocus. However, you can only use the center autofocus point, which limits composition.

You should also consider the 36-megapixel Nikon D800. The D800 has 15% better image quality, making it a better choice for landscape and commercial photographers. It's a better

Video: D810 vs. 5D Mark III
25:33 - *sdp.io/d810v5d3*

choice for most wildlife photographers, too, because the DX crop mode provides a 1.5X teleconverter effect while still capturing 24 megapixel images—more detail than even the 7D.

5D Mk IV ($2,500 new, $1,800 used)

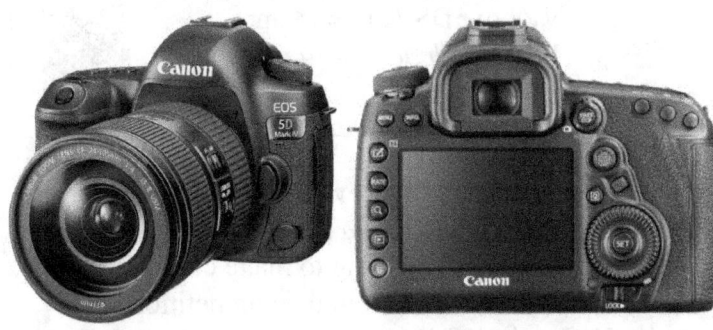

The 5D Mk IV is Canon's portrait and wedding professional camera body. The design has hardly changed from earlier models, but the autofocus system is greatly improved and will be worth the upgrade for most pros. The 30 megapixel anti-aliased sensor is a big step up from the 5D Mark III, but produces images with noticeably worse image quality than the competing Nikon D810, Sony a7R II, or even the 5DS and 5DS-R. Nonetheless, the image quality is more than sufficient for most working professionals.

While the 5D Mk IV technically supports 4K video, we found it unusable as a 4K video camera (and we've been filming at least two 4K videos per week since 2014). Canon chose to use an extremely inefficient video codec, motion JPG, which creates unusable large files with no benefit to image quality. Additionally, the 5D Mk IV has a severe crop factor of 1.74X (or more if you're producing content in 3840x2160). That means that you'll often need to change lenses before filming, and super-wide angle shots are simply impossible.

Canon advertised the Image Microadjust feature as being able to fix minor focusing problems in post. That feature would be game-changing for portrait and wedding photographers, but in our

Video: 5D Mark IV Review
25:33 - *sdp.io/5d4Review*

testing, it simply didn't work. Additionally, enabling it doubled the size of the files. As a result, we recommend 5D Mk IV photographers leave the option disabled.

While these heavily marketed features didn't fulfill their potential, we still recommend the 5D Mk IV for professional photographers with an investment in Canon.

5DS and 5DS-R ($1,600 used)

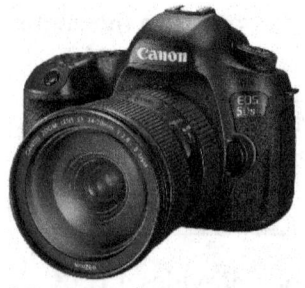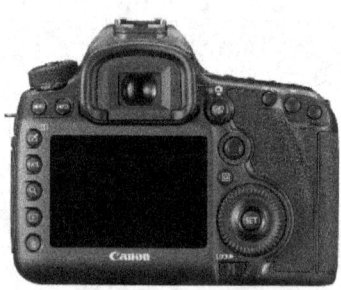

The 5DS and 5DS-R are Canon's answers to the high-resolution Sony a7R and Nikon D810. Both cameras are virtually identical to the 5D Mark III, with one major difference: they have an amazing 50 megapixel sensor.

According to our testing, no other 35mm-style camera provides anywhere near this level of detail. You don't need optically perfect lenses to benefit, either. In our landscape test, we found a visible improvement in detail with our standard recommendation of the Sigma 24-105 f/4 lens. Compared to the 5D Mark III, the 5DS-R's images look significantly better. They're only slightly better than the Nikon D810, but the difference is still obvious.

Many photographers hoped for significantly reduced noise and improved dynamic range. We did find some improvement to both noise and dynamic range, but Canon's newest cameras still

Video: 5DS-R Landscapes	
12:23 - *sdp.io/5dsrlandscapes*	

haven't quite caught up with the Nikon and Sony sensors. However, photographers can trade extra detail for reduced noise. Increasing noise reduction in post-processing reduces sharpness and noise simultaneously, and it's possible with editing to make clean, sharp images from the 5DS and 5DS-R that match the best images from the competition.

To help you take the sharpest pictures, the 5DS and 5DS-R include a mirror-dampening technology that reduces shake created by the movement of internal components when you take a picture. Additionally, the mirror-lockup mode now includes an automatically delayed shutter that will allow you to eliminate camera shake without using an external shutter release.

If you're a studio shooter, you'll also appreciate the USB 3.0 upgrade that improves the tethering performance. While most Canon cameras can be tethered to a computer to immediately copy pictures into post-processing software and allow you to preview the results on a bigger screen, USB 3.0 speeds up the process, allowing you to see images quicker.

Wildlife photographers will be tempted by the promise of more detail and the ability to crop even more tightly. Most wildlife photographers choose APS-C cameras, such as the 7D, 70D, or 7D Mark II,

Video: 5DS-R Studio Test	
4:13 - *sdp.io/5dsrStudio*	

because they have higher pixel densities—basically, more pixels crammed in a smaller space. Because wildlife photographers are often required to crop the image anyway, these APS-C cameras create more detailed wildlife photos.

With 50 megapixels, you can capture a full-frame image and get the exact same amount of detail you would get when taking pictures with the APS-C Canon 70D or 7D Mark II. Essentially, it's an effective digital zoom. You can use your full-frame wildlife lenses without any penalty, and decide later whether you want to crop.

There are a couple of serious problems that prevent us from recommending the 5DS and 5DS-R as general wildlife cameras. First, the 5 frames per second rate is usable for birds in flight, but it pales in comparison to the 10 frames per second the 7D Mark II is capable of. Also, the beta 5DS-R we tested began buffering after only 12 or 13 consecutive shots, meaning you have to be very careful not to take too many photos—something that makes wildlife photography much more challenging.

Ultimately, we recommend the 5DS-R only to very advanced wildlife photographers who spend the time getting very close to their subjects—close enough to fill the frame on their full-frame lenses.

| Video: 5DS-R Wildlife Test |
| **6:38** - *sdp.io/5dsrWildlife* |

For those photographers, it's simply unbeatable, especially when combined with Canon's incredibly sharp big primes, such as the 600mm f/4.

The 5DS includes an anti-aliasing filter designed to reduce the effects of moiré. The 5DS-R has a system to counteract those effects, improving sharpness. We recommend the 5DS-R over the 5DS to every type of photographer. In our experience with the 5DS-R and other high-resolution cameras lacking anti-aliasing filters, we've never seen moiré in real-world photos. You'll definitely see the effects when photographing test charts, however.

If you want the most detailed pictures possible, the 5DS-R simply cannot be beat by any other DSLR. If you're already shooting with the D810 or a7S, the difference is minimal in all but ideal conditions. In other words, you will see more sharpness out of mid-grade lenses, even when shooting handheld, but it's slight. When using sharp prime lenses, a tripod, and mirror-lockup, the difference is more distinct. Still, it won't be worth the trouble for most photographers to sell their existing cameras and lenses and switch to Canon.

There is one camera that produces sharper images: the Olympus E-M1 Mark II in high-res mode. However, it requires the use of a tripod. As an added benefit, the E-M1 Mark II has 4K video, an articulating touch screen, an electronic viewfinder, Wi-Fi, and other modern niceties.

1DX ($2,400 used)

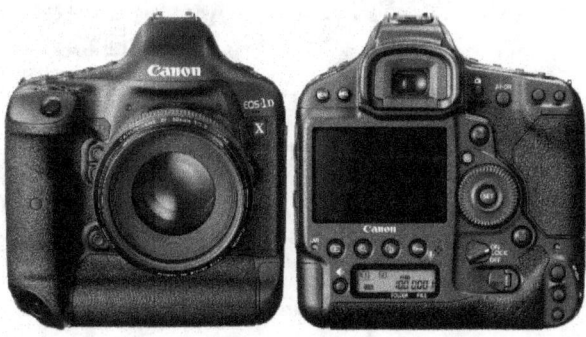

The 1-series Canons (currently the *1D X*) are intended for professionals who really abuse their camera bodies and don't mind carrying around the extra weight. With those cameras, you're paying thousands of dollars for durability, weather sealing, and longevity that very few people will need. Therefore, I almost never recommend them to people who ask which camera to buy. They're wonderful cameras, but if you need them, you probably already know, and wouldn't be seeking advice from me. The 1D X offers even stronger autofocusing than the 5D Mark III, and all autofocus points work up to f/8, allowing you to use a teleconverter with telephoto lenses. The amazing 14 frames per second greatly increases your chances of capturing the action at the perfect moment.

Image quality is also noticeably better than that of the lower-end full-frame cameras, despite having a lower megapixel count.

1DX II ($3,500 used)

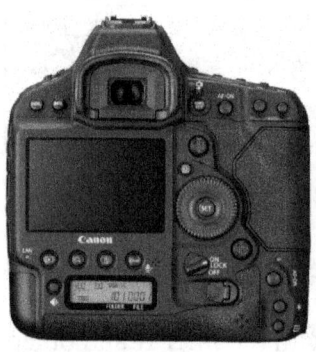 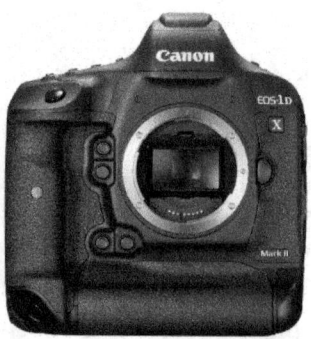

Like the 1DX, the primary purpose of the 1DX II is sports and photojournalism. Just because it's the most expensive Canon camera doesn't mean it's the best; most wildlife photographers will get better results with a 7D Mark II (for moving subjects) or a 5DS-R (for still subjects). Most other photographers, including landscape and portrait photographers, will make better images with a 5DS-R. The 1DX II is an incremental improvement to the 1DX. Some of the key improvements include:

- Higher framerates of 14 FPS (allowing you to see through the viewfinder) or 16 FPS (with the viewfinder dark)

Video: 1DX II Preview
33:56 - *sdp.io/1dx2preview*

- Improve live view and live view focusing. This is particularly useful when shooting at 16 FPS with the viewfinder dark; we've found you can do a reasonable job of tracking subjects using live view, making 16 FPS very practical.

- Dual-pixel AF, which provides better focusing while recording video or using live view

- 4K video at 60 FPS, offering four times the resolution of HD (1080p) video (but with a 1.4X crop). Note, however, that the

Video: 1DX II Video Quality
9:50 - *sdp.io/1dx2VQ*

  maximum framerate 4k video consumes 1 GB of storage for every 10 seconds of video. Therefore, you can consume a 128 GB CFast card in 12 minutes.

- HD video at 120 FPS, offering 4x slow motion (if you're rendering at 30 fps) or 5x slow motion (if you're rendering at 24 fps)

- A touch screen, which is useful for focusing with live view. Unlike most touch screens, however, it is not useful for quickly reviewing pictures. For some reason, Canon has provided more limited touch screen capabilities on this camera than on lower-end cameras.

- CFast and CF card slots. The CFast allows for the extended high framerate shooting and for 4K/60FPS video. However, CFast cards are currently much more expensive than CF cards, so factor that additional cost into your budget.

Compared to the Nikon D5, the 1DX II is overall faster, and a much more usable video camera. We generally prefer the usability of the Nikon D5, including the controls and better touch screen

Video: 1DX II Image Quality 3:52 - *sdp.io/1dx2IQ*	

functionality. However, most photographers won't notice a functional difference between the cameras. In most traditional sports and photojournalism scenarios, both cameras will produce indistinguishable results.

Compared to the Sony a9 ($4,500), the 1DX feels clunky and antiquated. The optical viewfinder doesn't preview your exposure and can't show you your histogram. The viewfinder blacks out with every exposure and makes loud noises. The 1DX II is overall slower. However, the 1DX II is time-tested and has native support for an amazing selection of Canon lenses. If you can choose Sony G-Master lenses for your needs, however, you should

Video: 1DX II Review 18:00 - *sdp.io/1dx2Review*	

definitely try an a9 before buying the 1DX Mark II.

1DX III ($6,500 new)

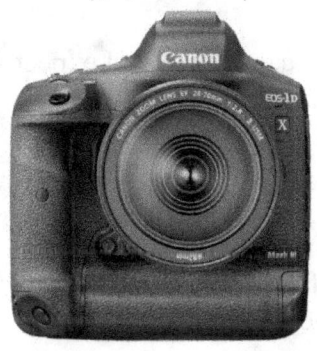

While the 2020 Summer Olympics were delayed, Canon's update to their top-end sports camera was not. The 1DX Mark III is a substantial upgrade to the 1DX Mark II, offering vastly more powerful video capabilities, HEIC files as an option to replace JPG with better compression and quality, better autofocus and more AF points, 16 FPS (or 20 FPS in live view), better image quality, GPS, and better wireless.

The 1DX Mark III is clearly superior to the 1DX Mark II. However, so many photographers have upgraded that they've driven down used prices on the previous-generation camera, which can now be had for about half the price of a new 1DX Mark III. Ultimately, for the professional photographer, you probably won't make more money shooting with the 1DX Mark III instead of the 1DX Mark II, but it will cost you more money.

On the other hand, the workflow benefits of the 1DX Mark III might make the expense worthwhile. These benefits includebeing able to shoot silently with more reliable autofocus, tagging your photos with GPS (improving your ability to organize the photos in Lightroom), reducing storage costs by using HEIC instead of JPG, and transferring files wirelessly.

Chapter 12: Canon EOS RF

The Canon RF mirrorless system will eventually replace the Canon EF DSLR system. While it's still young, Canon has brought many desirable professional lenses to the system, such as the 50mm f/1.2, the 85mm f/1.2, and the 28-70 f/2. Each of those lenses produce results that simply cannot be produced with any other autofocus lenses. Those lenses are the sole reason we have purchased 9 RF cameras for our video productions.

Canon also offers the amazing 600mm f/11 ($700) and 800mm f/11 ($900) wildlife lenses, which produce amazing results while being easily hand-holdable. No other system has light, inexpensive, and sharp wildlife lenses. As a result, we strongly recommend the RF system to wildlife photographers.

The Canon RF system supports adapters that allow you to attach Canon EF DSLR lenses with reliable autofocus. Therefore, if you're an existing Canon DSLR photographer, you can continue to use your existing lenses with your new mirrorless camera.

Canon EOS R10 ($1,000 new)

The Canon EOS R10 is Canon's entry-level RF-S camera. It has a smaller APS-C 24-megapixel sensor with a 1.6X crop factor. The R10 hasn't yet been officially released, but when it is, I'll update this buying guide with a full review.

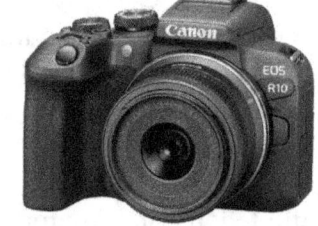

It's sure to be an extremely popular entry-level camera. We've reviewed so many similar cameras from Canon, including the M50 upon which this is heavily based, and I'm sure it's going to be a good all-around performer with amazing autofocus.

Like the Canon R7, it offers 15 frames per second with the mechanical shutter, so I expect it to be the best choice for parents shooting their kids sporting events. It also offers 23 FPS with the electronic shutter, but I'll need to test it for rolling shutter before I feel comfortable recommending that.

It records 4k video at 30FPS full-width, which should be more than enough for any budding vloggers and videographers. It records 4k/60 with a significant crop; if you want full-width 4k/60 video, you will need to spend an extra $500 on the Canon R7.

Before buying the R10, look closely at the full-frame Canon EOS RP. While the RP is older, it has a sensor more than twice the size, and fully utilizes the entire full-frame RF system. If you aren't shooting sports or wildlife and you don't need full-width 4k video, the RP will be the better choice for you.

Canon EOS R7 ($1,500 new)

The Canon EOS R7 is misnamed; Canon should have called it the Canon EOS R70. This distinction is important because the name R7 indicates it's a successor to the famous Canon 7D series of DSLRs. However, the 7D series was built more solidly, had a top LCD screen, and were generally intended more for professional use than the R70.

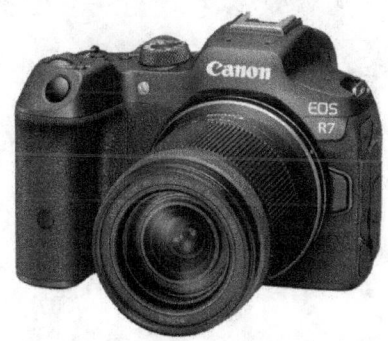

The R7 is, however, the successor to the excellent Canon 90D. It uses a similar 32-megapixel sensor and feels very similar in the hands. Like its namesake (the 7D series of cameras) it includes two card slots. However, they are both SD cards, and the buffer fills quickly.

If you look at Canon's advertising, you'll see that it offers an amazing 30 frames per second still photos. That sounds unbeatable for sports and wildlife photography, however, 30 frames per second is only available with the electronic shutter, and the R7's electronic shutter has a very slow readout speed. As a result, any moving subjects will have extreme rolling shutter (as discussed elsewhere in the book).

Here's evidence of that: Chelsea swinging a wiffle ball bat that gets so bent it looks like a banana. The picture is ruined.

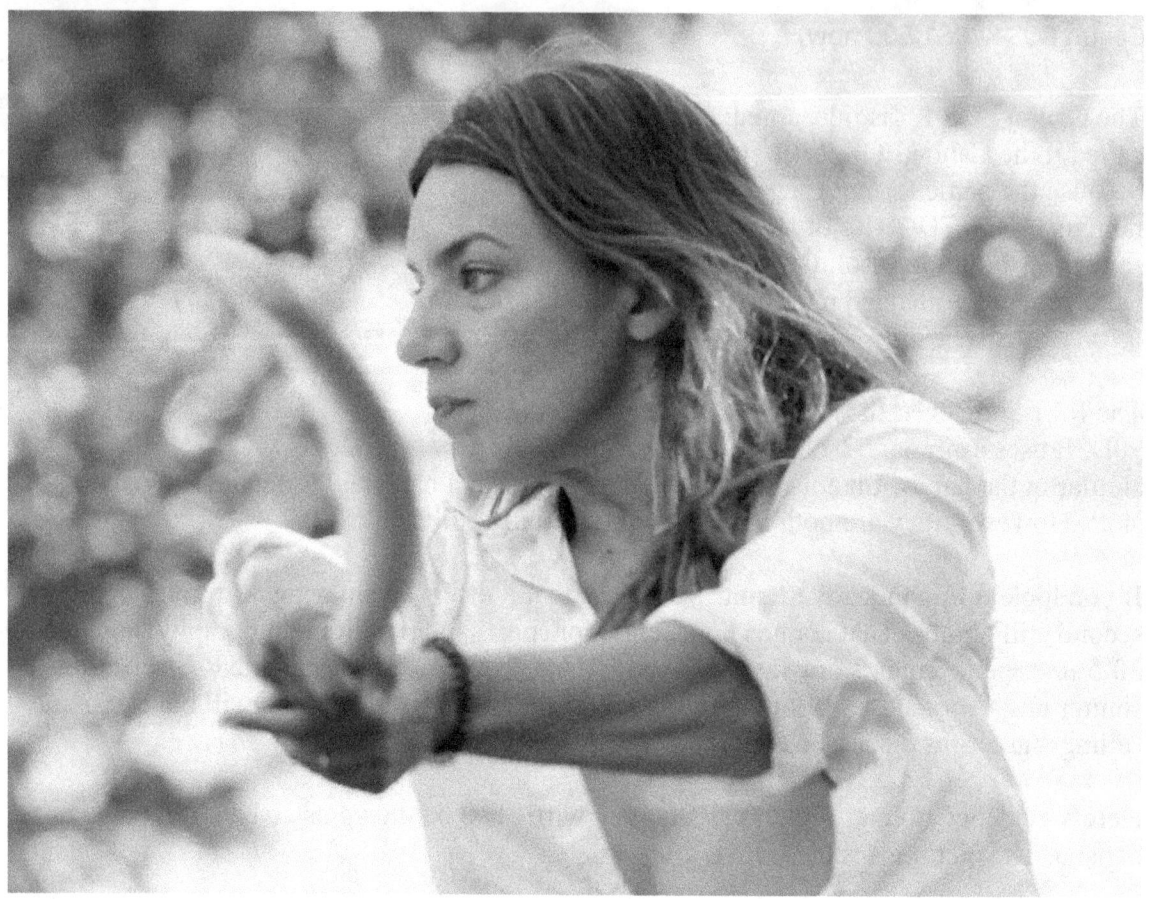

The answer is to not use the R7's electronic shutter for action, but instead to use the 15-frames per second mechanical shutter. At 15 FPS, the R7 performs amazingly, and though the frame rate is cut in half, it still outperforms the Fujifilm X-T4, the Canon 7D Mark II, the Canon 90D, and every other $1,500 camera we've ever tested.

While other cameras, such as the Olympus E-M1X, offer a higher FPS, the Canon R7 outperforms them with its stunning autofocus system. The R7 locks onto the eyes of athletes and animals with astounding accuracy. Ultimately, the R7 provides results in action scenarios that no other $1,500 camera can provide, and it doesn't require years of practice. An absolute beginner can pick it up and put it into shutter priority mode and get great sports photos.

With the Canon RF-S 18-150mm lens, the R7 is a good all-around camera for travel, portraits, and sports. However, if you're not shooting sports or wildlife, and you don't need the amazing autofocus or high frames per second, you'll get noticeably better images using the Canon EOS RP with the Canon RF 24-240mm full-frame lens. More importantly, the RP uses the entire image circle of Canon RF full-frame lenses; if you choose the R7, RF lenses will have a 1.6X crop factor.

Canon EOS RP ($1,000 new or $900 used)

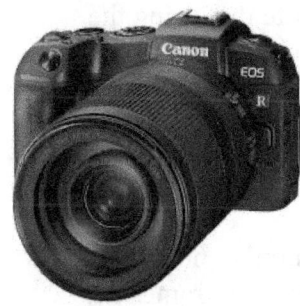

The Canon EOS RP is the best value in photography for non-professionals. It has a full-frame 26 megapixel sensor, good autofocus, and a flip-forward screen perfect for vlogging. If you're looking for a single lens for travel, landscapes, and general photography, buy the Canon RF 24-240.

The battery is a bit small, and it only offers 4k video with a heavy crop and limited autofocus. It has only a single SD card slot, and the autofocus system isn't ideal for shooting sports. Still, $1,000 for a functional full-frame camera makes it an excellent value and a good starting point into the powerful Canon EOS RF system. I routinely recommend it over more expensive APS-C cameras such as the Canon R7 and Fujifilm X-T4.

Incidentally, we currently use 5 Canon EOS RPs in our video studio. We definitely have more expensive cameras, but the RPs have been running for about 3 years now and there's no real reason to replace them. The 1080/60 video looks great and they've literally never missed focus even with super-shallow f/1.2 depth-of-field.

Canon EOS R ($1,600 new)

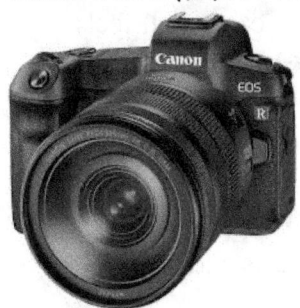

Canon's EOS R was the first RF camera. At launch time, it just wasn't ready; it was buggy and the autofocus system was extremely unreliable. Therefore, the reviews you'll see from 2018 and 2019 are often very negative (including our own).

However, since that launch date, Canon released a series of firmware updates that made it a much better camera. Today, we continue to use the R for vlogging because of its extremely reliable autofocus. Like the RP, it only records 4k video with a tight crop.

Canon EOS R6 ($2,500 new)

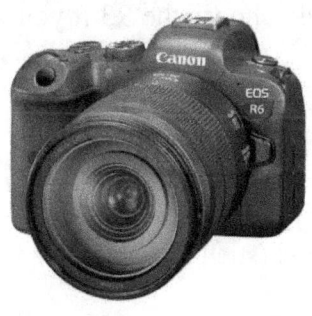

The Canon EOS R6 is a professional-grade mirrorless camera with 20 megapixels and two card slots. The autofocus system is amazing, providing human and animal eye AF corner-to-corner. You may never need to manually select an autofocus point. Unlike the R and RP, the R6 has two card slots, so professionals will be protected in the event a memory card fails. However, because its only 20 megapixels, your images won't be nearly as detailed as they would if you captured it with the older EOS R. If you plan to photograph sports, the EOS R6 is a better option.

The EOS R6 offers full-width 4k video at both 30 and 60 frames per second. However, it overheats very quickly. At 30 frames per second, it will record reliably for about 30 minutes. At 60 frames per second, it often overheats in less than 20 minutes. Full-width HD video, such as what the R and RP produce, does not overheat. Note that using the camera for stills will also contribute to the video overheating, meaning you might only be able to record 2 or 3 minutes of 4k video after shooting stills for an extended period.

Canon EOS R5 ($3,900 new)

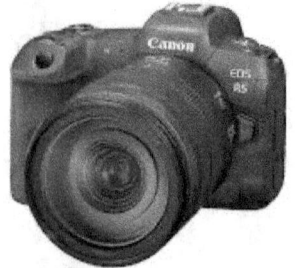

The Canon R5 is the flagship model for Canon's mirrorless RF lineup. It has 45 megapixels and reliably shoots 8 frames per second with the mechanical shutter or 18 with the electronic shutter. Like the R6, the R5's autofocus system is simply amazing.

Like the R6, it has amazing video capabilities that are limited by overheating. The standard 4k/30 mode will record indefinitely without any heat problems, but it's "line skipped", which means its not quite as sharp as the 4k/30 video produced by the R5. If you switch to high quality video mode, the video is as sharp as the R6's video, but it overheats in about 35 minutes of recording.

The R6 also offers 4k/60, 4k/120, and 8k/30 video. The 4k/60 and 4k/120 video are line-skipped. The 8k/30 video is amazing, but it only reliably records for 20 minutes, and if you switch to 8k after shooting stills you might only be able to record for 2-5 minutes.

Even with these caveats, the R5 might be my favorite general-purpose camera. It feels great in the hand; even better than the famous Canon SLRs. The low-light capabilities are better than any Sony, Nikon or Panasonic camera.

Canon EOS R3 ($6,000)

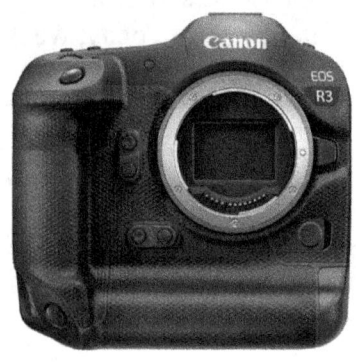

The Canon R3 is Canon's current flagship mirrorless camera, optimized for professional sports and action photography.

The R3 is the fastest camera Canon has ever produced with 30 FPS. However, it does not produce the best image quality because it has only a 24 megapixel sensor, less than half that of the Canon R5. You should choose the R3 if you meet any of these criteria:

- You shoot professional sports and action

- You use high shutter speeds, such as 1/2000, in artificial light indoors (such as photographing indoor sports)

- You need a silent shutter with more than 20 frames per second (such as photographing golf)

- You shoot very fast action, are concerned about rolling shutter, and you need more than 10 frames per second.

The R3 has an integrated vertical grip and a gigabit Ethernet network port, so it provides a direct upgrade path for DSLR photographers with a Canon 1DX-series camera.

One novel feature that we haven't seen since the Canon EOS-3 film camera: eye-controlled autofocus point selection. Whatever you look at the camera focuses on it. However, the accuracy of the eye-controlled autofocus depends on you recalibrating it at least daily. If we didn't recalibrate it every day, it was unreliable. The novelty of the eye-controlled system wore off in about a week and we have never used it since, instead relying on the multitude of other focus point selection methods. Please don't buy the R3 for the eye-controlled autofocus; it's a gimmick.

Chapter 13: Canon EOS M

Note: Canon does not seem committed to expanding the EOS M system of cameras and lenses. We were never satisfied with the selection of lenses and bodies for the EOS M lineup. If Canon is not planning to aggressively create new bodies and lenses for the system, we cannot recommend new buyers choose an EOS M camera.

Canon is the #1 manufacturer of DSLRs, but their effort to address the growing mirrorless interchangeable lens market came rather late. Some might even say their efforts are half-hearted, and that they'd prefer you buy a DSLR instead.

Perhaps their biggest selling point is compatibility with existing Canon EOS lenses using the EF lens adapter ($135). Indeed, the adapter does work, and you can autofocus with your EOS lenses. However, if you already have Canon lenses, you probably also already have a Canon DSLR, and I find it makes little sense to attach the M to a Canon EOS lens. You lose the size benefits the mirrorless system can bring, as well as the focusing and viewfinder benefits of a DSLR. In other words, attaching the M to an EOS lens gives you the worst of both worlds, in many ways.

If you take my advice and avoid adapting lenses, Canon offers you four native lenses:

- 11-22mm f/4-5.6 (equivalent to 17mm-35mm f/6.4-f/9)

- 22mm f/2 ($250) pancake lens (equivalent to 35mm f/3.2)

- 18-55mm f/3.5-5.6 IS ($300) (equivalent to 29-88mm f/5.6-f/9)

- 55-200mm f/4.5-5.6 (equivalent to 88-320mm f/7.2-f/9)

Also, Tamron offers an 18-200mm superzoom lens for those who don't want to change lenses.

As you might have inferred, I can't recommend the EOS M system to new buyers. There's nothing inherently wrong with the system, but the market for low-end interchangeable cameras has almost disappeared, and Canon doesn't seem to be investing further into the system. In fact, the M3 might be the very last EOS M camera.

Nonetheless, if you love Canon and the 35mm-equivalent focal length feels right to you, the M with the 22mm f/2 lens makes a great, compact, walking-around camera suitable for street photography, and the kit costs only $450. If you'd prefer a zoom lens, I'd direct you to the fixed-lens PowerShot series instead.

EOS M ($265 new, $240 used)

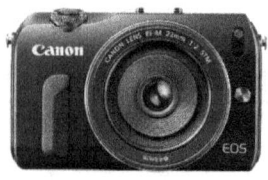 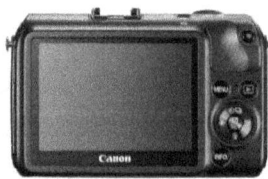

Released in 2012, Canon's first mirrorless interchangeable lens camera features an 18 megapixel APS-C sensor, 1080p video, hybrid focusing, and a touch screen. Functionally, it's quite similar to their lower-end APS-C DSLRs (with a 1.6X crop), and you can expect similar image quality (but in a smaller form factor).

It lacks a viewfinder, so you'll need to use the back of the camera to frame and focus your shot. The focusing system is capable for a mirrorless camera, though because mirrorless focusing is advancing quickly, it doesn't keep up with the latest generation of cameras, including the Sony a6000 and Fujifilm X-T1.

EOS M2 Mark II ($320 new kit, $270 used kit)

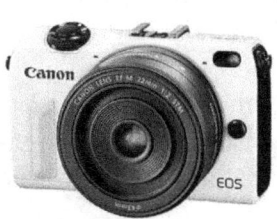

In December of 2013, Canon released an updated EOS M camera: the M2. However, they didn't initially release it in the United States, and instead continued to sell the original EOS M camera. Today, you can find copies of the M2 Mark II in the US, but you probably shouldn't bother.

The M2 offers significantly improved hybrid autofocus, and claims to focus 2.3 times faster than the original M. It's also about 10% smaller than the original M. Otherwise, the cameras are identical.

EOS M3 ($480 new, $410 used)

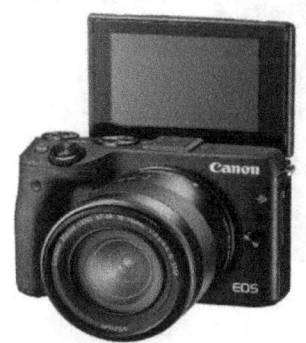 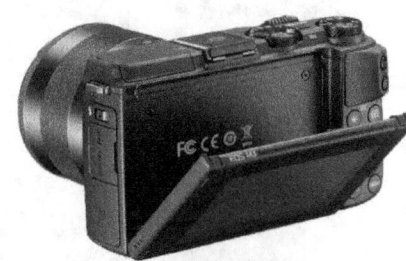

With a small size (but usable grip), 24 megapixels, good image quality, a fantastic focusing system, and a flippy touch screen that can be used for selfies, the M3 seems like the perfect DSLR companion.

Indeed, if you're already invested in the EOS M system, the M3 is the best camera in the lineup. However, the limited lens selection makes it difficult for us to recommend over cameras such as the Olympus E-M10 II and Sony a5100.

EOS M5 ($1,000 new, $925 used)

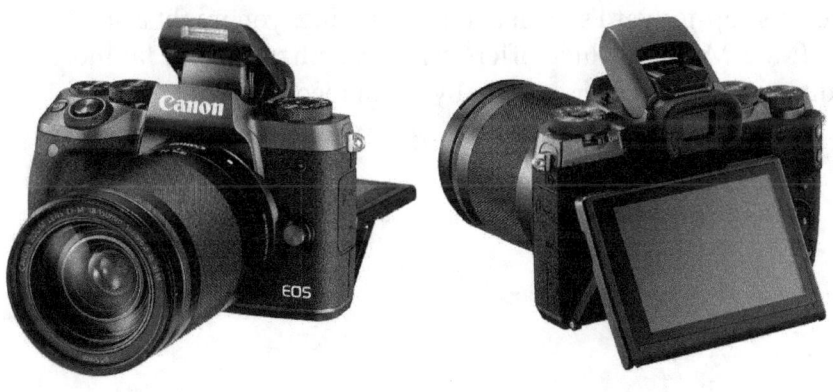

If you're looking for a Canon DSLR with an electronic viewfinder, you're in luck. Canon started with their 80D DSLR, shrunk it down, gave it an electronic viewfinder, and called it the M5.

It's a great camera! This was somewhat of a surprise to us, because we were disappointed with the earlier EOS M cameras. Focusing isn't as fast as Canon DSLRs, but it's good enough, and the touch-and-drag technique for changing the focusing point works great.

Unfortunately, Canon still doesn't make enough EOS M lenses… you're stuck with low-quality kit lenses or you need to add an adapter to use Canon EOS lenses. If you do use the adapter with DSLR lenses, you'll discover that the M5 is too small to balance those large lenses, and you'll just wish you had an 80D instead.

The flip screen flips down, rather than to the side. Thus, the screen can't face forward while the camera is attached to the tripod. The M6 uses a flip-up technique, making it a better choice than the M5 for vloggers you might use a tripod.

If you aren't yet invested in the EOS M system, take a serious look at the Fujifilm X-T20 instead. It's less expensive, offers 4K video, and many more native lenses (including high-quality lenses). The X-T20 also looks and feels better, produces better stills and video, and is generally more fun to use.

EOS M6 ($680 new, $550 used)

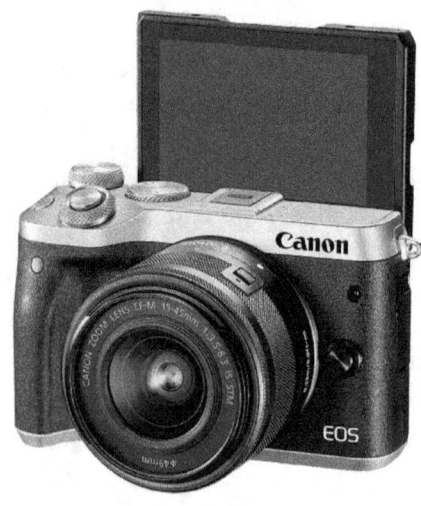
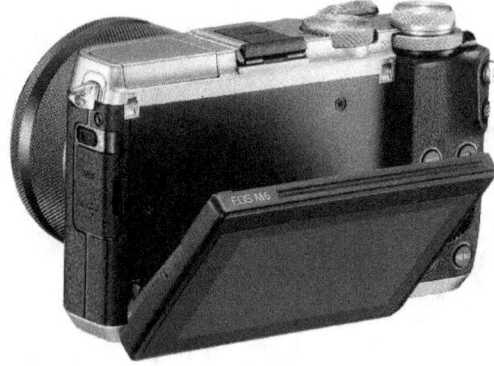

Finally, Canon make a retro-styled camera! The EOS M6 is a great all-around mirrorless camera, but it suffers from the same drawbacks as the M5: no 4K video, no sensor stabilization, and no native high-quality lenses. It also doesn't have a viewfinder, but if you prefer to use the rear screen, that won't matter to you.

If you're looking for a casual camera for snapshots and selfies, the M6 is a good choice. If you're drawn to the M6 for it's flip-forward screen and retro styling, you might also consider a silver Olympus E-M5 Mark II, which offers sensor stabilization and far more lenses. It doesn't focus quite as well, however, especially when recording video.

EOS M50 ($500 new)

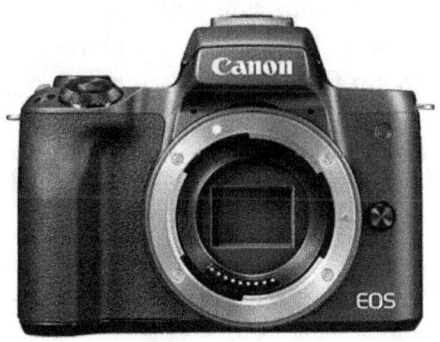 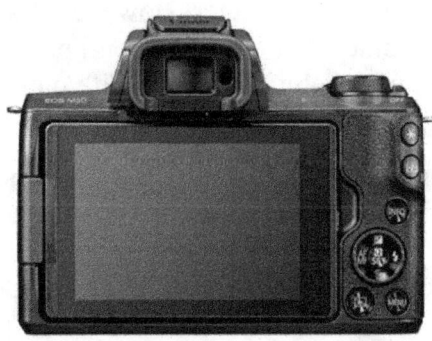

As of August 2020, the EOS M50 is the best $500 vlogging camera you can buy. It doesn't have the functionality of a Panasonic GH5 or the video quality of a Sony a7 III, but the small size, touchscreen that flips forward from the size, and amazing face detection autofocus means vloggers can reliably film themselves from a tripod or at arm's length and get everything in focus.

Canon has heavily marketed it as a 4k camera, and technically it's true: it does have 4k at 24p. However, in practice, the 4k isn't that useful. First, most YouTubers film in 30p, and 24p falls short of that. Second, the camera doesn't have Canon's famous dual pixel autofocus in 4k mode, meaning you can't reliably focus on your face or track moving subjects. Third, 4k video has an extra 1.6X crop, meaning you can't shoot 4k video at super-wide angles, you regularly have to change lenses when switching between stills and 4k video, and the effective sensor size is smaller than a micro four thirds camera.

In summary, you shouldn't get the M50 for its 4k capabilities. Like the other EOS M cameras, their native lens selection is extremely limited, so I can't recommend them for stills photography. However, as a vlogging camera, it's literally the best option available (but plan to publish in 1080/60p).

EOS M50 Mark II ($600 new)

The M50 Mark II is physically indistinguishable from its predecessor, and it includes a handful of updates for video creators:

- **Vertical video**. Yes, you can record vertical video by turning any camera sideways, but the M50 Mark II will put the vertical status into the video file's metadata. When you import it into a video editor, it'll automatically be vertical. It's a nice feature that saves you a couple of clicks if you're producing vertical video, but it's not a reason to upgrade or spend the extra $100.

- **Face and eye detect autofocus from further away**. The original M50 has great face and eye detect autofocus. This one detects you if you're smaller in the frame. If you're always close to the camera, as a vlogger would be, this won't make any difference.

- **Built-in WiFi and Bluetooth**. You can transfer your files wirelessly, but you probably don't want to. It's still a bit unreliable and video files can be quite slow to transfer. I still prefer using a memory card reader, even if I'm transferring the video to my smartphone.

- **YouTube streaming**. Technically you can stream directly to YouTube across WiFi without the need for a computer. However, it's a pain to setup, and you'll still need a computer or some other device to configure the stream. Plus, people watching your live stream probably won't mind if you just stream from your smartphone because that's what most people do.

- **USB webcam compatibility**. If you want to use a real camera for streaming from a PC (for example, if you're hosting yoga classes on Zoom), the updated camera can do that.

Otherwise, the results from the M50 Mark II will be identical to the original camera.

Chapter 14: Nikon F-mount DSLR

This chapter provides an overview of Nikon DSLRs, follow by recommendations for Nikon and third-party flashes and portrait equipment.

Note: Nikon does not seem committed to future support of the Nikon 1 system of mirrorless cameras and lenses. While we liked the Nikon 1 cameras that we used, if Nikon is not committed to releasing new bodies and lenses, we cannot recommend new buyers invest in the system. As a result, we have removed the Nikon 1 series from the buying guide. However, you can download that chapter at *sdp.io/n1guide*.

Nikon has the strongest lineup of camera bodies in the world, and one of the best lens and flash lineups. I routinely recommend Nikon equipment for everyone from beginners on a sub-$500 budget to professional landscape photographers with over $10,000 to spend. Elsewhere in this book, I've done my best to help with the difficult decision every DSLR buyer faces: Canon or Nikon. While I have found a handful of distinct benefits of the Canon infrastructure, in most ways, Nikon meets or exceeds the standard set by the #1 camera manufacturer. You almost can't go wrong by buying a Nikon.

If great video is more important to you than great still photos, you should consider Panasonic instead. Specifically, examine the GH2, GH3, and GH4 cameras (whichever new or used model fits your budget). They offer more video features, such as focus peaking and smarter focusing during recording. The newest cameras also offer 4k recording, which Nikon does not currently offer. Panasonic cameras are simply smaller, less expensive, and more capable than Nikon for most video work.

Nikon DSLR Buying Guide

This table shows the Nikon DSLRs currently available. In this table, FPS is Frames Per Second, the number of shots you can take in one second. IQ stands for Image Quality, and more stars is better. MP stands for Megapixels, which hardly makes a difference in the Canon lineup, because they're all pretty similar. Motor refers to whether the body has a built-in focusing motor, allowing it to work with older lenses that lack a built-in focusing motor.

Model	Price (new)	Price (used)	Arctic. Display	Wi-Fi	GPS	Full-frame	MP	FPS	Motor
D3100	$275	$165					14	3	No
D3200	$360 (kit)	$300					24	4	No
D3300	$450 (kit)	$350					24	5	No
D3400	$500 (kit)	$375		*			24	5	No
D5100		$300	*				16	4	No
D5200		$275	*				24	5	No
D5300		$400	*	*	*		24	5	No
D5500	$600	$450	*	*			24	5	No
D5600	$650	$530	*	*			24	5	No
D90		$165					12	4.5	Yes
D7000		$300					16	6	Yes
D7100	$700	$450					24	6	Yes
D7200	$1,000	$700					24	6	Yes
D7500	$1,250	$930	Tilt	*			21	8	Yes
D300S		$250					12	7	Yes
D500	$1,900	$1,400	Tilt				20	10	Yes

Model								
D600		$750			*	24	5.5	Yes
D610	$1,250	$850			*	24	6	Yes
D750	$1,800	$1,350	Tilt	*	*	24	6.5	Yes
D800		$900			*	36	4-6	Yes
D810	$2,800	$1,550			*	36	5-7	Yes
D850	$3,300				*			
Df	$2,750	$1,400			*	16	5.5	Yes
D3X		$1,500			*	24	5	Yes
D4		$1,800			*	16	10	Yes
D4S	$5,400	$2,700			*	16	11	Yes
D5	$6,500	$6,000			*	21	14	Yes

If you're investing in your first Nikon camera, I suggest taking your total budget, dividing it by two, and picking the Nikon camera that you can afford. Then, set aside the rest of your budget for lenses, flashes, a memory card, and a tripod.

Most photographers should choose one of the less expensive Nikon DX APS-C cameras (such as the D3x00, D5x00, D7x00). If you must make massive prints, or you need a full-frame FX lens for your style of photography, upgrade to one of the FX bodies (D6x0, D7x0, D8x0, or Df).

In order from least to most expensive:

- **D3x00**. They take fantastic pictures, and many photographers will be best served by choosing one of these entry-level bodies and saving more budget for lenses, flashes, and travel.

- **D5x00**. They take the same pictures as the D3x00 series cameras, but offer better usability: bracketing, an articulating screen, and a more powerful focusing system make the cameras easier and more fun.

- **D7x00**. They take slightly sharper pictures than the other cameras, they're more durable, and the focusing is faster. However, they give up the articulating screen of the D5x00 and they're heavier, making them less useful as all-around cameras. They're the best choice for most wildlife and sports photographers in the Nikon world, however.

- **D6x0**. The entry-level full-frame D6x0 series offers far better image quality than any of the previous cameras; noise is significantly reduced and pictures with full-frame lenses are much sharper. However, the focusing system isn't as solid as the D7x00 or D750.

- **D750**. The D750 offers similar image quality to the D610, but adds a faster focusing system and an articulating screen. The D750 is the only full-frame camera with an articulating screen, and that really improves its usability.

- **D8x0**. Perhaps the best full-frame studio and landscape cameras in the world, the D8x0 series captures clean, high-resolution images. You don't necessarily need more expensive glass to capture more detail with these cameras.

- **Df**. Nikon's retro camera has a low-resolution sensor but a stylish design. Oddly, it doesn't record video.

- **D3X/D4/D4S/D5**. If you need one of Nikon's top-end professional bodies, you probably already know it. These cameras actually take less sharp pictures than the D810, but they're extremely durable and fast.

The single-digit Nikons (currently the D3X, D4, D5) are intended for professionals who really abuse their camera bodies and don't mind carrying around the extra weight. With those cameras, you're paying thousands of dollars for durability, weather sealing, and longevity that very few people will need. Therefore, I almost never recommend them to people who ask which camera to buy. They're wonderful cameras, but if you need them, you probably already know, and wouldn't be seeking advice from me.

For that reason, the D810 is the highest-end camera that I recommend for amateurs and most pros. If you have an unlimited budget, get the D810 and several lenses. Otherwise, get the kit highlighted in the table that best suits your budget.

Within the professional bodies, the D3X and D4/D4S are very similar in price. The D3X is more of a general purpose professional camera body, while the D4 is specialized for low light, sports, and photojournalism.

Here are the unique aspects of each camera, and why you might want to choose them.

Nikon D3100/D3200/D3300/D3400/D3500 ($275-$500 new, $150+ used)

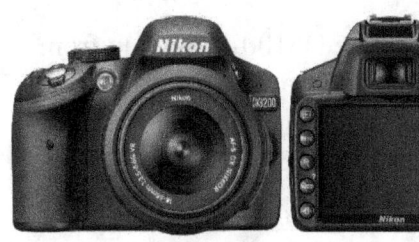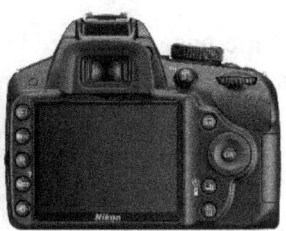

Nikon's entry-level DSLRs are perfect for beginners; saving money on the body lets you spend more on lenses, flashes, tripods, memory cards, and software, and those will have a bigger impact on your photography than buying a more expensive body.

The _D3100_ is available for outrageously low prices used; my target price is $200, but you can find them even cheaper if you're patient. If you're buying new, the _D3200_ is a better value. It has slightly better image quality and it takes pictures a bit faster. However, the D3200 creates much larger photos, which also make it much slower to copy and edit your photos.

The D3200 also adds a mic jack for recording external audio with your video. If you plan to record video and don't want to use the built-in mic (which is awful on all cameras), the mic jack is a must. The D3200 also jumps to 24 megapixels (the D3100 has 14 megapixels) providing much larger pictures. Those larger pictures require larger memory cards and more disk space, but only have slightly better image quality. The higher 4 frames per second (from 3 fps on the D3100) will help with action shots, but the buffer fills up too quickly for this to be a great sports camera.

The D3300 is about 9% smaller and lighter than the D3200 and D3100. Additionally, its updated AF-S 18-55mm kit lens is about 25% smaller and lighter than the kit lens included with the D3200 and D3100.

| Video: D3300 Review |
| **30:19** - _sdp.io/d3300Review_ |

For the D3300, Nikon continued using the 24 megapixel sensor, but removed the optical low-pass filter. This means that your pictures will be a bit sharper and more detailed, but you'll probably never notice the difference unless you use professional-quality lenses costing far more than the body itself. The D3300 also jumps to 5 frames per second, making it more useful for sports and action.

If this is your first camera, you might also consider the Canon T3/1100D. The cameras are equally functional; I'd buy whichever I found a better price on.

While we frequently recommended previous generations of Nikon's base-model DSLR, the D3400 ($500 new, $375 used) hasn't kept up with the competition. Nikon skimps on essential features, such as bracketing and a touchscreen. Basically the only new feature, SnapBridge, is universally hated. My advice: choose a D5500 instead and get an articulating screen (avoid the D5600's SnapBridge).

Nikon D5100/D5200/D5300/D5500/D5600 ($350-650 new, $275+ used)

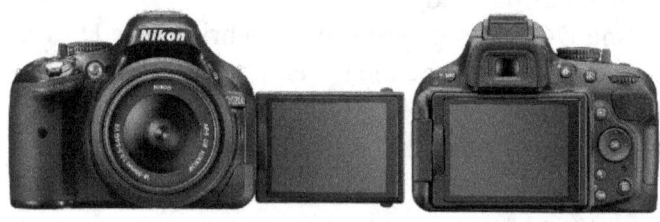

The _D5x00_ series of cameras offer similar image quality to the D3x00 series. Here, your extra money goes towards a very useful articulating screen, and depending on the specific model, extra features such as a touchscreen, Wi-Fi, and GPS.

The D5200 and D5300 provide a big megapixel increase over the D5100—jumping from 16 megapixels to 24 megapixels, and that megapixel increase does improve overall image quality.

The D5300 ($800 new, $600-$700 used) was the first camera in the Nikon lineup to offer Wi-Fi and GPS. The Wi-Fi can be useful for sharing images. The GPS is very useful for those who travel with their camera, because it

Video: D3300 vs D5300	
13:08 - _sdp.io/d3300vsd5300_	

helps you find your pictures by browsing them on a map (if you use Lightroom or another app that organizes photos by GPS data). For landscape or wildlife photographers, the GPS data can help you find your way back to a spot, so you can shoot the same location at a different time of year.

There are several good reasons to upgrade from the D5300 to the more expensive D7100 or D7200. The D7x00 cameras focus faster, which is great for sports and wildlife. The D7x00 cameras are also more durable and weatherproof. If you plan to shoot outdoor portraits with a flash, the D7x00 cameras support high-speed sync—an important feature that Nikon left off the D5300.

The D5500 ($500 new) offers very little over the D5300. In fact, one of the most notable changes is the removal of the

Video: D5500 (vs D3300, 70D & a6000)	
15:27 - _sdp.io/d5500Review_	

D5300's built-in GPS—if you'd like to see your pictures automatically overlaid on a map, you might save some money and buy the previous generation model.

The D5500 does offer a couple of new features when compared to the D5300. It now has a touch screen, which will make the smartphone generation more comfortable, and help all of us focus

Video: T6i & T6s vs D5500
23:14 - *sdp.io/T6iReview*

during video and quickly review our pictures. Nikon also removed the anti-aliasing filter from the D5500, so pictures will be slightly sharper. As a camera for hiking and travel, the D5500 might be the best DSLR available. Though it doesn't feel too small, it's extremely lightweight, making it easy to carry. It's also the first Nikon to include a touchscreen, which makes it easier to focus while using live view, and allows you to more quickly review photos.

The D5600 ($700 new) boasts one important new feature: SnapBridge, Nikon's latest and hated Wi-Fi implementation. Most people who upgrade prefer the Wi-Fi implementation in the D5300 and D5500; therefore, save yourself a few hundred bucks and get the old models.

If this is your first camera, you might also consider the Olympus E-M10 (either the original or the Mark II). The electronic viewfinder is useful for photographers of all skill levels, but beginning photographers will benefit the most from seeing their exposure before taking a photo, and the small size makes you more likely to carry it with you.

There are several good reasons to upgrade from the D5x00 series to the more expensive D7x00 series. The D7x00 cameras focus faster, which is great for sports and wildlife. The D7x00 is also more durable and weatherproof. If you plan to shoot outdoor portraits with a flash, the D7x00 supports high-speed sync—an important feature that Nikon left off the D5x00.

Nikon D90 ($150 used)

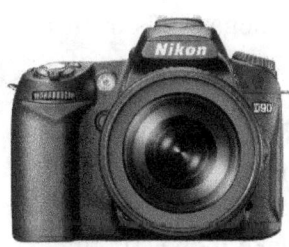 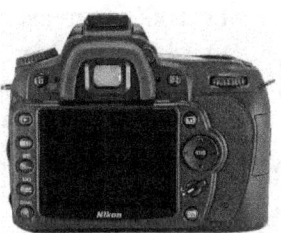

Since it's one of the older cameras discussed here (it was released mid-2008), the *D90* looks terrible on paper: 13 megapixels, 4.5 frames per second, and 720p video recording. However, it's a sturdy, capable camera that has one particularly compelling feature: a built-in focusing motor.

This built-in focusing motor allows you to use older Nikon "AF" lenses that lack a built-in focusing motor. These lenses rely on a mechanical coupling with the camera body to drive the lens focus.

You certainly don't *need* a focusing motor. You can simply choose from the wide variety of AF-S lenses that have built-in focusing motors. However, if you're the type who's always looking for used specialty lenses at amazing prices, such as older wildlife lenses, you'll get better lenses at lower prices if you can choose from those older AF lenses.

For that reason, the D90 is my recommendation for the eBay-savvy bargain-hunter. Everyone else should choose one of the newer Nikon bodies, however.

Nikon D7000 ($250 used)

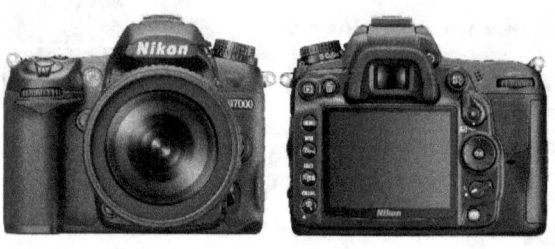

The *Nikon D7000* is an excellent camera for wildlife and sports photographers, with a fast 6 frames per second continuous shooting rate, a 16 megapixel DX sensor that gives you a 1.5x crop when using telephoto lenses, and a 39-point autofocus system.

There's one big, big weakness for those shooting action: a small buffer. It will work fine if you're shooting JPG, storing about 100 consecutive shots without slowing down. However, if you shoot raw (as most serious photographers will) you'll be limited to 10 or 11 consecutive frames before the camera slows down.

That means that you'll get less than two seconds of continuous shooting. That's not enough for your kid to make the run from third base to home, and it will be incredibly frustrating when shooting flying birds, which typically take 4-5 seconds in a single run.

If this is your first camera, you should also consider the Canon 70D and the 7D. The 70D has fantastic autofocus capabilities during video, and the 7D can be had used for $500-$750 (and it doesn't have the D7000's focusing issue). All these cameras have excellent autofocus systems, but the D7000 does have more focusing points (though that will probably never impact your photography) and much better image quality (which will impact your photography).

Nikon D7100 ($400 used)

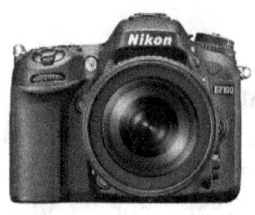 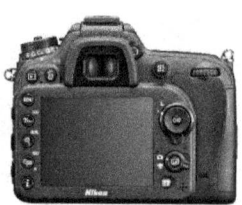

The D7100 would be an even better camera for wildlife photography than its predecessor, the D7000. However, the D7100's increase in megapixels from 16 to 24 further decreases the number of consecutive raw photos you can capture. At 6 frames per second, you can shoot for only one second before the camera slows down to one frame per second... and that's simply not long enough to adequately capture a bird in flight.

However, the combination of 24 megapixels and a 1.5x crop factor make the D7100 the most detailed wildlife camera available, though the requirement to shoot in raw definitely limits your image quality and your ability to recover blown out highlights (a common requirement for wildlife photographers).

If money is no object, consider upgrading to the $2800 D800, which has an amazing 36 megapixels and provides similar levels of detail in DX mode, which crops the full-frame sensor just like a D7100 (but without the buffer issue). If that's outside your budget, consider a used Canon 7D ($500-$1000).

Nikon D7200 ($700 used) and D7500 ($900 new, $550 used)

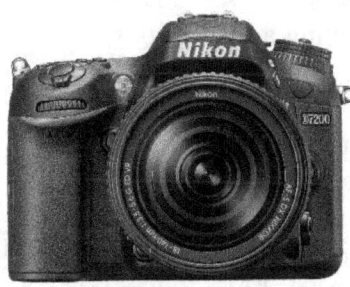 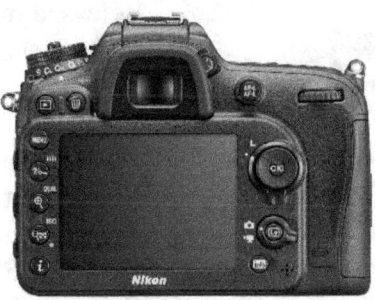

The D7200 is a very minor upgrade over the D7100, but it does offer one important improvement: a larger buffer. While the D7100's buffer filled up after about six shots (of 14-bit uncompressed raw photos), the D7200 can take 17 shots. That means you can hold the shutter down longer during sports and wildlife shooting, and you might be able to use raw instead of just JPG when shooting action.

The D7200 is the best Nikon camera for action photography where you can't fill the frame using a full-frame camera. In other words, if you plan to use a

Video: D7200 Technical Review
23:15 - *sdp.io/D7200TechReview*

70-200 lens but you can't get close enough to the action and you need to crop anyway, you'll get sharper results with the D7200 than you will with any other camera in the Nikon lineup, including the D810.

However, as a sports camera, the D7200 pales in comparison to the Canon 7D Mark II. The Canon takes ten pictures in a second, whereas the Nikon captures only six. That difference is huge to a sports or wildlife photographer—it

Video: D7200 Review
9:11 - *sdp.io/D7200Review*

increases the odds that you'll capture a shot the exact moment the athlete kicks the ball, or that you'll have perfect timing as a flying bird's wings hit their apex. The Nikon does have slightly better image quality than the D7200, but owning both, I always choose the 7D Mark II for sports and wildlife.

Nikon D7500 ($900 new)

In our tests of literally every camera on the market, the D7500 was the best sports and wildlife camera under $1,000. The autofocus is fast and reliable, the upgraded 8 FPS is very good for shooting action, and the handling is the best in class.

Compared to the D7200, the D7500 adds a tilting touchscreen, 4K video, 8 FPS, and (universally hated) SnapBridge Wi-Fi. It's generally a much better camera than the D7200, and a worthy upgrade. However, the D7500 drops from 24 megapixels to 21, so you can expect to see about 10% less detail. Regardless, for sports and

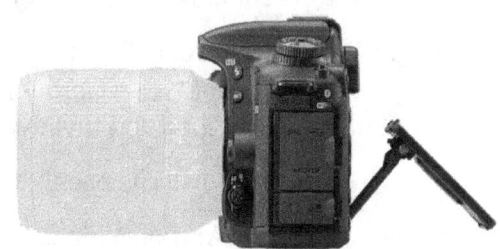

wildlife shooters who don't have the budget for a D500, the D7500 is the better choice.

Nikon D300S ($200 used)

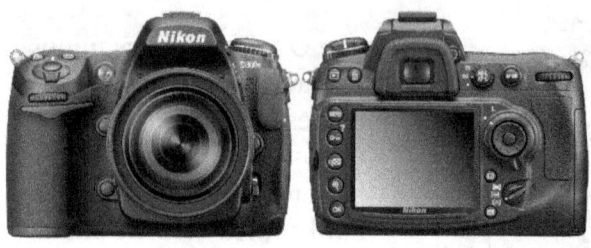

Released mid-2009, I would never recommend buying a new _D300s_. However, at around $200, a used D300S is an excellent option for a beginner camera and for wildlife photographers on a budget. It's only 12 megapixels, and Nikon's sensor was a bit dated even when the D300S was released. More importantly for wildlife photographers, it supports a stunning 7 frames per second (for up to 17 continuous raw frames) and 51-point autofocus. If you're a wildlife photographer not already invested in Nikon lenses, also consider a used Canon 7D, which costs about the same. The 7D has 50% more megapixels and can record 1080p video, instead of the D300S' 720p video. The 7D can also shoot at a slightly faster 8 frames per second, capturing 14-bit RAW files instead of the 12-bit RAW files that the D300S is limited to at higher speeds. Perhaps most importantly, you can use the incredible Canon 400mm f/5.6 with the Canon 7D. However, the 51 autofocus points of the D300S make it easier to track flying birds than the 7D, and the D300S has dual memory card slots.

Nikon D500 ($1,500 new, $950 used)

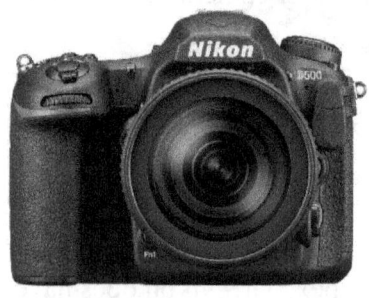 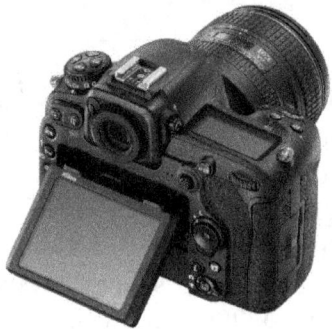

Finally, Nikon has a proper APS-C sports and wildlife body to compete with the Canon 7D Mark II. As a sports and wildlife body, it has these characteristics:

- **High frame rate**. 10 frames per second is outrageously fast. Most cameras have a rate of 5-7 frames per second.

- **Exceptional focusing**. The focus points cover almost the entire viewfinder. They focus faster than any camera we've ever tested, with the sole exception of the Canon 1DX Mark II.

- **Big buffer.** You can take 200 consecutive shots in raw, so you can keep up with non-stop action. Most cameras have a buffer of 10-40 shots.

- **Tilting touchscreen**. Similar to the full-frame D750, we found that the tilting touchscreen improves usability in a wide variety of scenarios, especially when working on a tripod and reviewing images.

- **Light-up buttons**. The backlit buttons made night photography much easier.

- **No AA filter**. Anti-aliasing filters noticeably reduce the sharpness of your images. In comparison tests, we got vastly sharper images using the D500 than we did using the Nikon D5, Canon 1DX Mark II, or Canon 7D Mark II.

- **SnapBridge.** SnapBridge uses Bluetooth to automatically transfer images to your smartphone, making it quick and easy to share images on social media. If you use Twitter, Instagram, or Facebook, and like to share images in real-time, this is a huge benefit.

It can also record 4k video with a tilting touch screen, making it a decently usable video camera. However, the 4k video recording includes an additional 1.5X crop, for a total crop

Video: D500 Review
20:06 - *sdp.io/d500review*

factor of 2.2X. As a result, the actual recorded image is very small, and is about the same as the Micro Four Thirds-mount Panasonic GH4.

After thorough testing, we've declared the D500 to be the best camera available for wildlife and sports, assuming you need the extra reach of the smaller APS-C sensor. If you aren't shooting action, the similarly priced Nikon D750 provides better image quality for portraits or landscapes.

Nikon D600/D610 ($700 used)

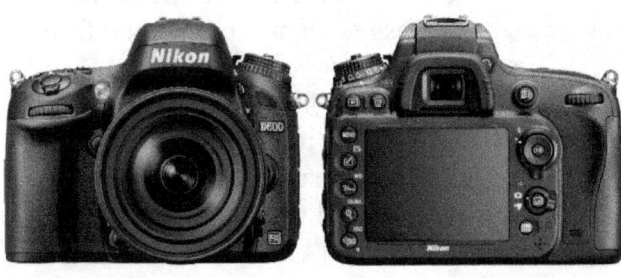

Nikon's low-end full-frame cameras, the 24 megapixel _D600_ and D610, are the answer for most photographers who can't tolerate the noisy low-light images of the DX bodies. The larger sensor captures more light, allowing you to get cleaner images when shooting indoors or in the evenings.

Both the D600 and D610 are great, well-rounded bodies.

While all the FX full-frame cameras offer substantially better image quality than the DX cameras, and 60% shallower depth-of-field (useful for portrait work), there are a couple of disadvantages:

- You'll need full-frame FX lenses to take advantage of the larger sensor, and these tend to be more expensive.

Video: D500 Preview
23:12 - *sdp.io/d500preview*

- Both the camera body and lenses are heavier (though few people complain).

In the studio, a DX camera can be passable for portrait work. Out of the studio, the D600 and D610 are ideal portrait cameras, allowing you to get full-frame background blur from your portrait lens with smooth, low-noise skin tones, even when working indoors or in the shade.

The D600 and D610 also make good choices for those who want to shoot indoor sports like volleyball, basketball, or hockey. While the focusing system isn't as robust as many of the other cameras, and the frames per second isn't as fast, the full frame sensor provides much cleaner images than any of the DX cameras, and these bodies are significantly less expensive than the higher-end bodies.

The D600 has become rather infamous for developing oil spots on the digital sensor. While photographers complain at length about minor problems with every camera body ever released, the oil spots on the D600 are a serious problem that shouldn't be underestimated. According to polls on the Internet, most D600 users have serious problems with oil spots that create visible spots on pictures. Unfortunately, I don't have a more reputable source than polls on the Internet, but I have spoken with several long-time D600 users who experienced the problem.

Fortunately, you can remove the oil spots by cleaning the sensor, as described in Chapter 5 of Stunning Digital Photography. Users also report that the oil spots go away after about 3,000 frames have been shot. Because of the bad reputation but fairly easy mitigation of the oil spot issue, a used D600 might be a particularly good bargain. Unfortunately, there's no hard data to allow me to guarantee that you won't see oil spots.

Video: D750 Preview
20:00 - *sdp.io/d750preview*

The introduction of the D610 in the fall of 2013 offered very little more than freeing Nikon from the D600's bad reputation. The sensor, processor, and autofocus systems are exactly the same as the D600. The D610 did add a quiet shutter mode (which is useful) and a 10% increase in frames per second from 5.5 to 6.

If you plan on shooting wildlife, you should choose a DX camera instead, such as a D7100 or a used D300S. If you plan to shoot sports and you don't expect to need to crop your pictures heavily, consider upgrading to the D750, which has a far superior focusing system. If you want the ultimate in image quality and an upgraded focusing system, check out the D810.

If you're not yet invested in the Nikon system, you should compare the D600 and D610 to the full-frame Canon 6D. The Canon camera can't quite match the Nikon's image quality, but it adds a GPS feature that I really enjoy and a somewhat useful Wi-Fi feature.

Nikon D750 ($1,500 new, $1,000 used)

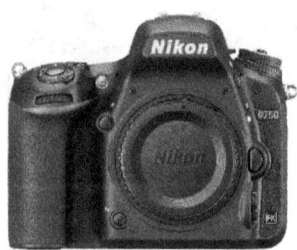 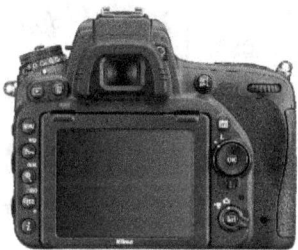

Fitting between the D610 and D810, the D750 combines the superior autofocus system from the D810 and D4S with the lower resolution 24 megapixel sensor of the D610. If you plan to take any sports or wildlife shots, it's a worthy upgrade from the D610. However, if you're a travel and landscape photographer, the D610 can save you $500 that might be better spent on lenses.

As the middle child in the Nikon FX lineup, the D750 is mostly made from parts of other cameras, but there is one very unique and important feature: a tilt screen. While dozens of cameras have a similar tilt screen, the D750 is the only full-frame camera to have one. They're immensely useful for creative composition, including shooting still subjects slow

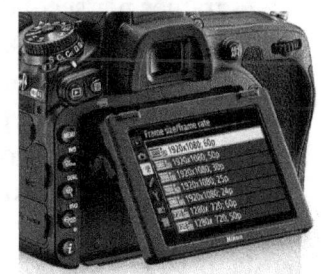

to the ground or when holding your camera over your head (for example, when shooting in crowds). However, the live view autofocusing is rather slow, so don't expect to chase wildlife or football players using the tilt screen.

If you plan to buy sharp, high quality lenses, upgrading to the D810 can provide you with more detailed images. However, if you choose to use the fairly unsharp 24-120mm f/4 kit lens, the D750 gives about as much detail as you'll be able to get.

Nikon D780 ($2,300 new)

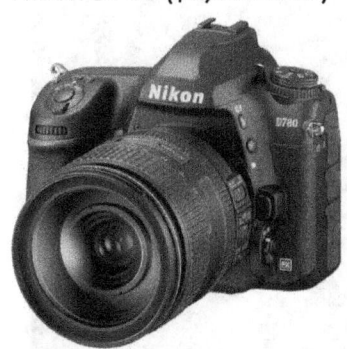

In January 2020, Nikon updated the D750 with technology developed for its mirrorless Z6. Essentially, the D750 DSLR becomes a Z6 when put into live view mode, which is incredibly useful for photographers who frequently use live view by holding the camera away from their eye, and any photographers who shoot video.

Benefits while using live view include drastically better autofocus and full-width 4k/30 video.

Unfortunately, little else was changed about the D780. The image quality is no better. When using the viewfinder, you would hardly notice a difference from using the D750. In a way, this is good news, because the D750 is an amazing camera that is available used or refurbished for less than half the price of a new D780.

As a result, the D750 remains the workhorse of portrait and wedding photographers for a good reason. The D780 tends to be a better choice for those professionals doing hybrid shooting where they might be expected to both shoot stills and record video.

Before buying a D780, definitely consider the less expensive Nikon Z50, which can adapt all the same lenses. The experience of using the cameras is very different, but the results are very similar. Also consider the Sony a7 III, which is slightly less expensive than the D780 (but more expensive than the Z50) and brings the benefits of mirrorless.

Nikon D800 and D800E ($450 used)

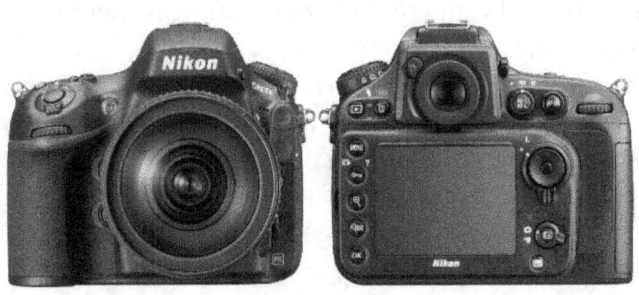

Though the D810 has now replaced them both, the Nikon _D800_ and D800E are incredible cameras, with better image quality than any other DSLR ever made—and that includes medium format cameras. The 36 megapixel full-frame sensor is simply incredible, making these models the ultimate choice for anyone primarily concerned with capturing detail.

In all other aspects, the D800 is a very capable camera, with 51-point autofocus (15 cross-type sensors), solid build quality, and good video capabilities. Off-the-shelf, the D800 is only 4 frames per second, making it a bit slow for wildlife and sports. If you enable DX crop mode (which crops the image 1.5X to a 24 megapixel image taken from the center of the frame), you get 5 frames per second. However, the option MB-D12 vertical battery grip speeds it up to 6 frames per second in DX crop mode.

The D800E is the same camera as the D800, but has the anti-aliasing (AA) filter removed. The AA filter simply blurs detail a tiny amount, helping to remove artifacts that can appear in fine details. However, it also removes a small amount of detail. Therefore, pictures from the standard D800 tend to look a little nicer to the eye, whereas the D800E technically captures more detail. This picture explains it much better than words can.

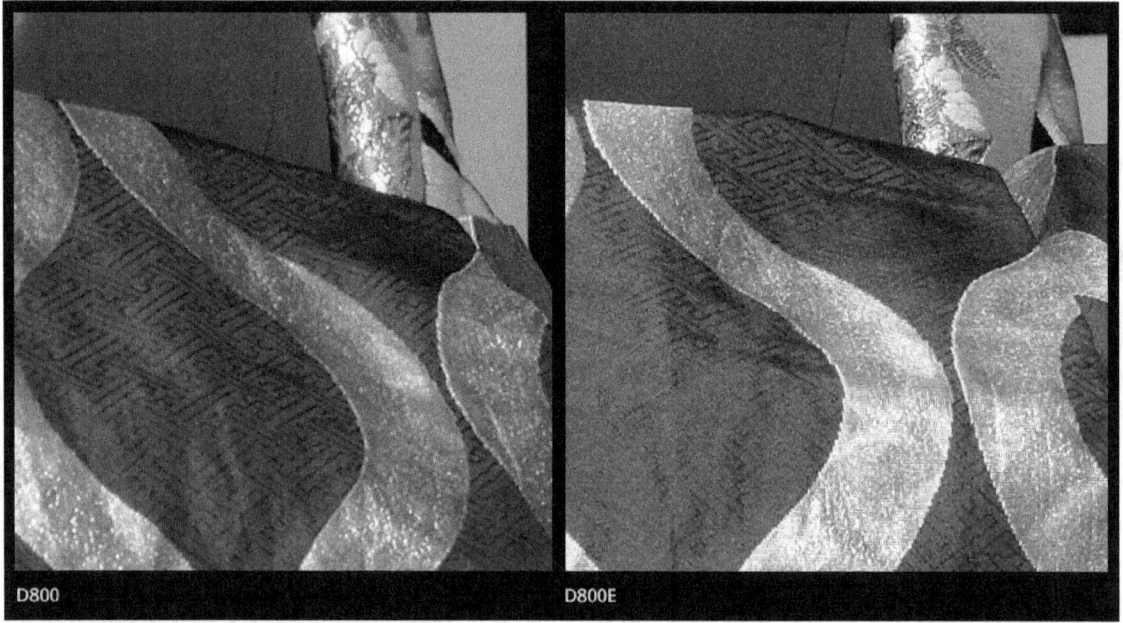

The standard D800 is the better choice for most users. If you're a commercial photographer who closely post-processes all their images and requires medium format levels of detail, choose the D800E.

If you zoom in tight on images from the D800, the images seem to be quite noisy compared to other full-frame cameras, even at ISO 100. The D800's high megapixel count means that each individual pixel captures less light, increasing noise. However, once you resize images to the same resolution as every other camera, the noise seems to disappear. In a nutshell, the D800 is noisier than the D600 and Canon's full-frame cameras, but in use, the D800's image quality is always superior.

The high megapixel count does cause some headaches, however. Having 50% more pixels than most full-frame cameras means that images take 50% longer to load, and your computer will need to work 50% harder when processing them. You'll also need 50% more storage space both in your camera and on your computer. If you're not using the sharpest lenses available, you won't even be able to take advantage of the detail, so be prepared to spend extra money on glass. Basically, the D800's huge images are a pain, and unless you enjoy or really need the detail, it's probably not worth it.

While I regularly recommend DX cameras for wildlife because they show more detail in distant subjects, the D800 offers a crop mode that gives you the best of both worlds. By cropping just the center of the massive 36 megapixel sensor, the D800's crop mode produces 24 megapixel images with a 1.5x crop, turning a 400mm lens into a 600mm lens while also providing a wide field of view that makes it easier to locate and track moving animals. For wildlife photographers with big budgets, the D800 is the ultimate body.

For most portrait photographers, the D600 or D610 is a better choice; save the extra $1,000 for lenses and lighting. However, wedding photographers will appreciate the D800's more powerful autofocus system.

Before buying the D800, check out the Canon 5D Mark III. The D800 has about 15% better image quality than the 5D Mark III, but the Canon's lower megapixel count increases the frames per second to 6, which is 50% faster when shooting full-frame images. The smaller images are also quicker to load and process. The Canon also offers faster autofocus, which will be important for sports and weddings.

Nikon D810 ($1,200 used)

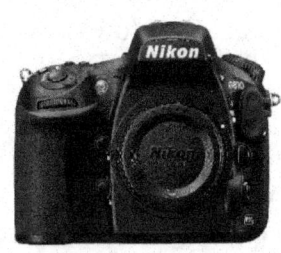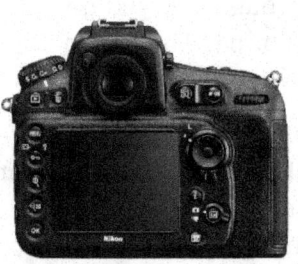

The D810 is, in my humble opinion, the greatest still photo camera ever made (as of October 2014). Sure, there are medium format cameras with bigger sensors and more pixels, and Nikon's own D4S is twice the price. But if I could only have one camera, I'd rather have the D810. The image quality is the best in the world, it has one of the best focusing systems ever made, it's smaller than a D4S or a medium format camera, and you can connect hundreds of Nikon-compatible lenses to it. If you can afford it, this is the camera for you.

Still, it's not the camera that I would recommend to most people. You can get similar image quality from a used D800E (this camera's predecessor), and that's the right choice for everyone who doesn't need the D810's greatly improved focusing system. Most portrait photographers should buy a D610 and

Video: D810 Preview
33:00 - *sdp.io/d810preview*

Video: D810 vs 5D Mark III
25:33 - *sdp.io/d810v5d3*

save the extra money for lenses and lighting; you won't miss the extra megapixels. Some hard-core professionals will need specific features of the D4S, such as the weatherproofing or the gigabit Ethernet.

But for wedding, wildlife, and sports photographers, or anyone who has the budget, the D810 is the greatest camera ever made.

Nikon D850 ($3,000 new, $2,100 used)

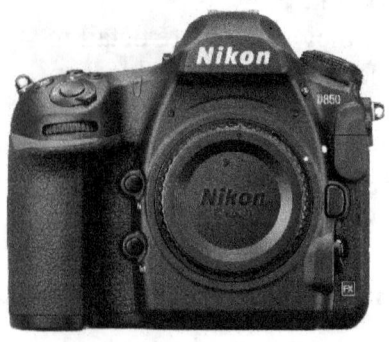 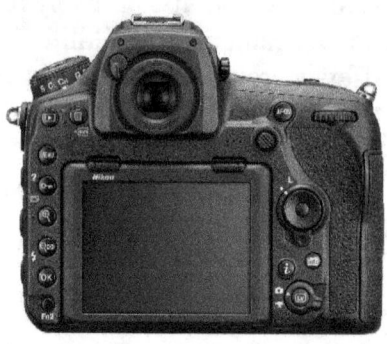

The Nikon D850 is a significant upgrade to the D810, offering sharper images, better autofocus, faster shooting, 4K video, and improved usability.

At the time of this writing, the D850 is the greatest all-around camera ever made. When Chelsea and I travel, we usually travel with two different cameras, because of our different preferences and goals. On our last trip, however, we travelled with a pair of Nikon D850s.

The key features are:

- **45 megapixel sensor with no anti-aliasing filter and low ISO 64**. Noise levels are similar to those of the Nikon D810, but the D850 produces sharper images, especially with very sharp lenses. Image quality is almost indistinguishable from the Sony a7R II and Sony a7R III, though the Nikon D850 has better raw dynamic range when shooting at its base ISO of 64. The Nikon's image quality blows away the Canon 5D Mark IV. The Nikon is also slightly better than the Canon 5DS-R, but after processing, it would be hard to distinguish the difference.

- **Better autofocus**. The D850 inherits the D5's amazing autofocus system, at least in theory. In testing, we found 3D tracking couldn't keep up with the D5 or D500. Nonetheless, the autofocusing is among the best in the world, beating the D810 and matching the Canon 5D Mark IV. Note, however, that the Sony a7R III as eye-detect autofocusing, which speeds workflow for portraits and some sports by not requiring the photographer to manually move the focusing point.

- **Proper 4K video**. Like the Sony a7R II and a7R II, the D850 allows you to record high-quality full-width 4K video, or you can crop it 1.5X to gain greater reach out of your lenses. At full-width, the video quality (especially in low-light) is better than the Sony a7R II and a7R III, a real feat. The tilting touchscreen makes it more usable than previous generations as a video camera. It's a vastly better video camera than the Canon 5D Mk IV, which lacks a tilting screen, requires a severe crop factor, and uses an unbearably inefficient motion JPG codec.

- **Higher framerate**. The D850 can shoot at a quick 7 FPS without a vertical grip, or at an amazing 9 FPS with the vertical grip.

- **Silent shooting**. The first SLR with proper silent shooting, you can use live view on the rear screen to take full-resolution images without disturbing clients. This allows greater discretion for wedding photographers shooting during the ceremony, press at sensitive events, and golf photographers.

- **Improved usability**. Don't underestimate the impact of a tilting screen on your work. Not having to lay on the ground when shooting low will increase the number of shots you take from ground-level, improving your composition. You'll also be able to hold the camera over your head and shoot over crowds with confidence.

- **Timelapses**. The D850 is the best timelapse camera we've ever used. It can shoot silently without moving the shutter, which means your 2,000 frame timelapse won't increment your shutter count. You can shoot 4k in-camera timelapses if you want to quickly add them to your blog, or full 45-megapixel 8k timelapses for higher quality. The amazing quality of the D850 raw files allows you to recover massive amounts of shadow and highlight detail for those sunrise and sunset shots.

It's the best DSLR ever made, but there's another contender: the mirrorless Sony a7R III. The Sony offers several big advantages over the D850:

- **Electronic viewfinder (EVF)**. The Sony's EVF previews your exposure, meaning you never make an exposure mistake. This eliminates the "chimping" process of shoot-examine-reshoot and speed's the photographer's workflow. You can see special effects in the viewfinder, allowing you to visualize black-and-white imagery. It also allows you to use the viewfinder while recording video.

- **Sensor stabilization**. Many of the sharpest and fastest lenses ever made lack built-in image stabilization, such as most 85mm f/1.4 prime portrait lenses. With high-megapixel cameras, camera shake becomes apparent even when following the reciprocal rule. The Sony's sensor stabilization has produced usable handheld images at slow shutter speeds (and lower ISOs) in many real-world situations for us, drastically improving image quality over what would be possible with an unstabilized DSLR.

- **Eye-detection autofocus**. As mentioned earlier, automatically focusing on the eye means less delay between shots in a portrait session, reducing workflow and minimizing the risk that you'll miss a split-second natural expression from a subject.

But the Nikon has far more lenses than the Sony, and the controls are more traditional. There is no right answer when choosing between the D850 and a7R III; you'll have to weigh the advantages of each and your own personal preferences for form factor. But if Sony offers all the native lenses you need, you should take a serious look at the a7R III before buying the D850.

Nikon Df ($2,750 new, $1,200 used)

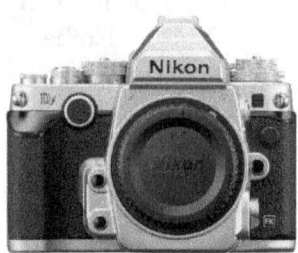 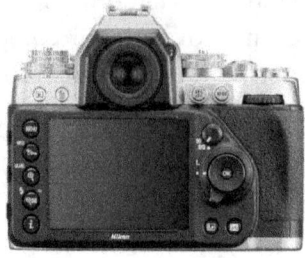

Here's a camera that doesn't fit in with any other DSLR made by Nikon or any other camera manufacturer—the Nikon Df. It's available in either silver or black. I wouldn't normally bother mentioning the finishes a camera body is available in, but the Df is all about style and fashion.

While most DSLRs have a "form follows function" design, the Df uses a "form over function" philosophy. It's a beautiful, retro camera modeled closely after the Nikon F3 from 1980; a year that many of us were young camera nerds, wishing we could play with the buttons and dials on our parents' or grandparents' cameras.

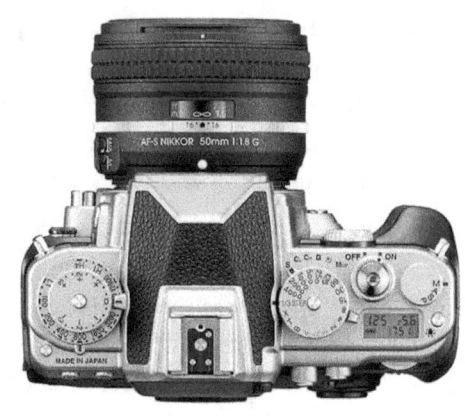

The buttons and dials on the Df are indeed cool. Instead of relying on pressing a button and spinning a dial to select your ISO, exposure compensation, shutter speed, and aperture, the Df provides dedicated dials resembling the mechanical dials of the early 1980's DSLRs. The dials aren't faster or more convenient than those found on other modern DSLRs; quite the opposite. However, they are infinitely cooler to use.

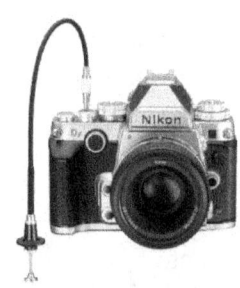

Functionally, there's only one other important nod to the past: an inclusion of a metering coupling lever that allows you to connect pre-Ai Nikon lenses made from 1959 to 1977. Basically, Ai lenses with an aperture ring have this little gear (the metering coupling lever) on the body mount that meshes with a matching gear on the body, and tells the camera what aperture you've dialed in. The camera uses this information to help with metering. Unfortunately, the metering coupling lever can cause some lenses designed before Nikon developed Ai to jam. On the Df (but not on any other DSLR), you can move this lever out of the way to connect these older lenses, or move it back in place for compatibility with Ai lenses.

The Df's firmware also provides the ability to configure pre-Ai lenses. Basically, you'll setup your pre-Ai lens in the firmware, flip the metering coupling lever out of the way, and then attach your lens. If you decide you want to use auto exposure, you can meter your scene with the camera, and then transfer the camera's aperture recommendation from the LCD display to the aperture ring on the lens. It's much like using an external meter.

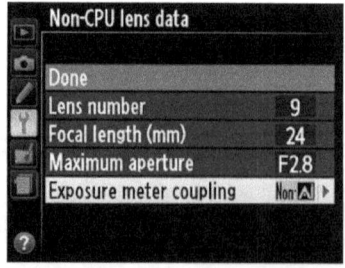

All Nikon bodies offer some level of compatibility with older lenses, however, so you don't necessarily need the Df to use an older lens; just those pre-Ai lenses.

There's also a PC Sync connection for connecting to older flashes, but that's still a fairly common feature on other modern DSLRs. It also supports a screw-in style remote cable release, which will definitely feel retro when your friend with the 6D is using his iPhone to trigger his camera. Of course, it has a focusing motor built into the body, like most of the higher-end Nikon cameras, so you can use lenses without autofocus motors.

A word of caution, however, from a photographer who spends almost as much time using old manual film cameras as modern DSLRs: using old lenses isn't going to be as much fun as you think. Yes, you can buy a used, manual Nikon 50mm f/1.4 Ai with a real aperture ring on it for under $100 (compared to $440 for the new equivalent). Attach it to your Df, and you can have a real, manual experience with plenty of mechanical dials to turn. However, the lack of split-prism or ground glass focusing screen in the Df makes manual focusing much more difficult than it was with the original F3. Further, older lenses simply weren't nearly as sharp as modern lenses, even when they're in perfect condition.

The Df includes most of the modern features you've come to expect: wonderful, low-noise image quality from the D4's 16-megapixel sensor, a 3.2 inch LCD screen, a modern autofocus with 39 autofocus points (9 cross-type, as shown in the following image), ISO 25,600 (expandable to 204,800), and even electronic dials if you decide not to use the mechanical-style dials on top of the camera.

Before buying it, however, you should be aware that it lacks several important modern features:

- **Video**. Seriously, it doesn't record video. It's an oddly selective nod towards retro-minimalism for a camera that includes so many other modern features, including live view and cheesy post-processing effects.

- **Ergonomics**. Consumer cameras from the 80s had smaller grips, as does the Df. The smaller grip slightly reduces the weight of the camera, but it also significantly reduces the comfort when using the camera. The dials were optimized for style, rather than usability, and you won't be adjusting those stylish dials without taking your eye from the viewfinder. However, their use is entirely optional, because the Df provides redundant digital controls. The shutter button is on top of the camera, rather than on the grip, where your index finger lands more naturally.

- **GPS and Wi-Fi**. Like most of the Nikon lineup, it lacks these niceties. Like many other modern Nikon bodies, you can connect the WU-1a wireless adapter ($46) or the GP-1/GP-1A ($300).

The Df is the most fun DSLR on the market today. If you're bored of your existing DSLR, it's a great choice. However, if you have $3,000 and want to make stunning images as efficiently as possible, choose a D800 instead.

Nikon D3X ($1,400 used)

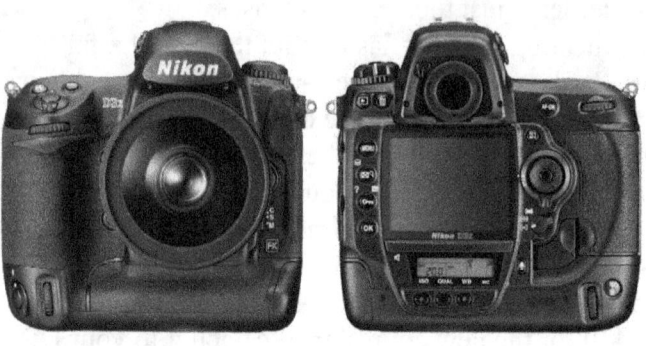

The 24 megapixel _D3x_ is Nikon's previous-generation professional camera (introduced near the end of 2008), designed for people who need their cameras to suffer substantial abuse and keep working. While all the Nikon cameras can handle a bit of rain, the D3x is designed to be used by photojournalists catching splashes from a fire hose, or paparazzi standing in a thunderstorm capturing Kanye and Kim fighting in the street. I tell every photographer to ditch the camera bag and expose their cameras to a little abuse, but the D3x loves the abuse, and is designed to handle being dropped over and over across its lifetime.

Because the D3x is designed to be so durable, used models hold up well. However, used models were probably owned by professionals, so they might show signs of wear from using the camera as it was intended.

Before buying the D3x, carefully consider a D800. The D800 feels flimsy compared to the D3X, it captures 20% fewer frames per second, and the autofocus system simply doesn't compare. However, the D800 has a 50% better megapixel count and about 7% better image quality. The D800 is significantly less expensive, too. In fact, if you don't need a bullet-proof professional-grade camera, the D800 is overall a superior camera to the D3X, and you'll have money left over for lenses and lights.

Nikon D4 ($1,400 used)

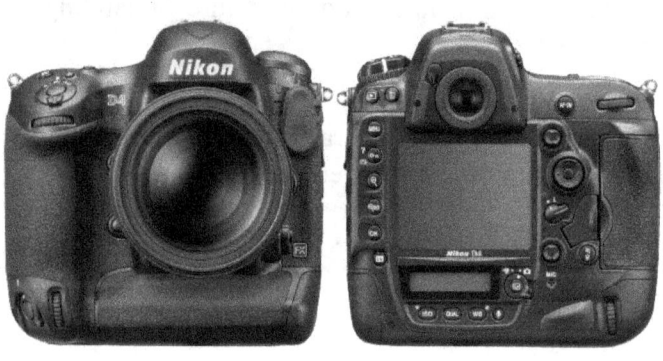

If it costs the most, it must be the best, right? Not necessarily. The _D4_ is Nikon's previous-generation professional-grade camera, and it originally cost $6,000. My target price for a used copy is now $3,500; allowing you to buy the top-end model for the price of the mid-range D810.

As with the D3x (the predecessor), the D4 is designed to be abused. If you're going to abuse your camera, you need the D4 or D4S. If you need to shoot 11 frames per second for 9 seconds, you need the D4. If you might make thousands of dollars by getting an action shot in focus, you need the D4. Otherwise, you're better off with one of the less expensive Nikons.

Surprisingly, the D4's image quality is almost identical to the D3x. Therefore, if you do need a professional camera but don't need the higher frames per second, you should consider buying a used D3x for about half the price.

Nikon D4S ($1,800 used)

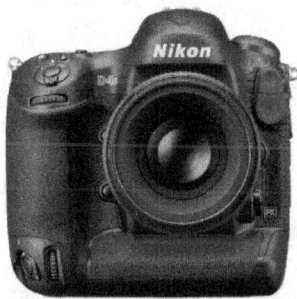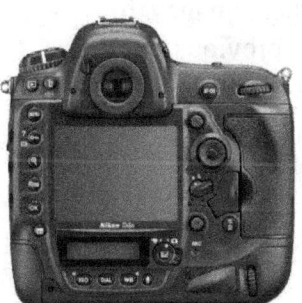

The D4S is Nikon's 2014 update to the professional-grade D4. It's a definite step up from the D4, with a very small jump in price, making it an excellent value for someone looking for a camera of this caliber. The most important improvements are:

- 11 frames per second with continuous autofocus. The D4 didn't allow continuous autofocus at the highest frame rate.

- Group AF, which causes multiple focus points to work together and can make tracking some types of moving subjects easier.

- Shorter blackout periods when shooting at a high frame rate, making it easier to track moving subjects through the viewfinder.

- 60 frames per second 1080p video, up from 30 frames per second.

- Gigabit Ethernet, allowing for faster tethering in a studio environment.

If none of these improvements are important to you, you might consider a used D4 instead. The image quality will similar, but you can find a used D4 for than $4,000.

> Video: D4S Preview
> **12:31** - *sdp.io/d4sPreview*
>

be
less

The 16 megapixel sensor seems to pale in comparison to the D810's 36 megapixel sensor, despite being more than twice as expensive. Indeed, the D810 does extract far more sharpness and detail out of professional lenses. The D810 also has better dynamic range, though the D4s has slightly less noise. The smaller files of the D4s are easier to manage, and make the high frame rates possible.

Nikon D5 ($5,500 new, $3,800 used)

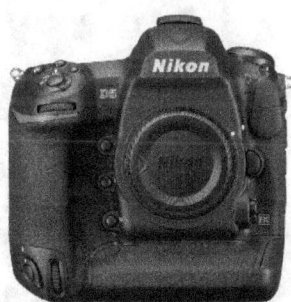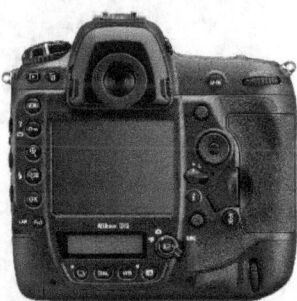

Nikon's top-end camera for professional sports photographers and photojournalists is amazing at a few things, and mediocre at everything else. Here's what it's great at:

- **Sports**. 12 frames per second, the best autofocus system we've ever tested, and the best low-light image quality we've ever seen. Indoors or outdoors, you simply can't do better than the D5.

- **Low-light photography**. When you have to use a high ISO (say, above ISO 3200) the D5 produces the cleanest, sharpest images. It soundly beats the Nikon D810, our previous low-light champion.

- **Durability**. It's built like a tank, designed to withstand unpleasant weather, and intended to last many years of hard use.

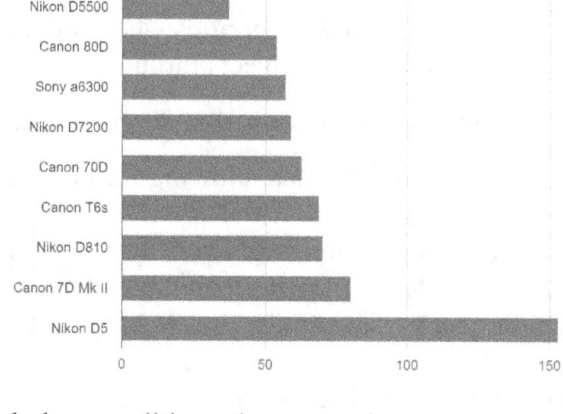

The autofocus system is simply amazing, getting almost 100% of shots in focus in any conditions. As an example, my daughter's indoor soccer games are poorly lit and require me to shoot through thick black netting. I had tried the 7D Mark II and D810, but gave up trying to photography her games, because they couldn't focus reliably through the net. The D5 has no problem.

As this chart shows, the Nikon D5 won our focusing test by a huge margin. In a series of tests taking pictures of a model walking towards the camera in indoor gym lighting conditions, the D5 took over twice as many sharp photos as the 7D Mark II. The combination of high frame rate and fantastic low-light focusing work together to make a camera that always gets the shot, and captures the best fraction of a second moment in high-action scenes.

In extremely low light conditions (like a dark bar), the D5 successfully autofocused in darker conditions than any other camera we've ever tested.

At high ISOs, the D5 produces sharper, cleaner images. Here's a low-light portrait of your author, with the D810 on the left and the D5 on the right:

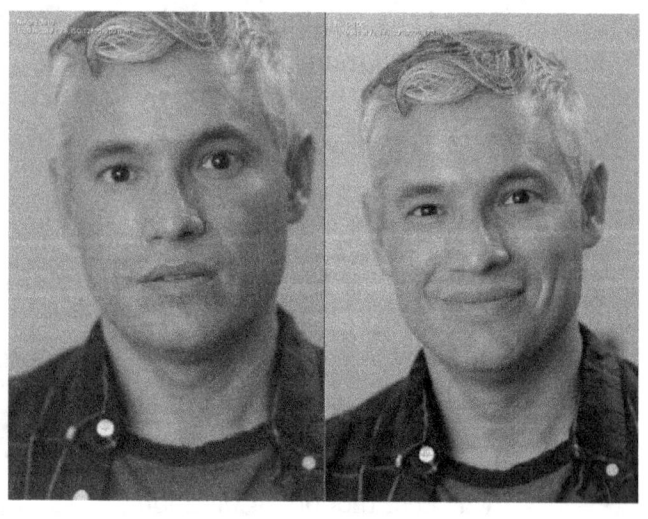

The D5's huge battery seems to last forever. We took over 1,500 photos and a full hour of 4k video, and it still had one bar left. On a Sony camera, that would have required about five battery changes. To a professional, it's vital to be able to go a full day without worrying about the battery quitting unexpectedly.

However, the D5 is not great at everything, despite being the most expensive camera:

- **Landscape photography**. The D810 produces sharper, cleaner images with more dynamic range. The D810 is also much smaller and lighter, making it easier to carry into the field.

- **Studio photography and portraits**. The D810's 36 megapixels and base ISO of 64 produces overall much better images than the D5's 20 megapixels and base ISO of 100.

- **Street photography**. There's nothing subtle about the massive camera. Though it has a silent mode, it only works in Live View, and the live view focusing system is almost unusable.

- **Video**. Though it has 4k video, at the moment, it only records for 3 minutes at a time. You can immediately restart recording, however, and we recorded a full hour without any problems. The 4k video has a severe crop factor of about 1.3x, so it's not full-frame. Video auto focusing is almost unusable. A Sony a7S II or Sony a7R II are far better choices for a full-frame 4k video camera.

Here's a side-by-side image, with the D5 on the left and the D810 on the right. The D810 shows vastly more detail in the eyelashes.
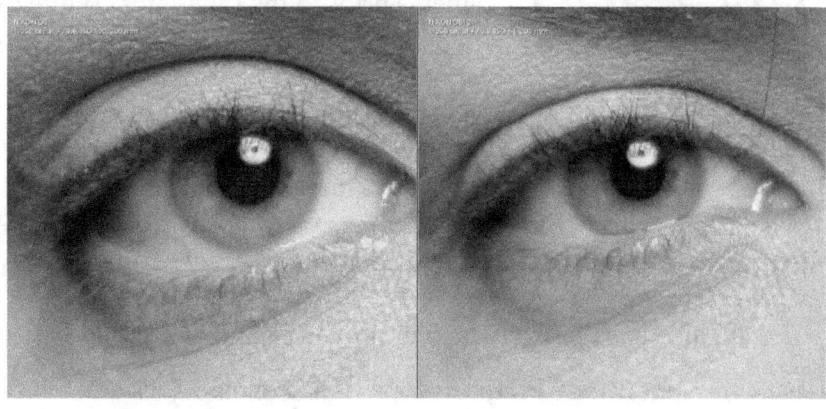
The D810 also has better dynamic range. In the closeup of the hair, shadows recovered in dark hair from a portrait shoot in the studio show far less noise on the D810 (on the right):

If you want the D5 for usability features, such as the vertical grip and Ethernet connection,
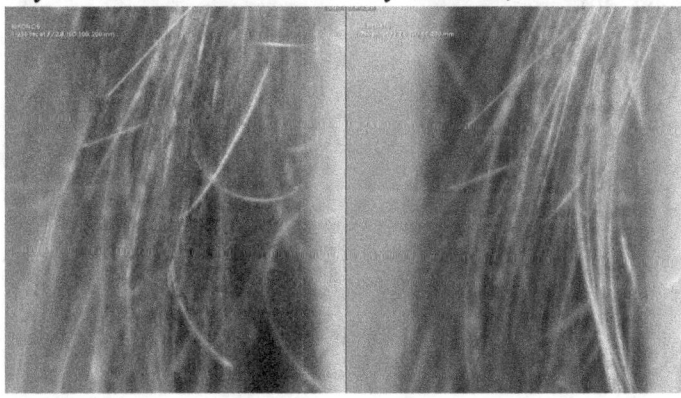
consider getting a battery grip for the D810 and an optical USB 3 cable instead of Ethernet. Compared to the 1DX Mark II ($6,000), we slightly prefer the D5. Nikon's 3D tracking system simply works better than Canon's version, and that feature has been really important in both sports and wildlife shooting. The video and live view focusing on the Canon is vastly better, but that's not a feature we often used.

Compared to the Sony a9 ($4,500), the D5 feels clunky and antiquated. The optical viewfinder doesn't preview your exposure and can't show you your histogram. The viewfinder blacks out with every exposure and makes loud noises.

Video: D5 Preview
34:39 - *sdp.io/d5Preview*

You can't tilt the screen on the Nikon for shooting over crowds at press events, and when you do use live view, the Nikon's focusing is awful. The Nikon is overall slower. However, the D5 is time-tested and has native support for an amazing selection of D5 lenses, and Nikon's 3D tracking works much better than Sony's alternative. If you can choose Sony G-Master lenses for your needs, however, you should definitely try an a9 before buying the D5.

Nikon D6 ($6,500 new)

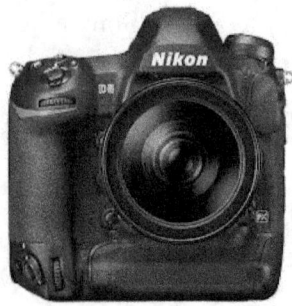

Nikon and Canon both tend to release updates to their top-end cameras prior to the Olympics. While the Summer 2020 Olympics were delayed, that didn't stop Nikon from releasing a very minor update to the Nikon D5.

The changes really were very minor. The D6 jumps from 12 to 14 FPS, but you'll only see that difference when not autofocusing. When tracking autofocus, as you would when tracking a moving subject, the camera needs to wait for the lens to autofocus before firing the shutter, and thus the difference between 12 and 14 FPS is minimal in the real world.

The D6 also has these new and improved features:

- Support for CFExpress type B as well as XQD
- A faster image processor
- A slightly improved autofocus system
- GPS
- WiFi
- Longer video recording times
- USB-C instead of USB 3.0

If both the D5 and D6 are the same price, definitely get the D6. However, if you can find a good price on a used D5, you might be happier saving a few dollars.

Which Portrait Lens Should You Buy?

Which lens you should buy depends on your budget. In order from least to most expensive (and least to most preferred):

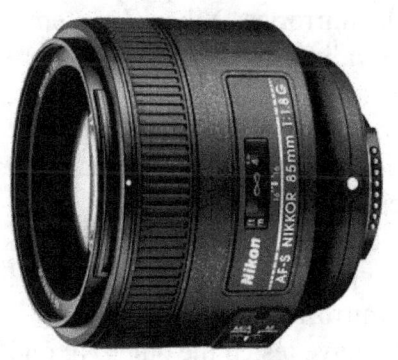

- $200 (DX or FX): *Nikon 50mm f/1.8G AF-S FX*

- $500 (DX or FX): *Nikon 85mm f/1.8G AF-S*

- $830 (DX or FX): *Tamron 70-200 f/2.8*

- $1,030 (DX only): *Sigma 50-150 f/2.8*

- $1,300 (DX or FX): *Tamron 70-200 f/2.8 VC G2*

- $2,800 (DX or FX): *Nikon 70-200 f/2.8 FL ED VR*

Chapter 15: Nikon Z

The mirrorless Nikon Z system seems to be replacing the Nikon F-mount DSLRs, providing smaller bodies, better video capabilities, and eye-detect autofocus. The Z-mount lenses are excellent and consistently sharper than the Nikon F-mount DSLR lenses. However, the Z-mount system is still very young. The native lens selection is limited, and several of the most important lenses are out-of-stock even though Nikon has technically released them to market.

All Nikon Z cameras can use Nikon F-mount lenses with the FTZ adapter. Therefore, if you have existing Nikon DSLR lenses, you can continue to use them with your new Nikon Z mirrorless camera.

We have tested the other Nikon Z cameras and have found that they're not quite as good as their biggest competitor: Nikon DSLRs. In our opinion, Nikon made the best DSLRs, and these mirrorless Z cameras have big shoes to fill. I would advise you to wait until Nikon releases their next-generation Z cameras before upgrading to this system.

If you're not already invested in Nikon gear, you might instead look to the Canon RF and Sony E-mount mirrorless cameras. Both systems are more mature with a wider variety of lenses and overall better results at a similar price point. Regardless, the Z system is capable of producing amazing results.

Nikon Z30 ($700 new)

The Nikon Z30 is Nikon's entry-level camera, intended as a successor to the D3x00-series of DSLRs. With the Z30, Nikon is targeting "creators", which in this context means young people recording videos.

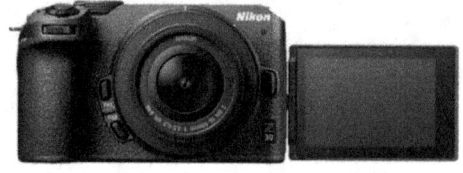

As a result, it has a flip-forward screen so people can see themselves when filming. It lacks an electronic viewfinder, but that's ok for the target audience who learned photography with a smartphone and rarely hold their eye to the viewfinder anyway. It also has a more sophisticated on-camera mic that promises better sound quality, but as a creator myself, I would still recommend using an external microphone simply because moving the microphone closer to the subject improves sound quality more than any other factor.

Unfortunately, it's going to be difficult for me to recommend this camera to young creators. While it advertises live streaming, that feature requires a USB cable attached to a computer, and young creators generally do not use a computer at all. Furthermore, streaming will be very limited: while you can stream to YouTube or Facebook, you cannot stream to Instagram or TikTok. For the most popular sites, you would still need to use a smartphone.

Ultimately, most young creators with a $700 budget will be better served by adding that $700 budget to their smartphone budget and buying an iPhone 13 Pro or Samsung S22

Ultra. In my testing of other Nikon APS-C cameras, such as the Nikon Z fc, a newish iPhone produced higher-quality stills AND video in all situations, but especially in low-light. The smartphones also allow creators to edit and publish their content directly on the device without fussing with Wi-Fi apps or memory cards.

Nikon Z fc ($950)

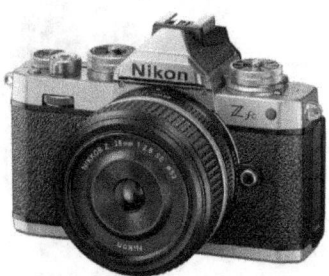

The Nikon Z fc is Nikon's answer to Fujifilm and Olympus: An attractive and capable retro-styled camera. It has a 20 megapixel sensor, 11 FPS (which will be closer to 5 FPS when shooting moving subjects in the real world) and 4k video. The Z fc is the most beautiful and fun Nikon camera of the last decade. However, it has one serious flaw: there's only one matching lens. It's available with a 28mm f/2.8 APS-C lens as part of a kit. You can attach other Nikon Z lenses with an adapter, but they won't match the aesthetic, and if you don't care about the aesthetic, you might be happier with a Nikon Z50.

By contrast, the Fujifilm X-T30 is $250 less expensive and offers similar looks and image quality, with even better handling (in my opinion). Fuji's X-mount offers dozens of designed-for-APS-C lenses that match the retro look, too.

This is sounding very negative, but I like the Z fc and I'm glad it exists. If you're a Nikon enthusiast looking for a camera to keep around your neck, you should definitely pickup a Z fc. However, if you're a retro camera enthusiast or you're looking to simulate the analog film experience, you'll be happier buying into Fujifilm's X-mount system.

Nikon Z50 ($900 new)

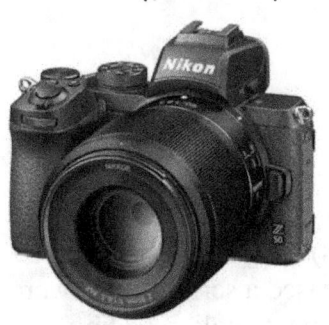

The 20 megapixel Nikon Z50 is Nikon's only APS-C mirrorless camera. While it allows Nikon to offer a price point below $1,000, it comes with a significant drawback: limited APS-C mirrorless lenses. At the current time, Nikon is only selling two APS-C mirrorless lenses, a kit lens and a telephoto zoom. If you're happy with those, the Nikon Z50 is a fine, general-purpose camera.

Note that you can attach Nikon's full-frame Z lenses. However, there will be a 1.5X crop factor. Also, those full-frame Z lenses are all quite expensive, betraying this budget-friendly camera. Note that you can use the Nikon FTZ adapter to attach Nikon AF-S lenses, which you might want to do if you're upgrading from a Nikon APS-C DSLR such as the D5500.

The Nikon has a flip-down screen, which is a rather odd choice. While the flip-down screen works well for handheld selfies, it is blocked by a tripod, making it awkward to use for filming yourself. You can attach a bracket to the camera to move it to the side so that you can see the screen.

For people looking for their first camera, it's hard to recommend the APS-C Z50 over the full-frame Canon EOS RP at the same price. However, for dedicated Nikon shooters who want to experience mirrorless at the lowest price possible, the Z50 is the best choice.

Nikon Z5 ($1,400 new)

At $1,400, the 24 megapixel sensor stabilized Nikon Z5 could be a breakthrough camera for professional photographers. It features Nikon's upgraded autofocus system, dual card slots, and sensor stabilization.

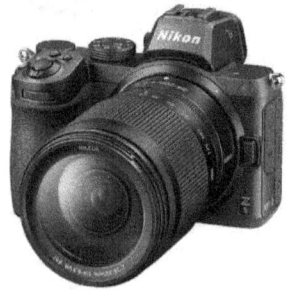

The Z5 offers good 4k/30 video. However, it lacks a flip-forward screen, so it won't be useful for filming yourself.

After several months of testing, we strongly recommend the Z5 for Nikon shooters looking for their first mirrorless camera. The Z5 takes advantage of Nikon's amazing mirrorless lenses, and if you upgrade to the amazing 24-70 f/2.8, you'll see a huge image quality improvement over Nikon's DSLR lens. While the Z5's eye autofocus system is frustrating and unreliable compared to recent Sony and Canon mirrorless cameras, DSLR users accustomed to manually selecting a focusing point will feel right at home, and that autofocus mode works as well as it does on Nikon DSLR cameras.

The closest competing cameras are the Canon RP ($900-$1,000) and the Canon R ($1,600). We can't recommend either Canon for professional use, because the both lack dual card slots, so the Nikon Z5 wins if you can't afford to lose a day of shooting. However, both the RP and the R are superior for video recording. Even though they don't offer full-width 4k, we found Canon's video autofocus system to be superior. Also, the two Canon cameras offer a flip-forward screen for people who need to film themselves.

Nikon Z6 ($1,600 new)

The 24 megapixel Nikon Z6 was one of the first two cameras launched with the Nikon Z system. In my opinion, Nikon launched the cameras before they were ready – the first firmware had serious problems with autofocus and auto white balance. As a result, many early reviews (including ours) cited serious problems. Since the launch, however, Nikon has released a series of firmware updates that greatly improved the overall functionality, adding important features such as eye-detect autofocus. While we'd still prefer the Canon EOS R at this price point, the Nikon Z6 is far better than it was at release date and is now a very capable camera. The price drop to $1,600 helps, too.

Like the Z5, it offers 4k/30 video but only with a tilt screen. While the Z6 has a nice top display, you should seriously consider purchasing a Z5 instead so that you can take advantage of the dual card slots. The Z6 has only a single XQD card slot, and while it's faster than SD cards (thus reducing buffering), you will lose all your pictures if that card fails. Many people have said that XQD cards do not fail, but in our poll of more than 4,000 photographers, a significant number reported XQD failures. Realistically, failures happen with any type of memory card.

Also like the Z5, the Z6 offers excellent sensor stabilization that will improve the sharpness of your handheld, slow shutter photos when using non-stabilized lenses.

Nikon Z6 II ($2,000)

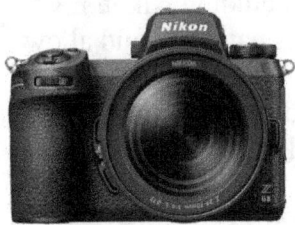

The Nikon Z6 II incrementally improves upon the original Z6 by offering two card slots, slightly better autofocus and 4k/60 FPS video with a 1.5X crop.

The closest competitor is the Panasonic S5, which matches it almost perfectly. The key differences are that the Nikon has a much better supply of mirrorless lenses and compatibility (via the FTZ adapter) with the massive array of Nikon F-mount DSLR lenses. The Panasonic offers a flip-forward screen, which might be a requirement for you or a drawback, depending on your shooting style.

Sony offers the a7 III at this price point, but it's quite outdated in 2021, and we'd rather use either the Panasonic S5 or the Nikon Z6 II.

If you're willing to spend $2,600, the Canon R6 is a much more powerful camera than the Z6 II. It offers full-width 4k/60 with the best video quality of any camera we've ever tested. It also has the most sophisticated focusing system we've ever used, capable of locking onto the eyes of athletes running straight at the camera or tracking birds, dogs, butterflies – anything. The R6 also offers 12 FPS with the mechanical shutter or 20 FPS with the electronic shutter, both with full autofocus, and we achieved those numbers in the real world (where the Z6 II slowed down significantly from the stated specs).

While the Canon R6 clearly has the greatest tech in this mid-range camera segment, the Z6 II is the best choice for photographers with an existing collection of Nikon lenses or a particular love for the brand.

Nikon Z7 ($2,500)

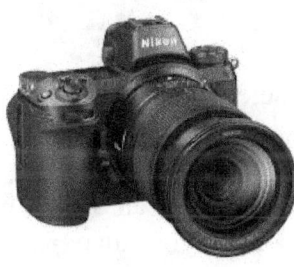

The 45-megapixel Nikon Z7 was originally marketed as a mirrorless version of our favorite DSLR of all time, the Nikon D850. Unfortunately, we found that not to be true. In real-world shooting tracking moving subjects, the frames per second dropped to 3-5, far below the 8 the D850 can do, or 10 when you attach a vertical grip to the D850.

Also, the Z7 lacks many of the D850's usability features, such buttons along the left side of the rear and lighted buttons. For these reasons, I often recommend buying a used Nikon D850 instead of a Nikon Z7.

The Nikon Z7 is significantly smaller than a Nikon D850, however, and it offers better video and eye-detect autofocus. Note, however, that even with the latest firmware update, we find the Z7's eye-detect autofocus to be too unreliable to use for shallow depth-of-field portraiture, and would still prefer to autofocus with a D850. If you're a hybrid shooter (creating both still photos and video), the Z7 is far superior to the D850 (which has a really clumsy video autofocus sytem).

The Z7 also connects to Nikon's superior Z lenses. In particular the Nikon 24-70 f/2.8 Z is far better than the Nikon 24-70 f/2.8 VR FX DSLR lens. That amazing Z lens might be a great reason for many shooters to upgrade to mirrorless, especially since you're choosing a high-megapixel body.

The Z7 also offers sensor stabilization, so your unstabilized lenses (including Nikon FX DSLR lenses with the FTZ adapter) will often produce sharper results handheld and allow you to use longer shutter speeds.

If you're looking for a high-resolution mirrorless camera, you should also consider the Sony a7R III. The previous-generation Sony has similar features and is $400 less expensive. The Sony also has many more available lenses.

Nikon Z7 II ($3,000)

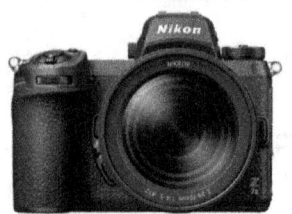

The Nikon Z7 II is an incremental upgrade to the original Z7. It offers two card slots and slightly improved autofocus. Honestly, after you apply all the firmware updates to the original Z7, there's not a lot of reason to upgrade unless you want the comfort of having an instant backup of your photos.

For us, the real question is whether Nikon D850 shooters should upgrade to mirrorless. We attempted to answer this question in a YouTube video (at *sdp.io/d850vsz7ii*), and the answer was a resounding, "Maybe."

If you're shooting landscapes, upgrading to the Z7 II is a no-brainer: the Z-mount lenses dramatically improve image quality. However, compared to the D850, be aware that you'll see banding in heavily recovered shadows thanks to the on-sensor phase-detect autofocus (PDAF).

If you're shooting video, again, the Z7 II is vastly better than the D850, which was hardly usable in live view mode at all.

For low-light work, the Z7 II's sensor stabilization lets you handhold shots with longer shutter speeds, even with unstabilized primes. The same applies for handheld video. Sensor stabilization could be reason enough to upgrade.

For portraits, the eye-detect AF won't produce better photos, but it WILL speed your workflow by finding the eye without having to move the focus point selector around manually.

For action, though, the Z7 II just isn't ready. The EVF lag makes it difficult to track erratically-moving subjects with a long telephoto lens: the optical viewfinder of an SLR still wins. The Z7 II's framerate really drops when shooting action, too, and you can expect to get about 5 FPS (compared to the 9-10 we get from the D850).

In a nutshell, if you're thinking about upgrading from a Nikon DSLR and you're shooting anything except action, the Z7 II is a worthy upgrade. If you're considering the Z7 II as your first camera, I would instead recommend a Sony a7R III ($2,300) because it has similar capabilities at a lower price. Also consider Canon R5 ($3,900), which has a more powerful autofocus system and is good enough for sports and wildlife. Canon and systems also offer a wider variety of mirrorless lenses.

Nikon Z9 ($5,500)

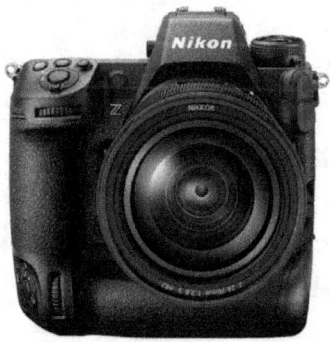

The Nikon Z9 is Nikon's flagship mirrorless camera and the spiritual successor to the Nikon D6 sports camera. It delivers.

45 megapixel 20 FPS raw, or 30 FPS JPG, is an amazing experience for any D6 owner. Both modes can be completely silent with sophisticated human- and animal-eye autofocus. It will even shoot 120 FPS with lower-quality 11 megapixel images. For sports or wildlife shooters, the Z9 is the ultimate Nikon. It's also amazing for landscapes and portraits, though the large form factor makes me prefer the smaller Z7 II for those tasks.

While the Z9 is the ultimate Nikon camera for sports and wildlife, after 6 months of testing, I've found that firmware 2.0 has some significant weaknesses compared to Sony and Canon's top-end cameras. Both Sony and Canon have superior autofocus systems that find and track subjects with far greater success.

I realize this is contrary to many other reviews you'll find. I can't explain other reviewers findings, but I'll say that we worked closely with two different Nikon shooters and they also preferred the autofocus from the Canon R5, Canon R3, and Sony a1. We worked with our Nikon rep to get the best performance possible from the Z9 and tested every single autofocus mode with both Z-mount and F-mount lenses. I'm confident, though disappointed, with the results.

The video specifications also over-promise and under-deliver. With firmware 2.0, the Z9 offers an amazing 8k @ 60 FPS raw. Indeed, the quality is amazing. However, it's limited to a proprietary raw format that isn't compatible with the most popular video editing apps, including Final Cut and Premiere Pro. In my real-world experience, video autofocus with the Z9 was so frustrating that I had to switch to manual focus, and unfortunately that doesn't work for our style of video creation. High resolution and high frame rates do not matter at all if a video jumps out of focus.

Still, the Z9 is a work-in-progress, and Nikon is promising more firmware updates to improve the performance. I know they listen closely to the criticisms we publish in our reviews, so I trust that the camera will improve with time.

Chapter 16: Sony E-mount

Note: Sony does not seem committed to future support of the Sony A-mount system of SLT cameras and lenses. As a result, we have removed the Sony A series from the buying guide. However, you can download that chapter at *sdp.io/saguide*.

Sony's E-mount cameras offer the image quality of an APS-C DSLR in a small, inexpensive package. Because of their relatively low price and amazing image quality, they're my standard recommendation for casual photographers.

In other words, if a friend asks me what camera they should buy, I'll steer them towards an a5100 or a6300, depending on their budget. But those friends aren't the type who would read this guide, or even care about megapixels; they're the average casual photographer, the mom or dad who wants pictures of their kids and their vacations. As a reader of this book, you might be happier with a more advanced camera.

Here's a list of Sony's APS-C bodies:

Model	Price (new)	Price (used)	Focusing	Tilt screen	Touch screen	Wi-Fi	Pop-up Flash	EVF	FPS	MP	Video
ILCE-QX1	$400	$350	OK							24	
NEX-3N	$230	$150	OK	Flip-up			*		4	16	1080/30p
NEX-5T	$350	$240	Good	Flip-up	*	*			10	16	1080/60p
a5100	$450	$350	Good	Flip-down	*	*	*		6	24	1080/60p
NEX-6	$480	$330	Good	Tilt	*	*	*	*	3	16	1080/60p
NEX-7		$350	Good	Tilt			*	*	10	24	1080/60p
a6000	$550	$400	Better	Tilt		*	*	*	11	24	1080/60p
a6300	$1,000		Best	Tilt		*	*	*	11	24	4K/30p

All the Sony APS-C mirrorless cameras are available as kits with the 16-50 f/3.5-5.6 lens, which I highly recommend. The kit lens is sharp, compact, and flexible.

They also allow charging over USB, which makes them easy to travel with. You can charge them from your laptop's USB port, or any plug-in or battery-powered USB phone charger. However, Sony does not provide a traditional battery charger with the cameras. If you want to charge your camera faster, you'll need to buy a separate charger.

The newest model, the a6300, offer better autofocus than the older models. This is an important consideration because even casual photographers struggled with the older models.

ILCE-QX1 ($400 new)

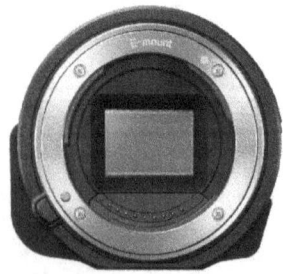 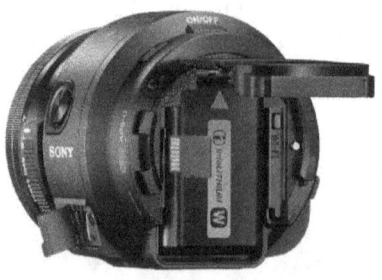

This unusual camera is little more than a lens mount, a sensor, and wireless communications. It lacks a display and common controls, requiring you to use a smartphone or tablet to control it.

Here's how Sony would rather have you think about the ILCE-

QX1: it's a way to upgrade your smartphone to an APS-C interchangeable lens camera, allowing you to take amazing pictures and instantly post them on Twitter and Facebook. Unfortunately, this design is so clumsy that most people don't consider it worthwhile. You need to use a plastic grip (included) to attach it to your smartphone, and it's awkward enough that it completely eliminates the benefits of portable smartphone photography. By relying on wireless communications for the viewfinder, the lag is excessive. It costs as much as a good quality mirrorless camera or DSLR, so you aren't saving much money by foregoing the camera body.

I do anticipate deep integration between our cameras and smartphones in the future. The QX1 and its non-interchangeable lens predecessors, the QX10 and QX100 are an interesting peek into the future. The technology is not ready for me to recommend, however.

NEX-3N ($230 new, $150 used)

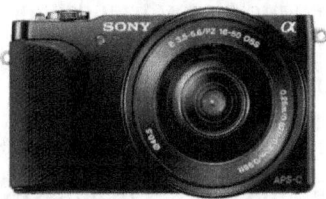 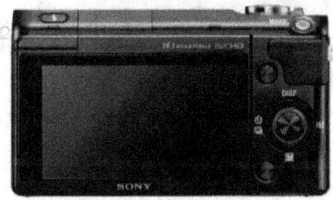

The NEX-3N is Sony's entry-level APS-C mirrorless body, and it's an amazing value. Other than being one of the least expensive new cameras you can buy, it's not the best at anything. Other cameras in the lineup offer 24 megapixels instead of the NEX-3N's 16 megapixels, but you won't notice the difference unless you also buy a lens that costs more than this body does new.

This camera does require some patience, however. The focusing is slow compared to the a5100 or a6000, even for still subjects. Low-light focusing can be downright frustrating. It doesn't have a viewfinder or a flash hot shoe, and the controls (for tasks such as manually setting the aperture, shutter speed, and exposure compensation) are limited. Therefore, you'll have to dig into menus if you don't like the camera's automatic settings.

You'll need to upgrade to the NEX-5T if you want a touchscreen, but the NEX-3N's screen does flip up for selfies. For the casual user, the NEX-3N offers incredible bang-for-the-buck.

NEX-5T ($350 new, $240 used)

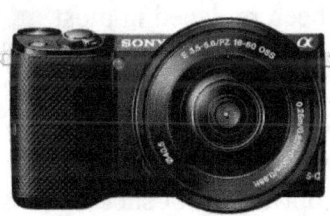 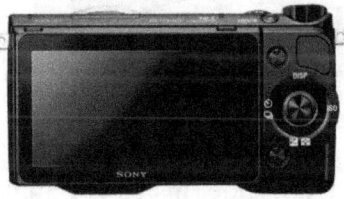

For an extra $100, the NEX-5T is a worthwhile upgrade over the NEX-3N for many. The image quality is unchanged, but this camera autofocuses faster and more reliably, leading to less frustration. It also offers better video quality, supporting up to 1080 60p. The touchscreen is a must-have feature for those of us addicted to our smartphones.

NEX-6 ($480 new, $330 used)

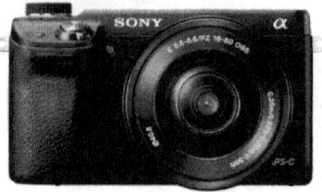 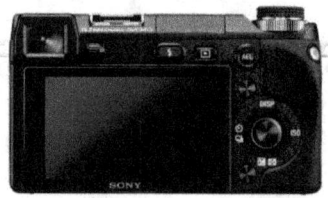

The least expensive camera in the Sony lineup to offer an electronic viewfinder and a flash hot shoe, the NEX-6 is a good choice for enthusiasts on a budget. Adding an external flash can do wonders for your indoor photography, so the flash hot shoe is an important feature. The only drawback to the less expensive cameras is that the screen doesn't flip 180 degrees for selfies.

a5100 ($450 new, $350 used)

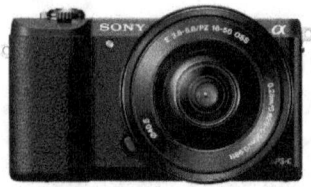 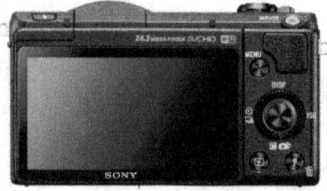

Very similar to its predecessor, the a5100 is the camera that I recommend to all my friends who want to take better pictures but don't want to actually learn photography. I wouldn't recommend it to friends who wanted to take sports or wildlife photos, however.

The a5100 is small, reasonable priced, and full of features. It's the little touches that make it such a great value: the touchscreen that flips backwards for selfies and the Wi-Fi/NFC for getting your pictures to your smartphone (and thus to your favorite social network).

I wouldn't recommend it to anyone more serious about photography. If someone even hoped to learn how f/stops or exposure compensation worked, I'd steer them to the a6000 instead for its electronic viewfinder

Video: a5100 Preview
12:21 - *sdp.io/a5100preview*

and better controls. However, the a6000 lacks the selfie touchscreen, which makes this a better choice for the casual photographer, regardless of the price.

NEX-7 ($350 used)

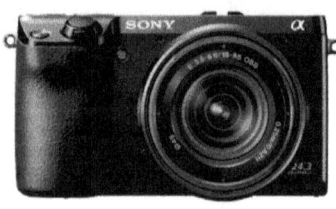 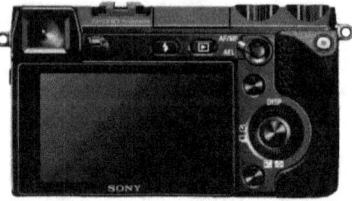

Formerly the king of Sony's E-mount lineup, the NEX-7 has been replaced in most ways by the more capable and less expensive a6000. I would never recommend buying a new NEX-7, since the a6000 focuses faster, has a standard flash hot shoe (instead of Sony's proprietary hot shoe), and is also less expensive new.

However, used copies can save you a few dollars compared to an a6000 and provide fantastic image quality and great manual controls.

a6000 ($550 new, $400 used)

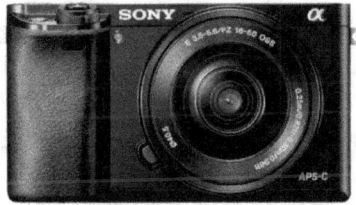 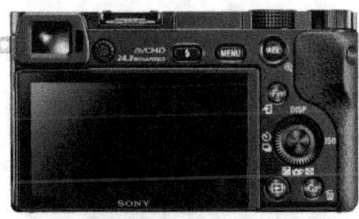

The top-end camera in Sony's E-mount lineup, the a6000 is a solid all-around camera for users who might want to manually adjust camera settings, including aperture, shutter speed, and exposure compensation. If I'm traveling light and want to take professional-grade landscape photos, I'll grab the a6000.

Video: a6000 Review
15:16 - *sdp.io/a6000review*

My one complaint about the a6000 is the lack of a touchscreen. Without a touchscreen, learning to use the camera is a little more difficult, and common tasks like reviewing pictures and focusing during

video are slower. If a touchscreen is a priority, choose the a5100 instead. Note that choosing the a5100 requires you to give up the electronic viewfinder, flash hot shoe, and some

Video: a6000 vs. E-M10
19:52 - *sdp.io/a6000vem10*

useful buttons and dials. Be sure to read the section on the a6100/a6400/a6600, which replace this generation of camera.

a6100 ($720) / a6400 ($900) / a6600 ($1,400)

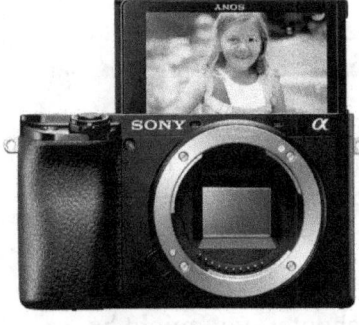

The a6100, a6400, and a6600 are Sony's new lineup of APS-C cameras, replacing the a6000, a6300, and a6500. They each feature a flip-up screen and an amazing autofocus system. They are all extremely compact and great choices for general-purpose family and travel cameras.

Most people will be happy with the entry-level a6100. If you upgrade to the a6400, your pictures will look the same, but you get these benefits:

- Sharper viewfinder

- Better weather sealing

- S-log video (if you don't know what that is, you don't need it)

Upgrade to the a6600, and you get these benefits as well:

- Higher FPS for sports and action

- Bigger battery

- Bigger grip

- Sensor stabilization for hand-holding in low-light conditions

- Eye AF when recording video

a6300 ($900 new)

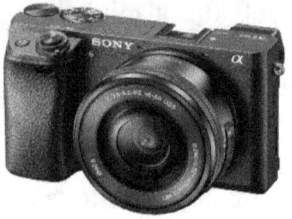 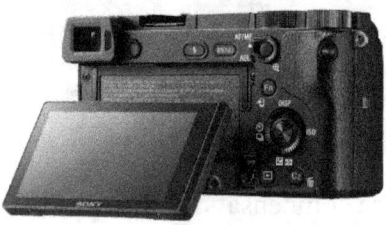

The $1,000 a6300 ($1,150 with a lens) is a compact camera that can do it all: landscapes, sports, low-light events, and portraits. In the Sony lineup, it's the best choice for sports, and it's the cheapest Sony to give you 4k and slow-motion video. It's the best all-around camera we've ever used, and it will be our top recommendation for beginning photographers who have the budget.

If you're a DSLR shooter who is reluctant to bring a big DSLR everywhere, the a6300 can get great images without the burden. It has class-leading focusing, video, and still image quality.

While the a6300 is a jack-of-all-trades, it's also a master of none. If you specialize in sports, landscapes, wildlife, portraits, or video, you might get more bang-for-your-buck with a different body. I'll make alternate suggestions throughout this review.

Here's a quick summary:

- **a6000 UPGRADE:** If you're an existing a6000 owner, you should upgrade only if you're struggling with focusing or if you want better quality video. The updated 2.35 megapixel viewfinder with 120 fps is twice as sharp and looks much better, and it's a huge advantage over DSLRs for most types of photography. The high ISO raw image quality of the a6300 was very slightly better than the a6000, but most people won't see a difference. if you want a noticeably cleaner and sharper images, save up for one of the full-frame a7 models.

- **LANDSCAPES:** If you're a dedicated landscapes shooter, the a6300 is the best APS-C camera you can buy. However, you could also buy a full-frame Sony a7 used for about $800, get half the noise, and fully utilize full-frame lenses.

- **SPORTS:** If you want a compact camera and occasionally shoot sports, the a6300 is the best mirrorless camera. If you're a serious sports shooter, you might be happier with a used original Canon 7D for half the price, because it has a dedicated focus point selector, a bigger buffer, no viewfinder lag, and more telephoto lens options. If budget were no concern, we'd grab a Canon 7D Mark II for sports.

- **LOW-LIGHT EVENTS:** The a6300 is a capable events camera. However, Sony doesn't currently offer any native f/2.8 or f/1.8 APS-C zooms.

- **PORTRAITS:** In good light, eye-detect autofocus worked great for casual portraits, even with shallow depth-of-field. However, once you factor in the cost of the lenses and flashes, you could more bang-for-your-buck from a Canon or Nikon DSLR.

- **WILDLIFE:** If you're hoping to shoot wildlife, there simply aren't any native Sony E-mount big telephoto lenses, and we found adapted lenses didn't autofocus well enough. You'll be happier with a DSLR for birding.

Video: a6300 Review
13:13 - *sdp.io/a6300review*

- **VIDEO**: The 4K video quality and 5x slow-motion blows away everything else we've tested except for the $3,000 a7S II. However, the ergonomics of the Panasonic GH4 are much better, and the Panasonic G7 is quite capable, $400 cheaper, and also has better ergonomics.

Image Quality

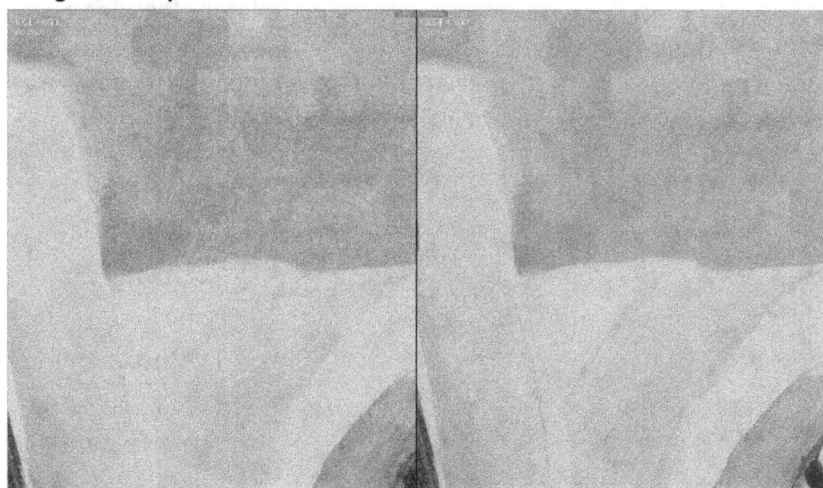

The a6000 had class-leading image quality, and the a6300 is basically the same. Throughout the ISO range, if you look really closely, you can see just a bit less noise in the a6300's raw files. Here's a 1:1 closeup (if you click the image) with the a6000 on the left and the a6300 on the right at ISO 25,600. These images were processed with Sony's raw processing software but they had the same amount of noise reduction applied. You can disregard the differences in color.

Every time a new generation of a camera is released, photographers hope to see 1-2 stops of image quality improvement. Based on our reviews and data from third parties, raw image quality improves by 1 stop every 8 years or so. Manufacturers often brag about bigger improvements, but they're generally referring to JPG quality… and if you care at all about image quality, you're probably shooting raw, anyway.

If you were hoping for a bigger jump in image quality, your best bet is to upgrade to a full-frame camera, like the Sony a7. If you have the budget, the a7R II has remarkable image quality.

For low-light shooting, you'll see the most improvement by using faster lenses. For example, switching to an f/1.8 lens from an f/5.6 lens. That'll give you about 3 stops of image quality improvement, and 8 times less noise in low-light environments.

Casual Photography

Casual photography is where the a6300 excels. It locks into focus fast in any conditions, including low light. The tilt screen means you can easily hold it low to the ground or over your head.

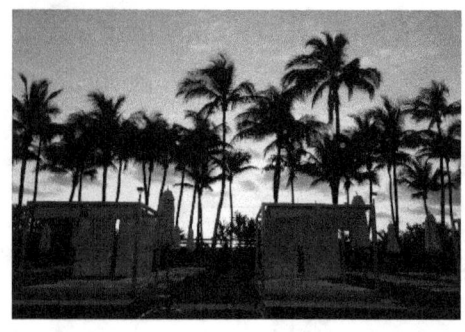 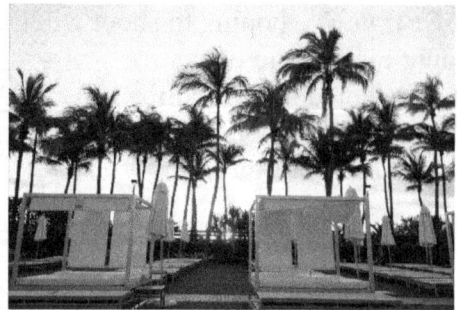 The raw files are as clean as any APS-C camera we've tested, and the dynamic range allows you to recover details in shadows and highlights, or fix exposure problems in post with very little penalty. The next pictures so the same shot straight out of camera, and after raising the exposure in post-processing. You can see that the raw file contains detail that you might think were completely lost. If you messed up your exposure settings, this can save your shot.

At night and in low light, like when you're with friends at a restaurant, detail drops but the noise is tolerable, especially for sharing pictures online. These next two pictures were taken handheld at ISO 6400.

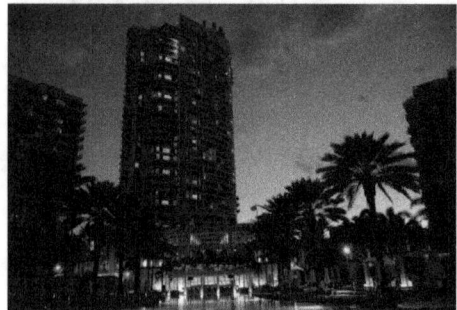 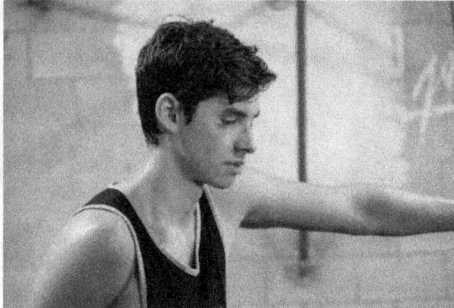 The best camera is the one you have with you, and the a6300 is the best camera to grab when you don't want something more cumbersome.

Sports

We never recommended the a6000 as a sports camera because we just got better results with DSLRs. The a6300 is the first mirrorless camera we can recommend for people who want to shoot sports; the focus tracking is great, the 8 frames per second is very fast, and

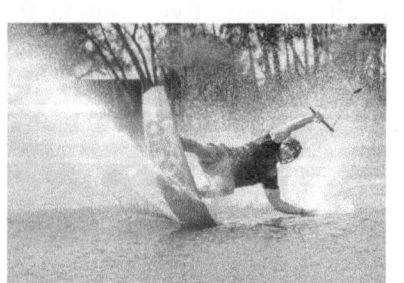 the viewfinder blackout and lag are much reduced.
The focusing in the a6300 is MUCH improved, thanks to its phase detect focusing system. While the focus tracking isn't as fast or accurate as big DSLRs like the 7d Mark II, At 8 frames per second, it's faster than comparably priced DSLRs.
For best results, just put the camera in Sports mode…that's the best way to ensure the fairly complex focusing system is correctly configured.

If you're serious about shooting sports, you'll still probably be happier with a Canon or Nikon DSLR. The optical viewfinder of a DSLR completely eliminates lag, and higher-end models have much bigger buffers that let you take more consecutive photos.

The high frame rate helps you capture that decisive moment. You can choose between 8 fps with continuous autofocus, or 11 without it. For the water skiing, we needed the autofocus to keep up with subjects moving towards us. For volleyball, or times when the subject was moving only side-to-side, the higher frame rate improved our odds of getting the perfect moment.

Check out this sequences of photos and notice how the last photos are so poorly composed.

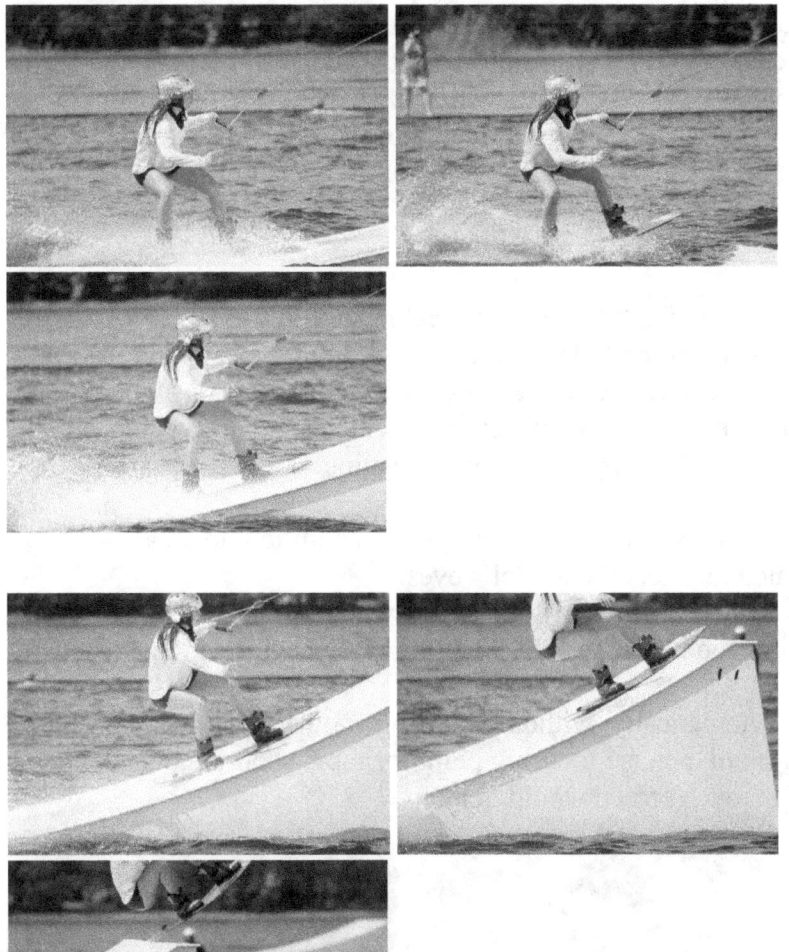

With practice, it's definitely possible to keep a fast-moving subject in frame. However, the display isn't quite real-time as you're shooting, so tracking a fast-moving subject is still more difficult than with an SLR.

As a result, I often lost track of the subject in the viewfinder after a long sequence of shots. With practice, you can learn to lead the subject.

Every reviewer was frustrated with the "Writing to Memory Card" message. You can't do anything with the camera until the entire buffer is written to the memory card, and when you're shooting rapidly, it never seems to stop writing. For some reason, Sony put the memory card write indicator on the bottom of the camera, where you can't easily see it. The buffer is too small to shoot action in RAW, so you'll need to use JPG, and even with the fastest memory card available, we missed shots because the buffer was full.

In sports, you often need to manually control the focusing point to stay focused on the key player. Often, you need to manually move the focusing point to the other side of the frame as the direction of the action changes. Manually controlling the focusing point on the a6300 is slower than on comparably priced DSLRs, which have dedicated thumbsticks. You'll need to carry multiple batteries to get you through most sporting events, or even a day of casual shooting. We really hope the next generation of Sony cameras adopts a bigger battery.

Portraits

The a6300 is very workable for portraits, but less than ideal in several ways.

Portraits often have very shallow depth-of-field, and that requires very fast and precise focusing on the model's closest eye. Face & eye detection are GREAT for this when they work, because you don't have to worry about selecting a focusing point on the eye. It requies an extra button press, which delays your shot by a fraction of a second–a critical amount of time in a fast-moving shoot.

But it doesn't always work… it totally depends on the lighting condition. In a backlit model shoot, it completely failed to detect the model's eyes.

Other times, it might be 80% or 90% accurate. If you shoot with a fast pace, it can be really frustrating. At slower paces, it's good, as is manual focusing with magnification & focus peaking.

We'd like eye-detect to work automatically, without requiring another button press. We'd also like a joystick to select a small, precise focusing point for rapid shooting, like every DSLR at this price point has. A touch screen would help, too.

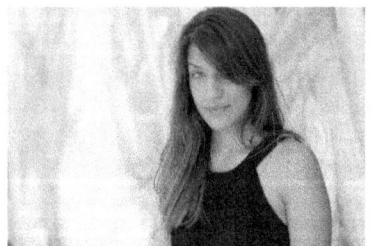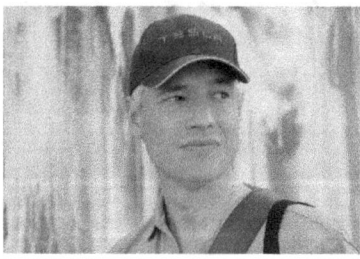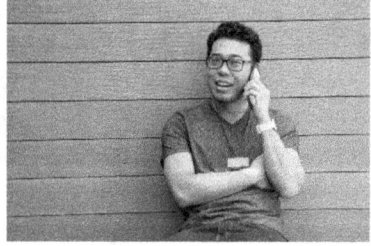

The a6300 only syncs with studio lights to 1/160th, which is slower than many cameras. In the studio, this means your subjects might have a bit more motion blur. In sunlight, you'll need to use an ND filter to shoot wide open if your flash doesn't support high-speed sync.

Street Photography

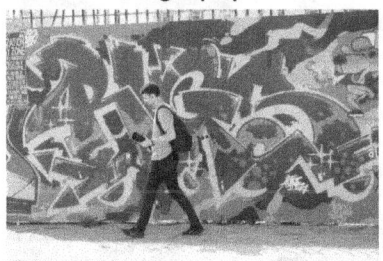 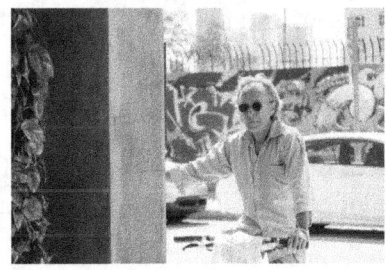

The a6300 is the perfect camera for street photography. It's small, discreet, and can be completely silent. When you're shooting blind, you can trust the autofocus system to lock on quickly.

Landscapes

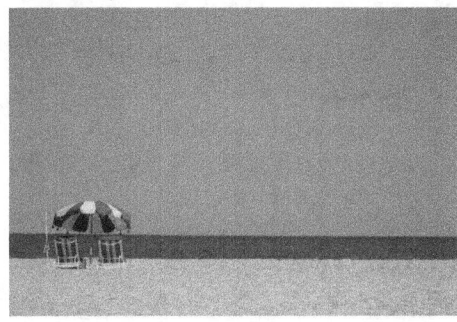 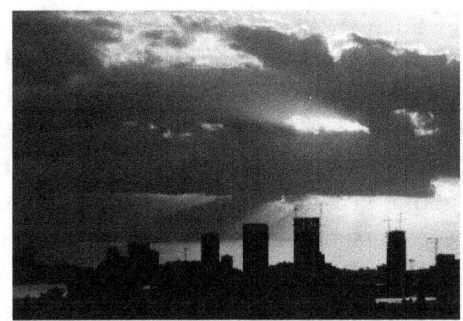

Like other Sony cameras, the image quality is just fantastic. It's unbeatable at this price point....but you're also within reach of the full frame a7, which offers much better image quality, especially with full frame lenses.

Wi-Fi

We used the WiFi constantly to post pics to Twitter & Instagram. This feature is critical to many modern photogs.

Sony's WiFi system is one of the better ones. Sharing still takes a minute or two, so we often just grab a snapshot with our phone, which has proper apps and a touchscreen. We'd like Sony to take their smartphone expertise and put full versions of the Instagram and Twitter apps on the camera, and add a touchscreen for tagging and labeling.

Video

The a6300 offers 4k video at 30 fps, and 1080p video at 120 fps. Both look gorgeous. S-Log3 provides unbeatable dynamic range for those experienced with color grading who are also shooting in extremely contrasty situations.

While the video quality is unbeatable at this price, the a6300 has some serious weaknesses as a video camera:

- It desperately needs a flippy touch screen. Many people, especially YouTubers, need to film themselves, and flipping a screen towards them is extremely convenient. You could, theoretically, accomplish something similar by connecting over WiFi from your smartphone, but that process is more time consuming.

- Continuous focus during video is still pretty useless for subjects that stop moving; it constantly hunts in and out and completely ruins the shot.

- The a6300 does have a mic jack, but it doesn't have a headphone jack! Therefore, you need to buy an external device to monitor your own sound. WTH?!

- You get the best IQ at 4k & 24p… the standard for film. If you jump up to 30p, the smoother standard for video, you'll see more noise at higher ISOs.

Our friend Max Yuryev tested the a6300 for video overheating with every frame rate, and it never overheated while outdoors in the shade at around 75-80F. Our a6300 did overheat in the hot Miami sun, however, and we weren't even shooting video, just stills. As long as you keep it out of direct sun, it should be fine.

Continuous autofocusing was often great at tracking moving subjects, but it often hunted or focused on the background, ruining the shot. With still subjects, the hunting made continuous autofocus unusable. The lack of a touch screen means you'll be manually pulling focus when you want to switch focus between subjects.

Nonetheless, at this price point, it's your best bet for getting 4k video of your kids sports. Our friend Jordan over at The Camera Store found the rolling shutter to be a problem, and it might be if you're shooting action, but it didn't hurt any of our shots.

If you're serious about 4k video, you might be happier with the GH4, which has a headphone jack, a touch screen that can flip forward, and access to the wide variety of nicely priced micro four-thirds lenses. We still prefer the much more expensive a7R II because it has a headphone jack, a stabilized sensor, and can switch between full-frame and Super 35 recording modes, giving us an option of crops. However, when I handed the footage to Justin, he thought it was from the a7R II. The a6300 is as good as the $3,200 a7R II.

Summary

The a6300 improves on a great camera. It's the best mirrorless sports camera ever, and a great, compact, all-around camera.

DSLRs still have some strengths… like the availability of f/2.8 and f/1.8 APS-C zooms, which Sony completely lacks. Yes, you can use the new full-frame f2.8 zooms, but we've found you always get sharper results when using lenses designed specifically for your sensor size. You could adapt DSLR lenses, but our tests show that adapted autofocus is much worse than with native lenses.

Our advice: get the a6300 if you shoot action or video, prefer mirrorless, and the Sony lens lineup has everything you might need. If you don't shoot action or video, the a6000 should be just fine, and it's half the price. Be sure to read the section on the a6100/a6400/a6600, which replace this generation of camera.

a6500 ($1,400 new, $1,300 used)

The a6500 is a significant step-up from the a6300, despite being released only about 6 months later. Most notably, the a6500 improves focusing speed and adds sensor stabilization, making it a better camera for both stills and video. It also reduces (but does not eliminate) the 4K overheating problem that plagued the a6300.

For detailed information, refer to the previous section reviewing the a6300, which is otherwise identical. Also, read the following section on the a6600, which replaced this camera.

Before investing in any of these cameras, you should consider one major drawback: Sony's APS-C lens lineup is rather basic. The APS-C lenses are fine, but they lack the professional-quality options that Fujifilm X-mount offers. Thus, you might consider a Fujifilm X-T30 or X-T4.

You might also consider full-frame cameras at similar price points. The Canon EOS RP is often available for the same price as the a6400. The Nikon Z5 is about the same price as the a6600. Sony's own a7 III can be found used for just a couple of hundred dollars more than the a6600, and I personally would much prefer a used Sony a7 III.

Sony Alpha/Full Frame Buying Guide

Just as Canon and Nikon offer both APS-C and full-frame DSLRs and allow you to swap lenses between the systems, Sony offers full-frame mirrorless cameras. The three models are all closely related:

- The standard 24-megapixel a7 II is the best value, and the best choice for most photographers investing in this system.

- The a7R II offers the greatest image quality, with 42 megapixels. It also has a much better autofocus system than the a7 II. It's the best all-around camera.

- The a7S II is a specialized camera primarily intended for low light video. Most stills photographers should avoid it because it offers only 12 megapixels of detail and a very primitive focusing system.

I believe full-frame mirrorless cameras will eventually replace full-frame DSLRs as the camera of choice for professional photographers and serious enthusiasts. The electronic viewfinders are fantastically useful, and every DSLR will seem bulky and heavy by comparison.

Today, they're the right choice for many, but not most. The most serious drawbacks, when compared to full-frame Canon and Nikon DSLRs, are:

- The native lens selection is quite limited, and lenses tend to be more expensive than their Canon and Nikon counterparts.

- While you can attach a wide variety of DSLR lenses using adapters, you give up much of the benefits of using a mirrorless camera, and if autofocus works at all, it will be much slower.

- Even with native full-frame E-mount lenses, tracking moving subjects with the autofocus is far less accurate than it is with similarly priced DSLRs. DSLRs are still the best choice for wildlife and sports.

For those satisfied with the current lens selection, and those primarily interested in still subjects, the a7 family is a perfect choice.

Here's a list of Sony's full-frame bodies. They all have almost exactly the same design, including a tilt screen (but without touch functionality), an EVF and Wi-Fi. They do not have a pop-up flash.

Model	Price (new)	Price (used)	Focusing	FPS	MP	Video (internal)
a7	$1,000	$800	Good	5	24	1080/6p
a7 II	$1,700	$1,350	Better	5	24	1080/60p
a7R	$1,900	$1,100	OK	4	36	1080/60p
a7R II	$3,200	$2,700	Best	5	42	4k/30fps
a7S	$2,200	$1,400	OK	5	12	1080/60p
a7S II	$3,000	$2,700	OK	5	12	4K/30fps

a7 ($1,000 new, $800 used)

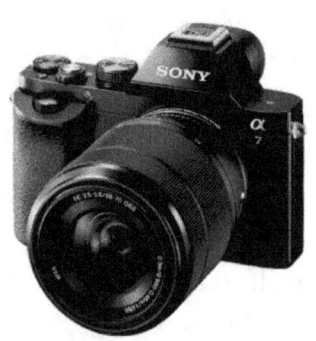 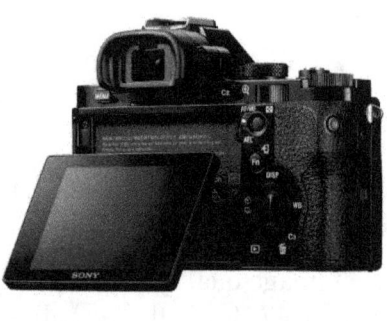

The a7 (and the other two a7 models) is an amazing camera and the future of photography, and its predecessors will almost certainly kill DSLRs. It offers full-frame 24-megapixel image quality, similar to a Nikon D610, but in a smaller, lighter package. The electronic viewfinder provides far more information than an optical viewfinder, and the ability to adapt almost any SLR lens provides incredible versatility for photographers capable of manual photography.

This is my current favorite full-frame mirrorless camera. If you want full-frame results, an electronic viewfinder, and the smallest camera possible, the a7 is the perfect choice. If you don't mind an optical viewfinder and a bigger body, the Nikon D610 has similar image quality, better focusing, and access to a much wider selection of native lenses.

a7 II ($1,700 new, $1,350 used)

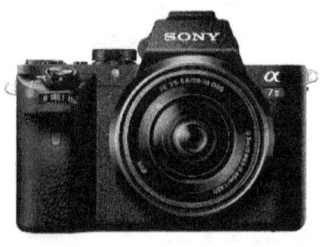 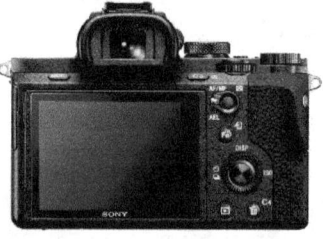

Compared to its predecessor, the Sony a7 II offers one important upgrade: Sony SteadShot Inside. This feature, also known as In-Body Image Stabilization (IBIS) stabilizes any lens against camera shake induced by handholding the camera at slow shutter speeds.

It's one small feature difference, but it's an important one that's made the Sony a7 II my favorite camera for artistic photography. SteadyShot Inside allows me to use slow shutter speeds with literally any lens, including adapted DSLR lenses, and fast prime lenses that don't have image stabilization built in. Now, you can attach that 55mm f/1.8 lens and hand-hold it at ¼ or 1/8ᵗʰ of a second and get clear shots, allowing you to use low ISOs even at night. No other camera in the world can do that with full-frame image quality.

Combined with the electronic viewfinder (standard on all mirrorless cameras) it means I can use my fast DSLR lenses with image stabilization, manual focus, focus peaking, magnification, and see a real-time preview with

Video: a7 II Preview
33:09 - *sdp.io/a7iiPreview*

a histogram and correct depth-of-field in the viewfinder. If I'm shooting black-and-white, I'll see the world in black and white when I look through the viewfinder.

If you plan to use zoom lenses that are already stabilized, or if you don't use slow shutter speeds handheld, choose the original a7 instead. At $1,000 new, the original a7 is the better value. The only other difference you might notice is

Video: a7 II Review
14:15 - *sdp.io/a7iiReview*

that the a7 II has an improved autofocus system (but neither camera autofocuses as well as comparably priced DSLRs).

If you have the budget, the newer a7 III is a worthwhile upgrade. Compared to the a7 II, the a7 III adds a much larger battery, far better controls, a vastly better autofocus system, and much better video quality.

a7 III ($2,000 new)

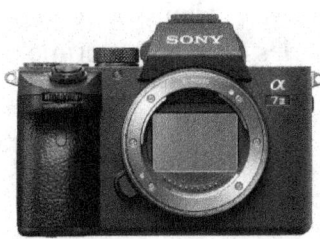
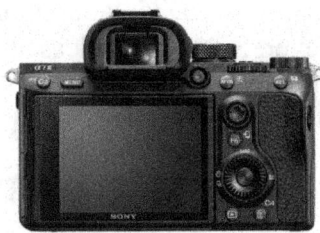

The a7 III is Sony's entry-level full-frame camera. There's nothing in it that the Sony a9 or Sony a7R III didn't have, but nonetheless, the a7 III is a historic camera that will change the future of all professional digital cameras.

The a7 III is revolutionary because of price: $2,000. That's about the same price as a new Nikon D750 or Canon 6D Mk II, but the a7 III is far, far more camera for the same money. 2017 was also the year that Sony offered a pretty complete lineup of enthusiast and professional-grade full-frame E-mount lenses. Finally, in 2018, there's a well-rounded, affordable Sony mirrorless camera that has no major drawbacks.

Here's a brief summary of our experiences with the Sony a7 III:

- Image quality is better than any other 24-megapixel camera we've ever tested. You'll still get more detail out of higher megapixel cameras like the a7R III, Nikon D850, or Canon 5DS-R when using sharp lenses and good technique.

- The focusing system is better than DSLRs. We hated the Sony a7 II's focusing system. The a7 III focuses accurately and instantly and is capable of focusing on eyes, something no DSLR can do. It also has focusing points that go to the edges of the frame, meaning photographers no longer have to use focus and recompose.

- The sensor stabilization works well. Handheld shots with unstabilized lenses (like primes) are sharper than with unstabilized cameras.

- Video quality is excellent, beating the Sony a7R III and Nikon D850 and rivaling even the Sony a7S II. It's better than any Canon camera. However, it lacks the flip-forward screen that most vloggers require.

- It has dual memory card slots, something the competing Canon 6D Mk II lacks. This is a requirement for us to recommend a camera for professional use. It can record both video to both slots, something most DSLRs can't do.

- The electronic viewfinder is great. It's not as sharp as the a7R III, but most people won't notice. In our experience, especially with beginning photographers, the EVF is vastly superior to the optical viewfinder found on all DSLRs because it can preview your exposure, show your histogram, and allow you to review shots in bright light where the rear screen is washed out.

The a7 III's $2,000 price tag doesn't reflect the total cost of the system. Basically, if you buy the Sony a7 III now, you need to use Sony's own lenses, which are more expensive than the third-party alternatives available for Canon and Nikon systems. Before you choose Sony over Canon or Nikon, be sure to add up the total cost of your camera body and all the lenses that you plan to purchase.

For example, as of July 2018, Tamron and Sigma still haven't released their third-party lenses "holy trinity" zooms for Sony E-mount, covering focal lengths from 16-200mm at f/2.8, you'll spend $9,000. If you buy the Nikon D750 and Tamron's holy trinity of zoom lenses, you'll spend only $5,500. However, Sigma has released a series of their Art prime lenses for E-mount, so the situation is improving.

I'd much rather shoot with an a7 III and the three Sony G Master lenses than a D750 and the Tamron lenses. However, if your total budget is $5,500, you'd only be able to buy the Sony a7 III and a single lens. A D750 with three lenses can produce much better results than an a7 III with only one lens.

a7C ($1,800 new)

Sony markets the a7C as the world's smallest full-frame camera… with interchangeable lenses, a viewfinder and sensor stabilization. That's a lot of adjectives, but in reality, the a7C is quite

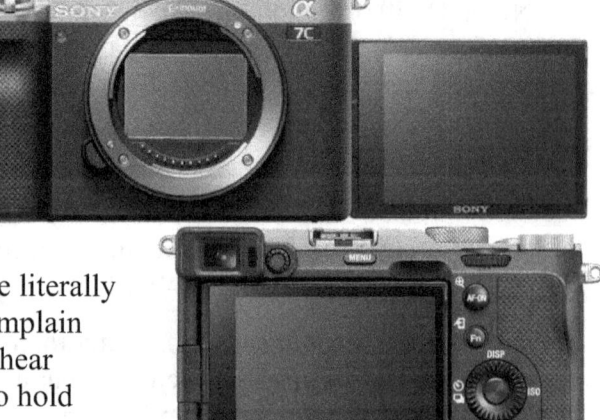

compact. This doesn't appeal to every buyer; I've literally never heard anyone pick up a Sony a7 III and complain that the camera was too big. In fact, I frequently hear people complain that the cameras are too small to hold comfortably.

Regardless, there is a market for compact, full-frame cameras, and some number of people who can afford the $1,800 price tag. Perhaps you're one of them.

The a7C is 95% similar to the a7 III. It has exactly the same sensor and image quality. Stills autofocus is very similar. Even in 2020, it has the same outdated menu system from 2018 and the rear screen only supports touch-to-focus; you can't touch to interact with the menus or settings. The exposure compensation dial still doesn't lock and is still prone to being accidentally bumped when not in use (or even turned off).

Compared to the $2,000 a7 III, it has several drawbacks:

- The grip is smaller and the a7C is less comfortable to hold (especially if you have large hands).

- The viewfinder is significantly smaller and feels lower quality; we found ourselves never using it, interacting only with the rear screen.

- The very useful thumb stick is gone; instead you'll have to click the flimsy directional pad to navigate menus or manually move your focusing point.

- The front control dial is gone, too, so if you're using manual mode, you'll need to use that same flimsy directional pad as a dial to adjust the secondary setting.

- The a7C has only a single card slot, so if your memory card fails, you've lost all your images.

- The sensor stabilization is decidedly worse, providing only about one stop of stabilization (and even that was unreliable).

As I write this, it begins to sound like the a7C is a decidedly worse camera than the 2-year old a7 III. It does, however, have a couple of advantages over the a7 III. The flip-screen is a must have for content creators, such as vloggers or other YouTubers. The continuous video autofocus is improved, though we still found it to be insufficient for anything but the simplest talking head videos. And, of course, it's smaller and cuter.

Indeed, those are the two most important traits of the a7C: It's small and cute. Most people reading this book won't care about those traits, but they will resonate with a few of you. In fact, I really like the a7C, especially in silver… Not enough to buy one at $1,800, but I like it.

a7R ($$900 used)

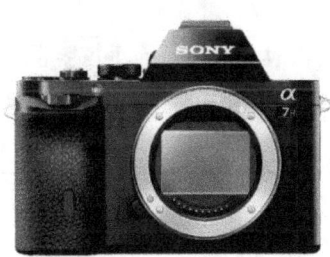 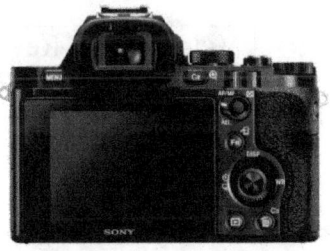

The a7R is Sony's top-end mirrorless camera, but I still recommend the standard a7 to most people. While the a7R is an amazing camera, the extra resolution (36 megapixels, compared to 24 on the a7) is mostly wasted, unfortunately.

The a7R has some flaws in the design that cause the shutter to vibrate the camera when taking a picture, reducing the effective resolution for many common shutter speeds, especially with telephoto lenses.

If you're considering the a7R because you want to create large, high-resolution images, you should also look at the Nikon D810 because the lenses available for the Nikon tend to be sharper at a similar price point. For example, DxOMark tested the a7R with the excellent Sony 24-70 f/4 ($1,200) lens and measured only 15 P-Mpix (Perceptual Megapixels). By comparison, the Nikon D800E (which has the same sensor), when paired with the Tamron 24-70 f/2.8 ($1,300), measures 23 P-Mpix at f/2.8. The Tamron also gathers half the light, allowing you to use a lower ISO and thus further improving image quality.

Similarly, the a7R combined with the 70-200 f/4 ($1,500) yields 23 P-Mpix. The Nikon D810 with the Nikon 70-200 f/4 ($1,400) yields 30 P-Mpix. Only the Sony 55mm f/1.8 prime proves to be sharp enough to take advantage of the a7R's sensor, with a P-Mpix rating of 30.

Of course, the D800E or newer D810 aren't the a7R. They're bigger and heavier, and they lack the amazing electronic viewfinder. If those benefits are worth giving up some sharpness, then this might be the camera for you. Otherwise, I'll steer you towards the much less expensive a7 or towards the D810, which eliminates shutter shake using an electronic shutter.

a7R II ($2,400 new, $1,600 used)

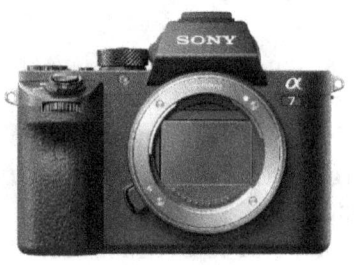 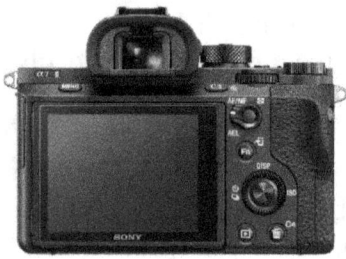

The a7R II is Sony's best all-around stills and video camera. As a stills camera, the 42-megapixel sensor has the best overall image quality of any camera we've used—the resolution is very high, there's not anti-aliasing filter, and the dynamic range is amazing. High ISO noise is noticeably less than other full-frame cameras.

As a video camera, it records in 4K either full-frame or with a 1.5X crop factor. We love the ability to add a 1.5X crop factor to zoom tighter than our lens would otherwise allow. The 1.5X crop factor (known as Super 35 mode) actually improves video quality by reducing noise by about half.

Video: a7R II Review (Part 1)
5:25 - *sdp.io/a7R2Review1*

Both stills and video benefit from SteadyShot Inside, which helps to stabilize handshake even if the lens doesn't have image stabilization built-in. That means you can use fast primes for even better low-light shooting. I was able to take handheld night shots in a city using a 24mm f/1.4 lens at ISO 100 and ¼ of a second—no tripod required.

Add to that the electronic viewfinder and tilting screen, and you have an incredibly versatile camera that has become our primary camera for most of our travel photography and video recording. However, it has some severe weaknesses compared to DSLRs:

Video: a7R II Review (Part 2)
6:32 - *sdp.io/a7R2Review2*

- **Slower, less accurate focusing.** The focusing on the a7R II is better than earlier mirrorless cameras, however, it still doesn't keep up with a DSLR.

- **Limited selection of high quality zooms**. Until recently, every test we did of Sony zooms produced the same result: extremely unsharp images compared to comparably priced Canon and Nikon zooms. Recently, we tested the Sony G-Master 24-70 f/2.8 lens, and that lens allowed the a7R II to produce images similar to the quality we expect from Canon and Nikon gear. However, the Sony version was heavier and more expensive. The 70-200 f/2.8 G-Master zoom should be available for testing soon. Sony still lacks a high-quality ultrawide lens (such as a 16-35mm); the existing 16-35mm lens tested as extremely unsharp compared to the less expensive Canon and Nikon versions.

- **Drawbacks to adapted lenses**. Because of the limited lens selection, we almost always use adapted Canon lenses on our a7R II. However, this comes with severe drawbacks. When you adapt lenses, focusing sometimes works, but it often fails. For consistency, we prefer manual focusing with adapted lenses. Manual focusing, even with magnification and focus peaking, is less precise than autofocusing, and if you miss focus by even a tiny amount on a 42-megapixel sensor, the entire image is less sharp… largely defeating the purpose of using a high-megapixel sensor and sharp glass. Additionally, adapting ultrawide angle lenses, such as the Canon 16-35, produces unsharp results at the wide end of the lens. This seems to be because of the design of Sony's E-mount and the requirements of the adapter; you simply can't get sharp 16mm pictures with this camera, using either native or adapted lenses.

These limitations prevent us from widely recommending the a7R II. Canon and Nikon DSLRs have a wider selection of higher-quality lenses at lower prices, and they tend to work more reliably. Nonetheless, those of us who require 4k video or who want a stabilized sensor don't currently have any other option than to use the a7R II and find ways to work around its limitations.

a7R III ($2,500 new)

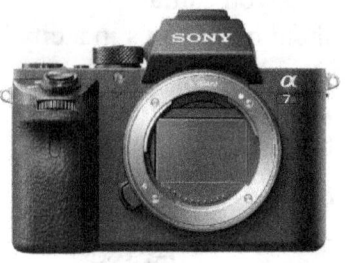 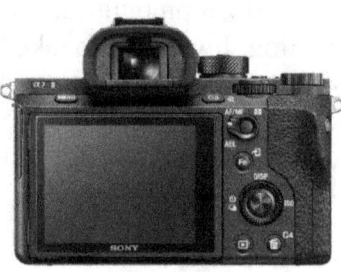

The a7R III is an incremental update to the incredibly successful a7R II. While image and video quality is mostly unchanged, the a7R III is worth the cost because of several key improvements:

- **Better autofocus**. Like the Sony a9, the a7R III has DSLR-like focusing speed and accuracy. The incredibly useful eye-detection autofocus is also significantly improved.

- **Better usability**. More buttons and bigger buttons mean you can now have conveniently-placed back-button focus and eye detection focus buttons. It has a thumb stick for quickly manually selecting a focusing point, and a touchscreen to select a focusing point when using the rear screen.

- **Better electronic viewfinder**. It inherits the a9's amazing viewfinder, offering better sharpness.

- **Higher framerate**. It can now shoot 10 frames per second, using either the mechanical shutter or the silent electronic shutter. That makes it fast enough for sports. Note, however, that it does not have the a9's fast readout, so the electronic shutter is less useful and will result in more flickering and rolling shutter. The mechanical shutter is the better option for most scenarios.

- **Bigger battery**. Battery life was a deal-breaker with the a7R II, requiring photographers to carry several batteries and constantly be charging them. The a7R III's battery can last about as long as a traditional DSLR, so you can make it through a typical day of shooting.

- **Dual card slots**. Finally, you can buy an a7 with two cards slots. A dead card won't mean you lose an entire shoot.

The hardest question to answer is whether to use the a7R III or the Nikon D850. For more information, refer to our D850 review in this book.

a7R IV ($3,200 new, $2,900 used)

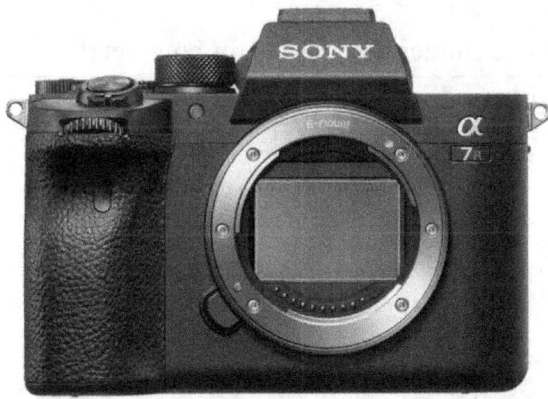

The Sony a7R IV is probably the greatest all-purpose camera ever made, and it's my personal choice for travel, landscapes, and portraits. But it's also overkill for most photographers (probably even myself). The 60 megapixel sensor doesn't offer significantly better image quality than the 42 megapixel sensor in the a7R III, according to our testing. It also doesn't significantly outperform cameras like the Nikon D850, Canon EOS R5, Nikon Z7, and Panasonic S1R. It does, however, create files that consume more storage space and slow down your post-processing. That was the biggest reason my wife, Chelsea, decided to continue using her previous-generation a7R III instead of upgrading.

Compared to the a7R III, the a7R IV's autofocus is a little more sophisticated. It's eye-detect is a little better. It has two UHS-II card slots, so buffering is less of an issue when writing raw photos to both slots (but it's still a major problem). It feels a little better in the hand. But will it every produce significantly better images than the older, less expensive a7R III? Probably not.

It offers a 240-megapixel mode which produces AMAZING pictures if you can get it to work. Unfortunately, in extensive testing and attempted real-world usage, I have only ever successfully captured photos in my own basement using a very heavy tripod. Even on higher floors of my house, the camera would move too much to create perfect images. For landscapes and architectural photography in the real world, I've never had a successful image after more than 100 attempts. You can dismiss the 240-megapixel mode unless you're building a space specifically for product photography or art reproductions.

a7S ($2,000 new, $1,200 used)

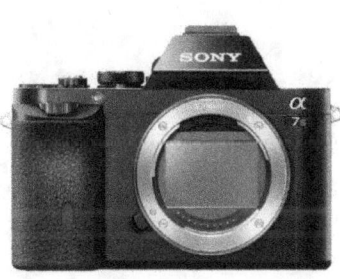 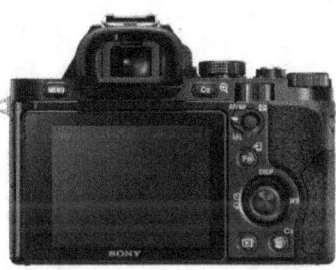

The Sony a7S is an amazing camera for a few, but it's not the right camera for most people.

The sensor separates the A7S from every other camera on the market. It's full-frame, but only 12 megapixels. That means that each of the 12,000,000 pixels is much larger than other full-frame cameras, and thus gathers much more light.

> Video: a7S Preview
> **12:03** - *sdp.io/a7sPreview*

Sony designed a 12-megapixel sensor to optimize it for recording HD and 4k video. As a result of that optimization and the large pixels, it's simply amazing for low-light video. If that's what you need, this is the camera for you.

If you want low-light stills, just about any other full-frame camera can give you similar results once you scale the images down to 12 megapixels.

Another benefit of the a7S is that it has an electronic shutter that allows for completely silent shooting; if you must be silent and work in low light, this is an ideal camera. The Panasonic GH4 also has a silent shutter option; however, the smaller Micro Four-Thirds sensor produces much more noise.

Video: a7S vs NX1 vs GH4
24:10 - *sdp.io/a7sNX1GH4*

a7S II ($1,900 used)

The Sony a7S II is Sony's full-frame, low-light camera. It's 12 megapixels, which means it doesn't provide the most detail for still images. It's true capabilities are seen once you begin shooting video, but we'll get to that a little later.

Video: a7S II Review
5:31 - *sdp.io/a7s2Review*

When compared to the less expensive, 36-megapixel D810, the Sony images are cleaner in low light when we scale the D810 images down to the same resolution.

This sample shows the images at ISO 3200, with the Nikon's scaled-down image on the left.

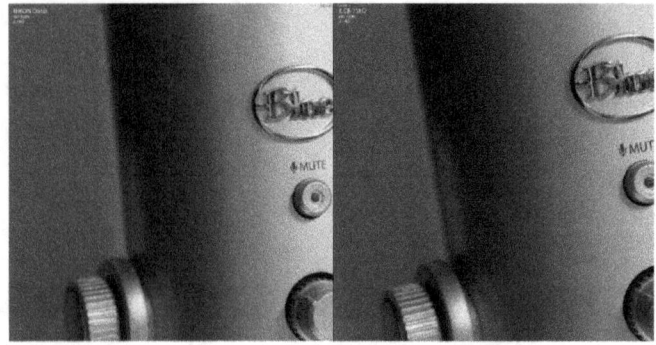

As you can see, the difference is subtle, and if we were to raise the noise reduction before scaling the Nikon's image down, any differences would disappear. For that reason, it's hard to recommend the a7S II just for stills.

Here's a side-by-side comparison at ISO 12,800, where the differences are more obvious:

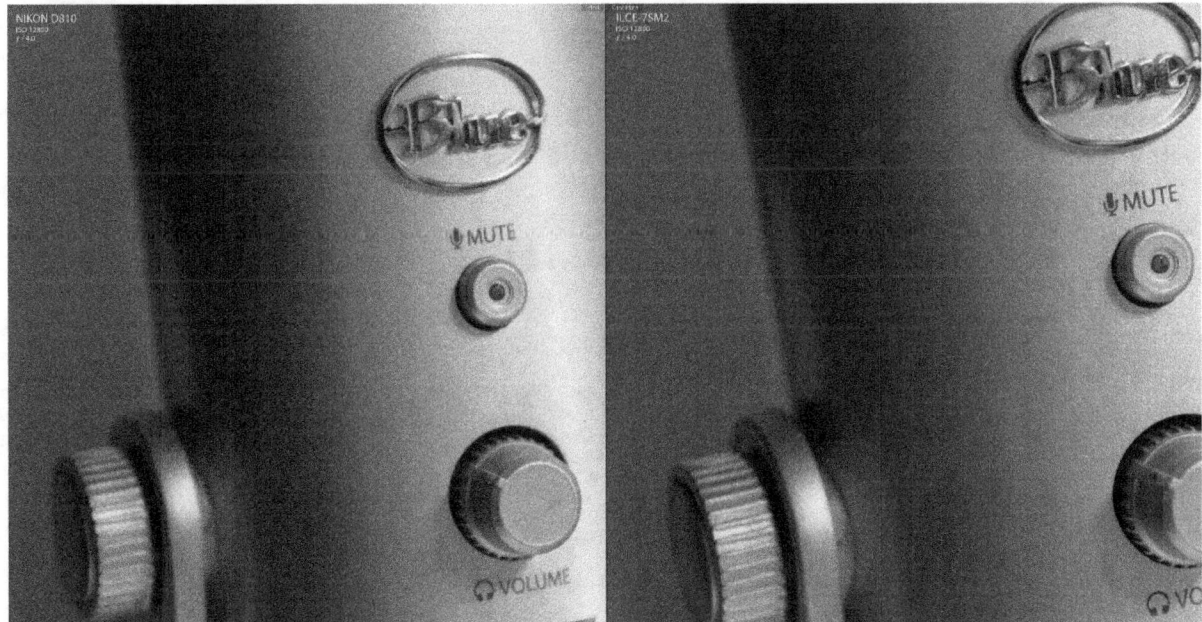

However, that's fairly easy to overcome with D810 raw files. Because they have so much more detail, we can raise the noise reduction to virtually eliminate noise, while still retaining more detail than the a7S II is capable of capturing. Additionally, the D810 has a vastly better autofocus system than the a7S II's contrast-based autofocus system, which is better suited to manual focus.

Where the a7S II shines is video. Indeed, the video quality is better than any camera we've ever tested, and the full-frame sensor means you get the shallowest depth-of-field from readily available 35mm lenses. It's remarkable. It can literally see in the dark, especially when paired with a fast prime lens.

Compared to the original a7S, the a7S II has several advantages for video:

- **Internal 4k recording**. With the a7S, you can only record 4k if you use an external recorder.

- **1080/120 FPS**. Recording 120 frames per second at full HD gives you the ability to do 4x slow motion at 30 frames per second. If you render your video at 24 frames per second, you can slow down to 5x slow motion.

- **Stabilized sensor**. The a7S II can reduce your handshake with any lens, including fast primes, making it excellent for handheld video.

- **Better autofocusing**. While the a7S II's autofocus system is still not the best on the market, or even better than the a7R II, it has improved since the original a7S.

All of the above technical specifications are nice to know, but the true test of a camera is whether or not you want to shoot with it. Tony and I took our copy out on the town to test it's low-light capabilities and overall usability and, overall, were really pleased with the a7SII performance.

Check out these pictures from our night out:

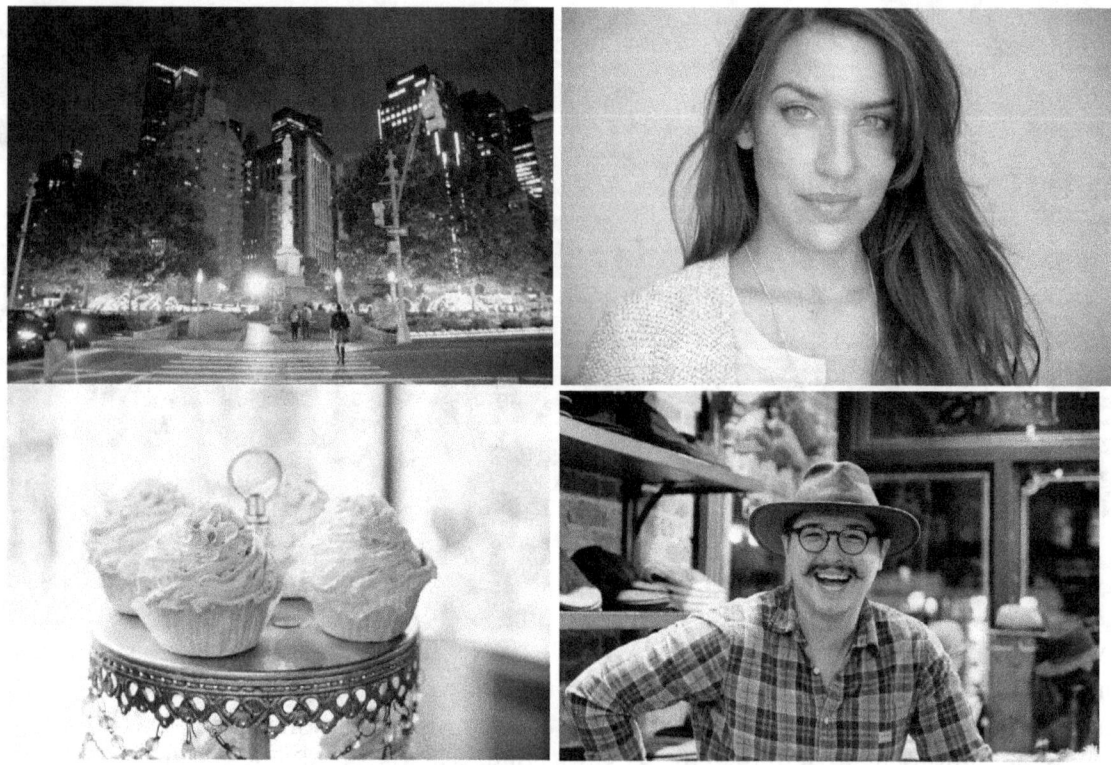

a7S III ($3,500)

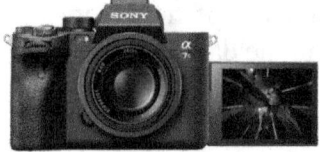

The Sony a7S III continues the a7S-series legacy of being a dedicated video camera in the form factor of a mirrorless stills camera. It uses the same sensor as the a7S II, but offers several important improvements:

- Dramatically improved autofocus capable of keeping a moving subject's eye in focus

- Quadrupled framerates, from 4k/30 to 4k/120, with little to no overheating or recording limits

- Improved menu and touchscreen

- Amazing, huge, high-resolution electronic viewfinder

- Flip-forward screen

- Dual card slots (with recording to both) supporting either SD cards or faster and more expensive CFExpress Type A.

While expensive, the a7S III is the best video camera at this price point for professional videographers, vloggers, and YouTubers who need autofocus. The closest competitors are two Canon RF cameras:

- **Canon EOS R6 ($2,500)**. This offers a flip screen, 4k/60 (not 120), and great video autofocus. The low-light video is actually better than the Sony a7S III.

However, both 4k/30 and 4k/60 overheat quickly. They might record for a full 30 minutes, but if you take some time setting up the shot, or even just using the menus or taking stills, recording time can be significantly shorter. It also doesn't record video to two card slots. Therefore, it's impossible to recommend it as video camera to professional shooters.

- **Canon EOS R5 ($3,900)**. This outperforms the a7S III in one big way – it offers 8k/30. The 45 megapixel sensor also makes the R5 a far superior stills camera, so it seems like the perfect professional hybrid rig. However, like the R6, it has severe overheating problems when shooting video, and it only records video to a single card slot. The R5 is a stills camera that can be used to record an occasional video clip, but we cannot recommend it as a professional video camera.

Because of these factors, we chose the a7S III for our own video recording. Why buy an a7S III instead of a "proper" video camera? There really is no proper video camera with full-frame 4k/120 and autofocus tracking. If one existed, it would probably cost over $10,000.

a9 ($3,500 new, $2,500 used)

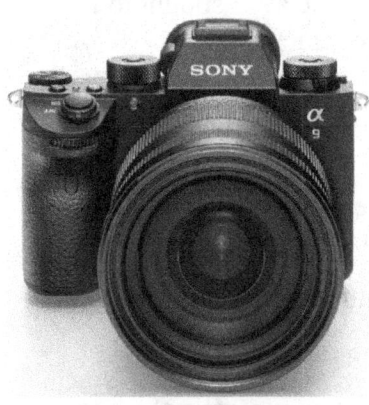 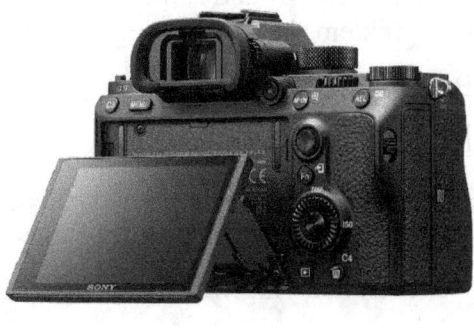

The Sony a9 is a historic camera. When the first reviewers used the camera, skeptical portions of the photographic community assumed their positive reviews must be the result of fraud, payola, or incompetence. But, as a long-term owner of the Sony a9, I can testify that it's a huge leap forward from any other camera.

The Sony a9 is a great all-around camera, but it's specialty is sports. As a sports camera, it's significantly more productive than any other camera ever made, including the Nikon D5 and Canon 1DX Mark II. The most significant feature is the total lack of viewfinder blackout. With other cameras, taking a picture briefly interrupts the viewfinder, either making it black (with DSLRs) or pausing live view (with mirrorless cameras). With the a9, your view of the action is never interrupted, and that made it much easier to track subjects. The a9 also captures 20 FPS (in ideal circumstances). The top cameras from Nikon and Canon capture 12 and 14, respectively. All cameras capture fewer frames when they're tracking focus, but the Sony still captured more images than any other camera. Optionally, it can capture those pictures silently, which is perfect for golf, weddings, and press conferences. The Sony images are better quality than the Nikon and Canon cameras, too. Autofocusing is perfect, even when tracking fast-moving subjects. This was always a weak point in the Sony lineup; focusing never quite matched DSLRs of the same price point.

The a9 blows every other camera away. The focusing points extend to the edge of the frame, so you're free to compose the picture however you want. Eye-detection focusing snaps the nearest eye into focus, even through glasses, making it amazing for shooting portraiture.

With that said, the Nikon D5 and D500 have 3D tracking, which tracks a subject as it moves around the frame. The Sony a9 has a similar feature, but the Nikon cameras simply worked better. The a9 will shift focus between multiple subjects when it shouldn't, and the Nikons stay locked onto the subject you select.

The Sony a9 also greatly improved the usability by adding a thumbstick and slightly better buttons. I still prefer the feel of a Canon or Nikon DSLR, however.

Compared to other Sony cameras, the a9 supports two memory cards for redundancy. However, only one card supports the faster UHS-2 standard, so if you want to avoid buffering, you'll want to configure RAW files to write to card 1, and a JPG backup to write to card 2.

a9 II ($4,500 new)

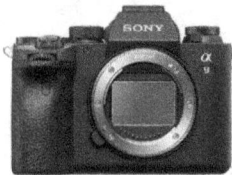

The original a9 was extremely well-received, but we and other reviews quickly found some serious flaws. The a9 II sought to fix those problems by introducing these improvements:

- Better weather sealing

- Bigger grip and buttons

- 10 FPS mechanical shutter (up from 5), which is required when using a flash or to prevent banding when shooting with artificial lights at fast shutter speeds

- Better control over anti-flickering with the electronic shutter

- Dual UHS-II card slots to reduce buffering when writing raw to both cards

- Improved autofocus (though we didn't detect a difference in our testing)

- Gigabit Ethernet port (up from 100 Mbps)

- Support for secure FTP transfers

- 2.4Ghz and 5Ghz wireless (the original a9 supported only 2.4Ghz)

- Better eye-detect autofocus when shooting video

- Voice memos (required for some professional workflows)

These improvements will definitely be worth the increased cost for those who need to use the mechanical shutter, including basically all sports photographers who need to use the mechanical shutter while shooting sports in artificial light with a high shutter speed. However, in general usage, it was impossible for us to notice a difference compared to the original a9.

a1 ($6,500 new)

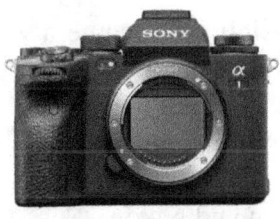

The Sony a1 is Sony's most powerful and most expensive camera. It shoots at up to 30 frames per second (though we frequently counted 15-20 FPS), 8k video, and offers 50 megapixels. 50 megapixels has become common, and the a1's image quality isn't significantly better than the $2,300 Sony a7R III. To be clear: your $6,500 does not get you better image quality.

While the a7R III isn't suitable for sports and wildlife, while the a1 excels at it, and that's what you're paying for. The ability to shoot 30 FPS silently with no viewfinder blackout is pretty remarkable.

Notice, however, there should be a huge asterisk (*) next to the FPS specification, because it only achieves 30 FPS under ideal circumstances:

- Using a lens on *Sony's list of supported lenses*.

- Shooting JPG or lossy compressed raw – not HEIC, and not lossless compressed raw. That's a big drawback, because those new formats would help reduce storage space and transfer time. Choosing the lighter file formats drops the FPS to 20.

- Using a shutter speed of 1/125 or shorter. When photographing motorsports at 1/30, the a1 achieved only 6-9 FPS (less than that of the slower Canon R5).

- Not putting too much pressure on the autofocus system – tracking an athlete up-close, for example, will cause it to drop to 15-20 FPS.

Still, when 30 FPS works, it's amazing. You can truly choose the perfect wing position of a bird in flight, or that split-second moment an athlete's fingertips touch the ball.

There's more to it than simply getting the FPS right. The electronic viewfinder (EVF) has no noticeable lag and we had no problem tracking erratically-moving subjects like fast moving birds through an 840mm lens.

Banding under artificial lights is significantly reduced on this camera, too. While banding can still occur, we found it to be no worse than fast mechanical shutters.

And the shutter is VERY fast on this camera. This gives you the ability to sync your flash at 1/400 – with either the mechanical or electronic shutter. Shorter sync speeds mean sharper pictures, less need for HSS, and more power to overcome ambient light.

Chapter 17: Panasonic L-mount

The L-mount is an alliance between Panasonic, Leica, Sigma, and possibly other camera and lens manufacturers. Full-frame mirrorless cameras and lenses made for the L-mount should work interchangeably, though in our experience, you'll get the best results by using Panasonic lenses on Panasonic bodies, Leica lenses on Leica bodies, and Sigma lenses on Sigma bodies.

The L-mount launched in February of 2019, several months after Canon and Nikon launched their full-frame mirrorless camera mounts and many years after the Sony E-mount. While creating an alliance allowed L-mount cameras to use Leica's existing lenses, those lenses were limited in number and extremely overpriced. Sigma was able to introduce several L-mount lenses, but most of their work has simply been adapting DSLR lenses to the mount, so the lenses do not benefit from the shorter flange distance mirrorless cameras make possible. In other words, the existing lenses are bulky, expensive, and limited in variety.

Competition from the big three camera manufacturers has been fierce: the L-mount has gathered very little market share. While Panasonic, Sigma, and Leica have each found narrow niches and loyal followers, the actual interest in the cameras is very low. As the market for interchangeable cameras continues to shrink, I have concerns about the long-term future of the L-mount. Thus, buyers considering their first L-mount camera should seriously consider looking instead at the Canon RF and Sony E mounts, which have much larger market shares and likely better long-term outlooks.

This chapter will focus primarily on the Panasonic full-frame mirrorless lineup because the Leica and Sigma options haven't developed a significant following.

S5 ($2,000)

The Panasonic S5 is really the S1 Mark II – a more compact and capable successor to their first-generation camera. The only real disadvantage it has when compared to the S1 is the lack of the top screen, which is somewhat helpful, even though Panasonic used an extremely outdated LCD display.

The S5 seems to have been designed to compete with the popular Sony a7 III, offering a stabilized full-frame sensor, 24 megapixels and two cards slots. It also offers a flip-forward screen. In our direct comparison to the a7 III, we found the S5 to be equal or superior to the Sony. The eye-detect AF was excellent in both AF-S and AF-C. As a video camera, the S5 is capable if used with static subjects. The full-width 4k/30 looks great, and you can also record 4k/60 with a 1.5X crop, which is useful for slow-motion B-roll. The video autofocus was still frustrating. With still subjects, Panasonic's Depth-from-Defocus system pulses slightly (albeit less than previous generations). With slowly moving subjects, it failed to keep up with them, leaving shots out-of-focus. Regardless, if you're video style uses only manual focus, the S5 is an excellent hybrid camera.

S1 ($2,500)

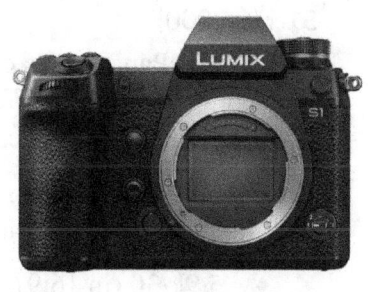

The S1 is Panasonic's camera for professional wedding and portrait photographers, as well as for general-purpose stills and video. It offers a stabilized 24 megapixel sensor, a tilting (but not flip-forward) screen, and a bulky, heavy, rugged body. Unfortunately, it feels like Panasonic rushed the S1 and S1R to market early to compete against Canon and Nikon's new full-frame mirrorless cameras. At the current price, the S1 offers generally inferior performance to the Sony a7 III, Nikon Z6, and Canon R6, especially in terms of autofocus. The S1 has a very primitive contrast-based focusing system which is suitable only for still subjects in AF-S. Therefore, it's fine for still portraits, but photographers shooting a bride walking up the aisle will be frustrated with it. Even still portraits of animals or young kids (who tend to move quickly) will be difficult if they have shallow depth-of-field.

Still, if you're a Panasonic loyalist or if you feel that most mirrorless cameras are too small, you might appreciate the S1 as a sturdy, weather-sealed, bulky mirrorless camera. It's certainly capable of producing good results, but most photographers should wait for the next generation of Panasonic cameras.

S1R ($3,700)

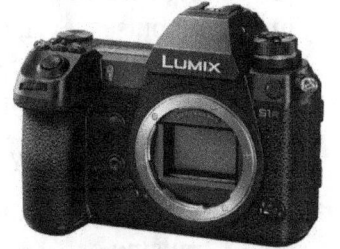

The S1R is Panasonic's camera for stills photographers who require 45 megapixels of high resolution. While Panasonic has a loyal following, the S1R has failed to capture significant market share because of its comparatively high price, limited native lens selection, and frustrating contrast-based autofocus system.

At the time of its release, it was aimed at the popular Sony a7R III. The primary benefits of the S1R compared to that camera are better weather sealing, a bigger body, a top LCD screen, and a more versatile tilting screen. However, the Sony a7R III is now priced 30% less than the S1R and it has a significantly bigger selection of native lenses and vastly superior autofocus.

The current-generation Sony a7R IV is still slightly less expensive than the S1R, but offers 60 megapixels, a special 240-megapixel mode, and vastly superior autofocus.

The 45 megapixel Canon EOS R5 also competes in this segment, adding a flip-forward screen, truly amazing autofocus, limited 8k/30 and 4k/120 video, and many unique professional lenses.

As a result, it's impossible for us to recommend the S1R over its competition. If you're loyal to Panasonic, we suggest waiting for a successor.

S1H ($4,000)

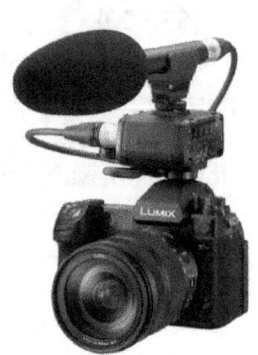

The S1H is Panasonic's camera for serious videographers. It has a 24 megapixel sensor and offers a wide variety of tools that professionals need (but amateurs and enthusiasts probably won't use):

- 6k/24 (but in a strange 3:2 format that you would always need to crop the top and bottom from)

- 5.9k/30 (in 16:9), allowing cropping while still publishing in native 4k

- Manual control over dual native ISO

- 12-bit raw video to an external recorder

- Vectors, waveforms, timecodes, and other tools professional videographers require

- Certified for recording Netflix content

- A screen that both tilts and flips forward

The S1H is a camera for use by professional videographers, not by vloggers. While it does have a flip screen, the contrast-based autofocus (known as DfD or Depth-from-Defocus) hunts and pulses, even on still subjects. Therefore, most will not be happy with the continuous autofocus capabilities.

The S1H's closest competitor is the $3,500 Sony a7S III, which has several advantages:

- A wider variety of native E-mount lenses

- Phase-detect autofocus for perfectly tracking moving subjects without hunting or pulsing

- Full-width 4k/60 (the S1H requires a crop)

- 4k/120 for better slow-motion

- 16-bit raw video to an external recorder

- 1.5 lbs, far lighter than the S1H's 2.3 lbs

- Digital microphone connection through the flash hotshoe

If you're a serious videographer using manual focus and publishing in 4k/24 or 4k/30, both cameras will produce similar results. If you're shooting run-and-gun, sports, interviews, or filming yourself, the a7S III is the superior choice thanks to its better autofocus.

Chapter 18: Fujifilm X-mount

Fuji makes remarkable, unusual, and quirky mirrorless cameras. In some ways, they're the most high-tech mirrorless cameras available. For example, the X-T1 is better at tracking quickly moving subjects than any other mirrorless camera we've ever tested, and it has the best viewfinder ever made.

In other ways, they're the lowest-tech mirrorless cameras. They certainly look retro, some of them have traditional optical viewfinders, and they lack features like HDR or decent bracketing.

Like Leicas, Fuji cameras aren't right for most people, but they're perfect for a select few. You'll first need to determine whether you might be a Fuji person. Fuji people:

- Are tech-savvy, but love retro design

- Love buttons, dials, complexity, and tweaks

- Happily spend hours reading a manual and weeks mastering a camera's quirks

Fujifilm has one of the better selections of lenses, but it still pales in comparison to the lenses available for Micro Four-Thirds or most DSLR systems. Of the available lenses, many of them (especially the first models released in 2012) have serious quirks, like unreliable autofocus. Nonetheless, Fuji has an excellent assortment of high-quality prime lenses that I recommend over the zooms: 14mm, 18mm, 23mm, 27mm, 35mm, 56mm, and 60mm. You also have five solid zoom lenses covering the usual focal ranges.

Fujifilm X-mount lenses have an aperture ring on the lens itself, which was common on cameras 30 years ago, but is quite unique in the world of modern cameras. You can also control the aperture from the camera, or use the lenses in fully automatic mode. Most lenses have the aperture numbers written on the lens itself, allowing you to view or select your aperture without looking at the LCD. However, lenses with variable apertures (including the kit lens) do not have the aperture numbers written on the lens, requiring you to look at the display to view or adjust the aperture.

All the Fujifilm X-mount cameras have similar, excellent 16-megapixel image quality. Rather than a traditional Bayer red/blue/green filter, the Fuji cameras (except the X-A1) use a slightly different pixel arrangement that, in my opinion, produces slightly better-looking noise. None have a touch screen.

All Fujifilm cameras share some of the same quirks. For example, selecting a specific focusing point is slower and clumsier than on other cameras. You'll also need to change focusing modes more frequently than with other cameras. You can train yourself to do this fairly quickly, but the (relatively) clumsy interface and need to change modes will cause you to lose some candid shots during the first few weeks of use (and maybe even after that). For example:

- You can't directly change the focus point with the directional buttons, but must first push a button. Sony A-mount cameras work the same way.

- With most lenses, including the otherwise excellent kit lens, you can't manually focus when the camera is in autofocus mode.

- Regardless of the mode, you can't manually focus if you have the shutter half-depressed.

- If you enable face detection, the camera always uses the center autofocus point when it doesn't detect a face, and you can't change it.

Another quirk is that they only offer bracketing at +-1 stop. Most cameras offer at least +-3 stops, and many offer +-5 or +-7 stops. +-1 stop bracketing is almost useless, and as a result, you'll need to manually adjust the exposure compensation if you need to bracket more stops. That makes bracketing a clumsy process, and HDR is difficult to do well without a tripod.

If you read this description and thought, "I'd rather concentrate on the composition and lighting than choosing camera modes," then you're not a Fuji person, and I'd steer you towards Micro Four-Thirds or Sony instead. If you thought, "That sounds like fun; I love a challenge," then you're a Fuji person.

Fujifilm Lenses

Fujifilm's lens selection can't compete with Canon and Nikon, and the lenses are priced higher than the mainstream equivalents. Nonetheless, they have the lenses that most photographers will need.

While I recommend zooms for most casual and professional photographers, in the Fujifilm world, I prefer using their primes. The reason for uniquely Fuji: the prime lenses (and fixed aperture zooms) have an aperture ring on the lens with the aperture settings marked. Therefore, you can turn the aperture dial on a prime lens to select automatic aperture, f/1.4, f/2.8, or whatever aperture you need.

The variable aperture zooms have an aperture dial, too, but they don't have the f/stops marked on them because they would be different depending on where you were zoomed to. Therefore, to select an aperture, you need to turn the dial while looking at the LCD screen.

That works, but the extra step takes away from the mechanical mood that I love about the Fujifilm bodies.

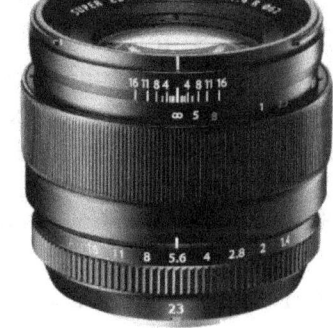

My favorite walking-around lens is the 23mm f/1.4 ($800, pictured), roughly equivalent to a full-frame 35mm f/2. It's a wide-angle lens, which isn't as traditional as a normal lens, but it's easier to crop when you can't get close enough to your subject. If you prefer a normal lens, the 35mm f/1.4 ($500) is roughly equivalent to a 50mm f/2.

More casual photographers should opt for one of Fuji's image-stabilized OIS zoom lenses. Fuji actually offers three normal zooms:

- **16-50mm f/3.5-f/5.6 ($230)**. Equivalent to a full-frame 24-76mm f/5.3-f/8.4, the wide-angle end of this lens makes it the best choice for casual and event photography. It's not the sharpest lens in the lineup, and focusing can be slow.

- **18-55 f/2.8-f/4 ($600)**. Equivalent to a full-frame 27-84mm f/4.2-f/6, this is Fuji's fastest zoom. It's the best choice for low light work, and it's substantially sharper than the 16-50mm.

- **18-135mm f/3.5-f/5.6 ($900)**. Equivalent to a full-frame 27-206mm f/5.3-f/8.4, this super-zoom is the right choice when you don't ever want to change lenses. Image quality is slightly compromised in favor of versatility.

Fuji offers a solid variety of portrait lenses:

- **60mm f/2.4 ($550)**. Equivalent to a full-frame 90mm f/3.6, this isn't the most powerful portrait lens, and the focusing speed is awful. If you don't mind manually focusing, it's your best bet for getting a headshot under $1,000.

- **56mm f/1.2 ($1,000)**. Equivalent to the popular 85mm f/1.8 full-frame lens, this sharp lens offers great background blur and pleasing features for portraits. Unfortunately, it costs twice as much as an actual full-frame 85mm f/1.8.

- **56mm f/1.2 APD ($1,500)**. An upgraded version of the previous lens, this version offers nicer bokeh for portraits. Basically, it removes the rings that you see around specular highlights in the background. It also removes compatibility with phase detect autofocus systems, meaning you'll have to use the much slower contrast detection autofocus, slowing down your portrait sessions and reducing the number of keeper shots. It's hard to justify for most buyers at this price, but wealthy Fuji fans will adore it.

- **50-140mm f/2.8 OIS ($1,600)**.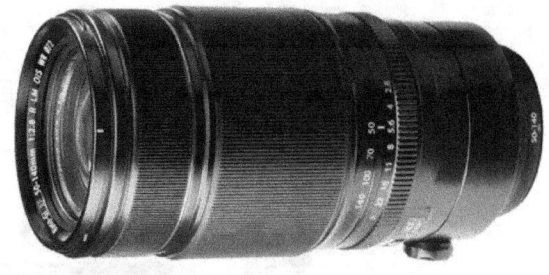
Pictured next, this is Fuji's equivalent of the full-frame 70-200 f/4 lens, with image stabilization. While prime lenses are fun for portraiture, a zoom is almost a requirement for professional portrait work because the zoom allows you to quickly switch between headshots and wider angles, capturing expressions and changing compositions without moving yourself or your subject. As with most Fuji lenses, this lens is expensive, costing twice as much as the full-frame equivalent from Canon and priced higher than the excellent Tamron 70-200 f/2.8, which offers better background blur. If you're invested in the Fuji world, this is your best option for a working portrait lens.

Fujifilm Sensors

Fuji's X-Trans sensors are one of the camera lineup's most exciting features. The X-trans sensor uses a more random-seeming red, blue, and green color filter that can improve the appearance of noise at high ISOs.

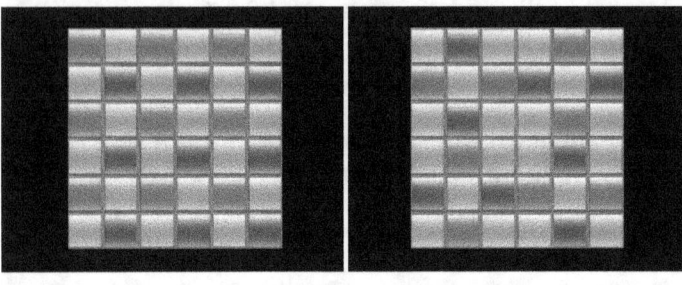

Honestly, most people won't notice the difference between the two sensors. Almost every digital camera in the world uses a Bayer sensor, but pixel peepers will appreciate the better-looking shadows when compared to pictures taken with other APS-C cameras. This figure shows a traditional Bayer filter on the left and the X-trans filter on the right.

Fujifilm Bodies

The following table, and the sections that follow, provide an overview of the current Fujifilm camera lineup.

Model	Price (new)	Price (used)	Focusing	Tilt screen	Touch	Wi-Fi	Pop-up flash	EVF	FPS	Video
X-A1	$300	$200	Good	*		*	*		5.6	1080/30p
X-M1		$450	Good	*					5.6	1080/30p
X-E1		$250	Good				*	*	6	1080/30p
X-E2	$700	$450	Better			*	*	*	7	1080/60p
X-Pro1		$500	Good					*	6	1080/30p
X-Pro2	$1,700	$1,350	Great					*	8	1080/60p
X-T10	$800	$400	Great	*		*		*	8	1080/60p
X-T20	$900	$800	Great	*	*	*		*	5	4K/30p
X-T1	$800	$550	Great	*		*	External included	*	8	1080/60p
X-T2	$1,600	$1,300	Best	*				*	11	4k/30p

X-A1 ($300 new, $200 used)

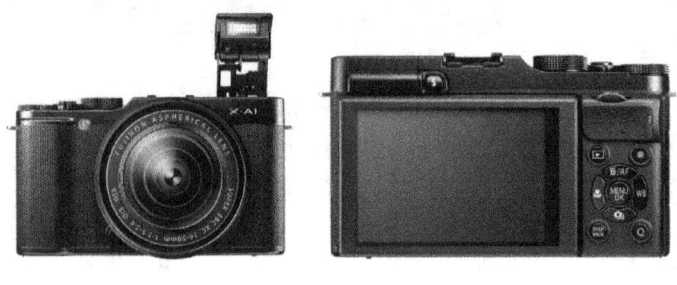

Fujifilm's entry-level camera provides about the same image quality as all the other bodies in the X-mount lineup, in a small, sturdy, and good-looking package. It's so good, in fact, that I recommend it over the X-M1 to most photographers.

You might consider upgrading to the X-M1 for its X-trans sensor, which can improve the appearance of noise at high ISOs. The X-M1 feels a bit better in the hands, too. The differences are fairly minor, however.

X-M1 ($450 used)

The X-M1 offers a better look-and-

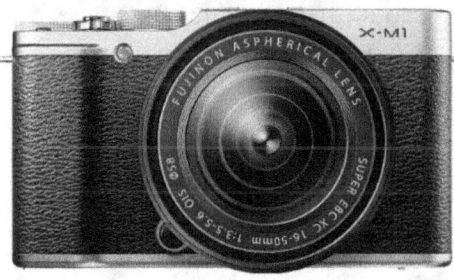

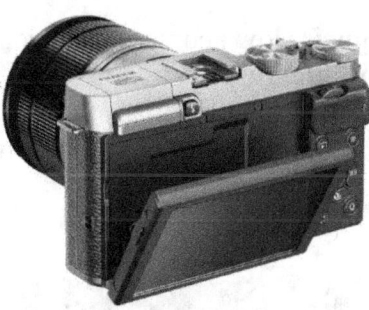

feel than the X-A1 and also replaces the X-A1's standard Bayer sensor with Fujifilm's proprietary X-trans sensor. Other than the difference in the filters, the X-M1 feels better than the X-A1. If you want a viewfinder and don't mind giving up the articulating display, upgrade to the X-E2.

X-E2 ($700 new, $500 used)

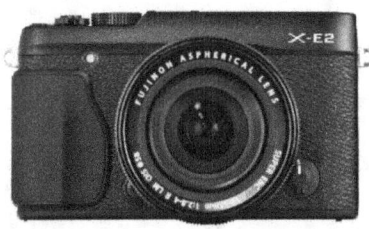

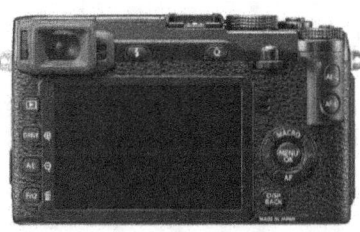

Compared to the X-M1, the X-E2 adds an electronic viewfinder and several buttons, simplifying manual control. The X-E2 also provides 1080p video at 60 fps, instead of just 30 fps. However, it loses the X-A1's articulating display, making the X-A1 a better choice for the casual photographer.

The EVF on the X-E2 is excellent, but it will be a bit difficult to see in bright sunlight.

X-Pro1 ($500 used)

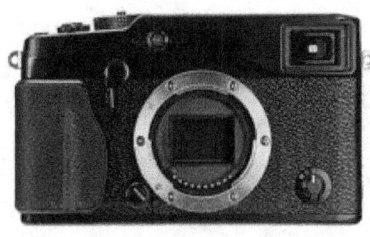

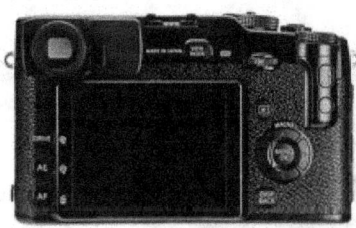

The X-Pro1 is Fujifilm's top-of-the-line viewfinder-style camera. In both look and feel, the X-Pro1 closely resembles one of my favorite medium-format film cameras of all time: the GW690 (and its sequels).

The X-E2 has a unique viewfinder that can switch between a traditional optical viewfinder and an electronic viewfinder. The optical viewfinder behaves like older viewfinder film cameras; if you zoom in or choose a telephoto lens, the viewfinder doesn't change. Instead, the camera shows crop lines in the viewfinder marking the edges of your picture. If you miss your old viewfinder film camera, you'll love it. Otherwise, I'll steer you to the X-E2. Most people will prefer the electronic viewfinder. As an EVF, however, it's rather slow and grainy compared to newer Fujifilm cameras, including the X-E2 and X-T1. That won't matter much for the types of photography this camera is best at: travel and street photography.

Likewise, the autofocusing is slow and unreliable compared to that of some newer mirrorless cameras at this price, especially the X-T1. If you buy a used X-Pro1, be sure to install the latest firmware update; it greatly improved focusing speed with several important lenses, which makes the camera much more usable.

X-Pro2 ($1,700 new, $1,350 used)

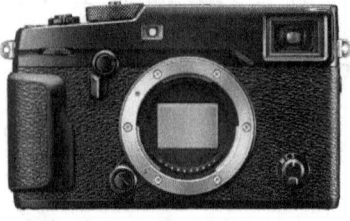 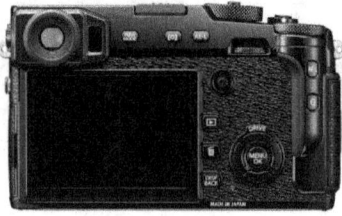

Like the X-Pro1, the X-Pro2 has a novel viewfinder that combines the best of an optical and electronic viewfinders. Compared to the X-Pro1, it offers these improvements:

- 24 megapixels (up from 16)
- Vastly improved focusing
- Wi-Fi
- 1080/60p video (for smoother video)
- An external mic jack
- A higher resolution display

X-Pro3 ($1,800-$2,000)

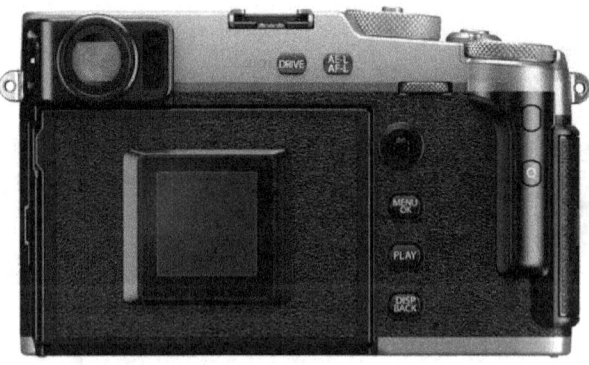

Like it's predecessors, the X-Pro3 is a camera that's more about the experience of taking pictures than the results it produces. I personally own and use an X-Pro3 as my "fun" camera.

I exclusively use the optical viewfinder because it gives me an analog feel. I love that eye-detect autofocus works so well even with the optical viewfinder, however, I wish that manual focus didn't require using a digital display. Unfortunately, the design of the X-mount system doesn't allow for a solution to that problem, so I prefer to use only autofocus lenses with the X-Pro3.

The biggest upgrade over the X-Pro2 is a very controversial feature: the rear screen is hidden by default. Instead of a conventional screen, there's a small screen that displays your camera settings and resembles the top of a film box. It's strange.

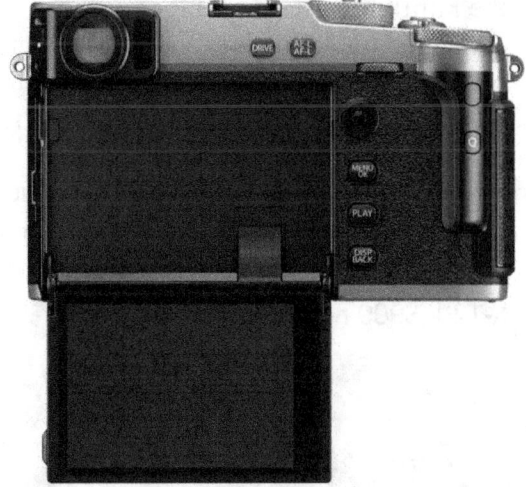

You can flip the screen down to access the menus and shoot with the camera held at arm's length, so it's not completely removing the digital experience like my Leica M10-D.

My favorite thing about the X-Pro3 is that it gives me the option of an analog experience. My least favorite thing is that it doesn't fully commit to that analog experience. I'm seeking a screen-free photography experience, and the X-Pro3 almost has that, but the presence of the digital screens still tempts me to review my pictures, to adjust settings, or to film a video.

With that said, it is still the most analog digital camera you can buy for under $2,000. The only superior option are the digital Leica's without a rear screen, such as the M10-D, which will cost you about $6,000-$7,000 on the used market. For the extra $4,000, you pay for the privilege of truly analog manual focus, no screens at all, and no video recording. While the Leica is three times more expensive, it isn't likely to lose much value at all, and you'll likely be able to get most of your money back even if you use it for a full decade. At least, that's my hope.

X-T10 ($800 new, $400 used)

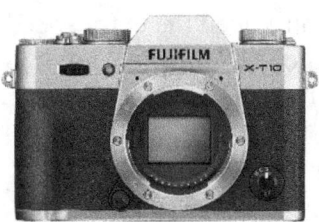 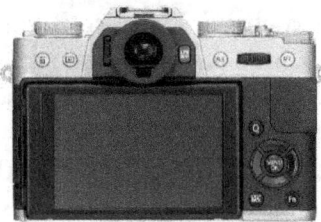

For those longing for the look and feel of a film DSLR, but who don't want the size or price tag of the bigger X-T1, the X-T10 is a great choice. We love the look of the shutter speed and exposure compensation dials, though changing them is a very deliberate and slow process.

Because the dials are on top of the camera and have the settings physically written on them, you need to look at the top of the camera to make adjustments. However, with the kit lenses, you need to look at the back of the camera to adjust the aperture. That makes the process of changing settings a bit awkward. To overcome this, use fixed-aperture lenses, such as the Fujinon 16-55mm f2.8 ($1,200) or the 35mm f1.4 ($400). These lenses have the aperture marking on the lens, so you don't have to look at the display to know which f/stop you're choosing.

To be clear, none of the Fujifilm cameras offer the best bang-for-the-buck when compared to DSLRs—mostly because the Fujifilm lenses are a bit overpriced compared to alternatives from Canon and

Video: X-T10 vs E-M5 II vs a6000
7:33 - *sdp.io/xt10review*

Nikon. Canon and Nikon don't give you the same experience, however. While Canon and Nikon might be the most efficient way to get results, the X-T10 is the most fun way to get results.

Consider upgrading to the X-T20 for the touchscreen, improved focusing, and better quality video and stills. The X-T20 is definitely a worthwhile upgrade.

X-T20 ($900 new, $800 used)

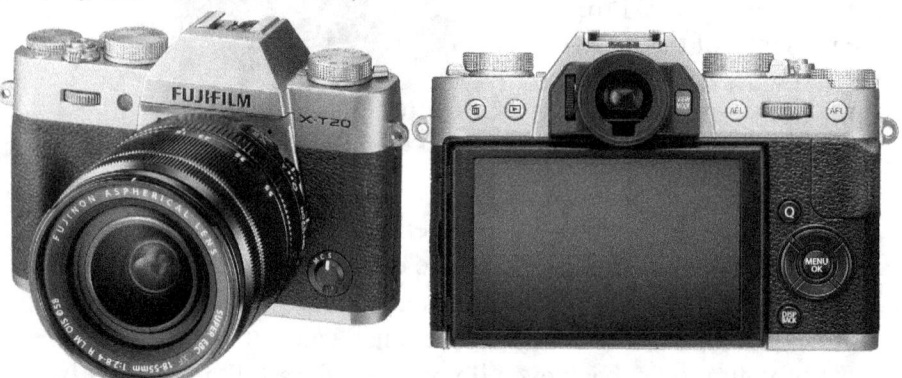

The X-T20 is the best value in mirrorless cameras. It's stylish and cool, especially in silver, so you never feel embarrassed. The analog controls feel great and make it fun to use (though you'll want to upgrade to the 16-55 f/2.8 or prime lenses to get totally analog aperture controls).

Image quality is the same as the more expensive X-T2, so if you don't need the extra speed or the bigger grip, the X-T20 is perfect. Stills and 4k video are gorgeous. Unfortunately, the screen doesn't flip forward, and it lacks a headphone jack, so if you're serious about video you'll be happier with a Panasonic G7 at this price point.

Though the specs don't match that of the Sony a6300 and a6500, I find the X-T20 to be much nicer to use. It produces better images overall, too, because Fujifilm's APS-C lenses are generally sharper and faster than Sony's limited APS-C lens lineup.

X-T30 ($1,300 new)

Compared to the X-T20, the X-T30 adds the newest 26 megapixel sensor, though we found virtually no difference in image quality. The newer generation also improved autofocus and battery life. The sports finder view is worthwhile for those shooting any sort of action sports; it works well and reduces EVF lag.

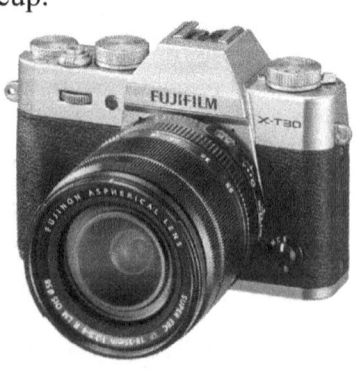

X-T1 ($800 new, $550 used)

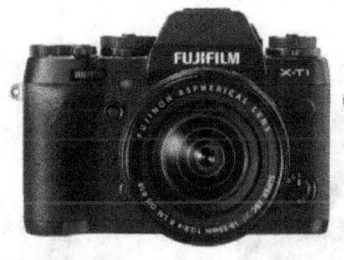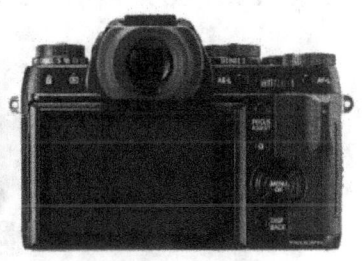

The X-T1 is Fuji's top-end camera, and the improved technology is a relatively small part of what separates it from the rest of the lineup. More significantly, it's bigger, heavier, weatherproofed, and is designed more like an SLR than a viewfinder. That means the viewfinder is in the center of the camera, aligned with the lens, rather than the upper-left corner. Most photographers who use their right eye find the rangefinder design of the other cameras to be more comfortable, but SLR users will be comfortable with the X-T1.

Since our initial review of the X-T1, Fujifilm did something remarkable—they released a series of firmware updates that fixed just about every single concern we raised. In that review, you'll hear us say that the camera seems like a

> Video: E-M1 vs. X-T1
> **41:03 -** *sdp.io/em1vxt1*

beta product. Indeed, it was (unofficially), and the firmware is now at version 4. While we consider it a mistake to release the camera before the software was polished, we applaud Fujifilm for fixing problems and adding new features like no other camera manufacturer ever has. As a result, I recommend you ignore the criticisms in our current X-T1 review. We hope to film an updated review after version 4 of the X-T1's firmware is officially released.

As with the X-T10, the X-T1 isn't the most efficient or cost-effective way to take a picture. Choose the X-T1 because you enjoy the process of manual photography, but you don't want to deal with film. When paired with a fixed-aperture lens, you can select your shutter speed, ISO, and f/stop without looking at a digital display, a process we really enjoy. When using the X-T1, we embrace that manual photography philosophy, and prefer using fast prime lenses, such as the Fujinon 23mm f/1.4, 35mm f/1.4, and 56mm f/1.2 (for portraits).

The X-T1 makes some claims that might make you think it's an excellent sports camera, however, our tests showed the autofocus system didn't compare to similarly priced DSLRs, such as the Canon 70D, 7D Mark II, and Nikon D7200. If you want to shoot sports and prefer mirrorless cameras, consider the X-T2 which is significantly better.

You might also want to upgrade to the X-T2 for dual memory cards and overall faster shooting.

If being cool is important to you, you might check out the limited edition graphite X-T1, shown next. It's about $200 more expensive than the base X-T1 (despite being functionally identical), and used copies are hard to find, but we find it to be a more handsome camera.

X-T2 ($1,600 new, $1,300 used)

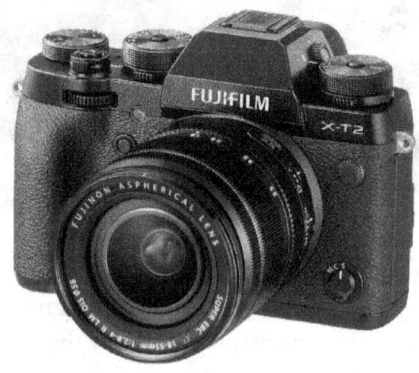 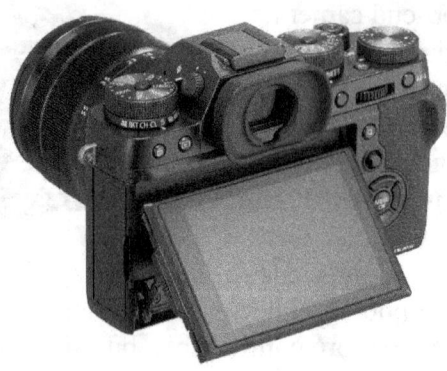

The X-T1 and X-T2 taught me an important lesson: Fujifilm listens. Our initial review of the X-T1 was (deservedly) fairly scathing: the camera was downright buggy and there were virtually no lenses available for it. Fuji did something no other camera manufacturer had ever done before: they fixed every single bug with firmware updates and added dozens of new features to address every single one of our concerns that could be fixed with software. For those concerns that required hardware changes, Fujifilm made the X-T2.

It's my favorite camera of all time to actually use. The analog dials make photography fun. It feels great and looks great. The lenses are vastly better than Canon, Nikon, and Sony APS-C lenses, giving you full-frame sharpness in an APS-C package.

While we love the X-T2, we still have some pretty important complaints. It lacks a touchscreen, which severely hampers the usefulness its beautiful 4k video. It also lacks sensor stabilization, which Sony, Pentax, Panasonic, and Olympus all offer. Fujifilm offers several great stabilized lenses, but my favorite lens is the 16-55 f/2.8 and the primes, which are incredibly sharp but not stabilized. In the real world, this resulted in a large number of shots lost to camera shake, even when following the reciprocal rule.

The X-T2 is fun to use, but discovering large numbers of shots with camera shake is not at all fun. Fujifilm listens, however, so I'm confident the X-T3 will have sensor stabilization and a touchscreen.

Focusing is excellent; noticeably better than the X-T1. It's not my favorite camera for sports because focus tracking isn't as perfect as it is with a similarly-priced Nikon D500. Nonetheless, an X-T2 and 50-140mm lens are one of the better combinations for shooting indoor and outdoor sports.

The same combination is great for shooting portraits. However, the eye detection focusing isn't 100% reliable, and when it misses focus, it can be frustrating. Ultimately I reverted back to using a single autofocus point and manually placing it on the eye.

X-T3 ($1,000)

The X-T3 was a significant upgrade to the X-T2, offering improved battery life, better autofocus and powerful 4k/60 video capabilities. Fortunately, it kept everything that I loved about the X-T2.

I immediately purchased the X-T3 and began using it as my primary "fun" camera, for times when I didn't need full-frame results. However, I stopped using it for two reasons:

- The battery life was still frustrating, and I don't like carrying extra batteries. If you are the responsible type who packs extras and remembers to charge everything, this isn't a problem for you.

- The lack of stabilization on the sensor and my favorite lenses meant that a high percentage of my photos had camera shake. Yes, I'm familiar with the reciprocal rule, but with sharp lenses, you'll see some amount of shakiness even when shooting at the equivalent focal length.

While I still always loved shooting with the X-T3, the X-T4 was a welcome and practical improvement. If you can stretch your budget to pick up the new model, you'll be glad you did.

X-T4 ($1,700 new)

The X-T4 is Fujifilm's flagship camera of 2020, and it's proof that Fujifilm listens carefully to our complaints. It addressed almost every concern we had:

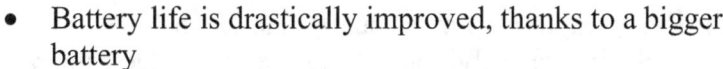

- Battery life is drastically improved, thanks to a bigger battery

- Sensor stabilization reduces camera shake with unstabilized lenses

- It adds a flip-screen for users who want to film themselves

However, the X-T series hasn't kept up with rapid improvements to mirrorless autofocus systems by Canon and Sony. As a result, we were frustrated by the inaccuracy of eye-detect autofocus for portraits. A full-frame Canon RP focused better and produced cleaner results at close to half the price.

The sensor stabilization is good, but it only offers about 2 stops of improvement. That's dramatically less effective than the 5 stops of improvement offered by Sony cameras and the 8+ stops of improvement offered by the latest Canon cameras.

Introducing sensor stabilization also introduced overheating when recording 4k/60 video. In full sun, our camera overheated after about 13 minutes of recording, though other users found it overheated either faster or slower. Your results may vary, but the overheating seriously limits its use as a hybrid or vlogging camera.

Additionally, Fujifilm's lens selection is showing its age. Important lenses, like the 50-140 f/2.8 and 56mm f/1.2, now have outdated autofocus systems that can't show the potential of the X-T4. Fujifilm does offer some newer lenses with better autofocus, but if those are the focal lengths and apertures you want, you might be disappointed with the results.

Still, the X-T4 remains my top recommendation for enthusiasts who want an fun, sexy camera with an analog feel. However, in 2020, it's falling behind the competition.

X-H1 ($800 used)

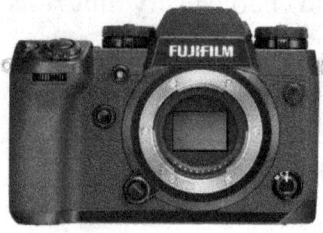 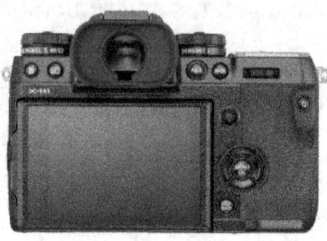

The Fujifilm X-H1 is Fuji's top-end X-Mount camera, and it's excellent, but also flawed. At launch time, Fujifilm priced this camera at $1,900, which proved to be far too high; they quickly dropped the price.

The most noteworthy new feature is sensor stabilization, which is a first for Fuji cameras, but has long been standard in other mirrorless cameras and Pentax DSLRs. Other improvements include a deeper grip (which we find really helpful for handling larger lenses) and some video improvements.

Fujifilm is a company that listens to customer complaints and fixes them. It's clear that Fujifilm had been hearing three complaints about three things with the X-T2:

- "The grip is too small for handling big lenses."

- "Your best lenses aren't stabilized and I'm getting shaky shots."

- "I need higher bitrates if I'm going to do professional-grade video."

And the X-H1 addresses those complaints perfectly. Indeed, it's the best video camera in Fujifilm's lineup. And, technically, it's the best X-mount stills camera, too.

For me personally, however, I'm torn between the X-T2 and the X-H1. The X-T2 just feels better in my hands; it feels more solid. I love that the X-T2's controls are all analog; the X-H1 replaced the exposure compensation dial with a very nice digital top LCD.

But the X-H1 adds sensor stabilization, a feature that most modern cameras from Sony, Panasonic, Olympus, and Pentax have. It's much needed for serious photographers in the Fujifilm lineup, too, because the excellent 16-55 f/2.8 lens (the top quality walking-around lens) is very sharp but unstabilized. Sharp, unstabilized lenses leads to excessive camera shake, even when following the reciprocal rule. The X-H1's sensor stabilization solves that problem, allowing you to handhold pictures at slower speeds, and generally saving a high percentage of pictures.

So, my ideal Fujifilm camera wouldn't be an X-H1, but rather an X-T2 with sensor stabilization. That's not an option, so for the time being, I'm picking up the X-H1 and I'm still loving it.

With that said, the X-H1 inherits the biggest problem of the X-T2: an underpowered battery. Historically, mirrorless cameras were designed to be small and lightweight, so having a small battery made sense. However, the X-H1 is very large; bigger even than some DSLRs, and it still uses the same tiny battery. It simply isn't enough power; if we go out for a day of casual photography (for example, on vacation) I need two or even three batteries. That's OK, but that also means I need to charge two or three batteries every night. It's not impossible to live with, especially since the X-H1 charges via USB, but it's a nuisance. Sony's latest generation mirrorless cameras use a larger battery, and Fuji should have made the same choice.

You can partially solve the battery life problem by purchasing the vertical grip for the X-H1; it provides access to a total of three batteries. It's still a pain to keep all three batteries charged while traveling, however.

The vertical grip also increases the maximum frames per second of the camera to 11 FPS, adds a headphone jack for monitoring sound, and allows 4k shooting for 30 minutes, so it's a requirement for shooting action or video. Unfortunately, the grip is an extra $330 (plus $135 for the extra batteries) for functionality that's included for free in other cameras. The grip also makes the X-H1 excessively large and less comfortable to carry on a strap. If you have big hands, or you often shoot vertically, you might prefer using the vertical grip, however.

The X-H1 is an amazing camera, especially with the vertical grip and extra batteries. However, that brings the price to $2,365. That means it's more expensive than the $2,000 Sony a7 III, and the Sony has several advantages:

- A full frame sensor with more than 1 stop less noise for raw still images, or 2+ stops less noise for video

- Access to fast full-frame lenses without crop, providing the option for more background blur.

- A better selection of autofocus adapters for Canon lenses

- Autofocus points that go to the edges of the frame

- Eye autofocus that works with continuous AF

- A much larger buffer for uninterrupted continuous shooting

- A smaller size with better battery life

- Access to professional support options (at the time of this writing, Fujifilm only offers professional support to GFX 50S owners)

Nonetheless, there are thing I love about the Fuji that the Sony doesn't have. I love the Fujifilm lenses with the aperture ring. Fujifilm has a better menu system.

Note: As of 2020, Fujifilm seems to have unofficially discontinued the X-1H. It has a strange history, being launched at the same time as the similarly priced but far more capable Sony a7 III, then seeing a dramatic and rapid price drop, and then being replaced as the flagship camera by the overall more useful and fun X-T4.

Chapter 19: Fujifilm GFX-mount

Let's talk about the sensor size. GFX is frequently referred to as "medium format." For many photographers, that term conjures the mental image of elite commercial photographers using top-end Hasselblads to shoot covers for fashion magazines, or of famous landscape photographers of yore like Ansel Adams.

That's not what the GFX mount is. GFX has a sensor size 1.7X bigger than full-frame (which gives it a crop factor of 0.79X). My own medium format film camera, a Mamiya RB67, has a 6x7 "sensor" size about 4.3X bigger than full-frame. While GFX is called medium format because it's bigger than full frame, it's only slightly bigger. It's still far smaller than medium format film cameras, and even calling it medium format seems intended to confuse potential buyers.

When a photographer upgrades from APS-C to full-frame, they enjoy an increase in sensor area of 2.25X (for most cameras) or 2.56X (for Canon cameras). When a photographer upgrades from full-frame to GFX, the increase is significantly less – only 1.7X. So, there is a benefit, but it's not as significant as many believe it would be.

Photographers often think of the "medium format look" that was defined by brands like Hasselblad in the film era. At that time, medium format cameras were the only way to achieve shallow depth-of-field and minimal grain (noise). In the digital era, however, this relationship functions very differently.

As you might have read earlier in this book, you can use crop factor calculations to exactly understand the depth-of-field you'll get when attaching lenses to different sensor sizes. Those calculations quickly reveal the deficiencies of the current GFX lens system.

The GFX portrait prime is 110mm f/2 ($2,800), which produces full-frame results equivalent to a 87mm f/1.6. That means that the common 85mm f/1.4 full-frame lens (about $1,500) produces more background blur and cleaner low-light images. In this comparison, the full-frame equipment would produce better results, in a smaller package, for less money.

That's just one example, but if you examine the entire GFX lens lineup, you will not find a single example of a lens capable of producing unique results that cannot be accomplished with any of the full-frame mirrorless systems. The GFX 32-64mm f/4 ($2,300) is equivalent to a 25-50mm f/3.2, which doesn't even compare with the common 24-70 f/2.8 ($1,000-$2,000), much less the Canon RF 28-70 f/2.0 ($2,900).

The "medium format look" isn't entirely about depth-of-field or low-light capability, though that seems to be most of it. If either of those two elements are what you're seeking, you can get the best results using the Canon RF full-frame system and choosing fast lenses such as the 50mm f/1.2, 85mm f/1.2, and 28-70 f/2.0.

Another element of the "medium format look" is dynamic range, which is currently controlled almost entirely by sensor size (though computational photography will hopefully change this). This is where the GFX series shines: the GFX cameras have about a full stop more dynamic range than any full-frame camera. This only applies when shooting at the base ISO, so, when shooting in bright conditions or using a tripod. For landscape photography, I always bracket my photos so I have the option to combine them and extend the dynamic range. However, that only works well with still subjects such as landscapes. Thus, if you are shooting in a situation that requires the ultimate dynamic range with a moving subject, such as a portrait, a GFX camera would produce unique results.

Besides depth-of-field, noise, and dynamic range, a fourth frequently cited factor in the "medium format look" is "compression." Compression is 100% determined by the lens angle-of-view, which you can calculate using crop factor. The idea that medium format has magical compression is completely a myth; we (and many other reviewers) have done side-by-side blind tests showing APS-C, full-frame, and medium format photos taken with equivalent settings and the results are clear: there is nothing magical about the "medium format look."

Another factor to consider is support. While I personally have never used Fujifilm support for GFX cameras, I have had multiple photographers reach out to me with horror stories of recurring problems that could not be fixed and of waiting multiple months for their gear to be returned. With that said, I've heard support horror stories from every camera system, and support varies from one country to the next, but it's something to be aware of.

So, should you invest in the GFX system? Perhaps. The GFX50S and GFX50R offer resolutions that match the top-end Sony, Nikon, Canon, and Panasonic full-frame cameras, but they also offer better dynamic range. The GFX100 offers more resolution AND more dynamic range. The GFX50S and GFX100 also offer tilting viewfinders, which are incredibly useful for both landscapes and portraits and make it much easier to get lower perspectives. If those unique traits are what you're looking for, than the GFX system might be for you.

GFX50S ($5,500 new, $3,700 used)

In February 2017, Fujifilm launched their mirrorless medium format platform with the 50-megapixel GFX50S. In 2020, it feels a bit primitive, even at its used price. It's large and slow. The contrast-based

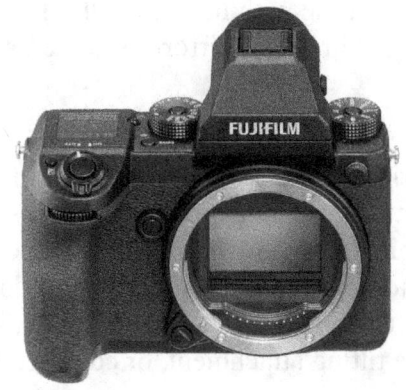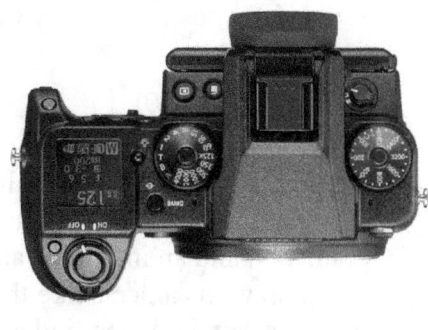

autofocus system is rather unreliable, even for still subjects, and essentially unusable for moving subjects.

The electronic viewfinder is removable, which is cool. You can use the EVF-TL1 adapter ($570) to let it tilt to the side or up and down, which is useful for many types of photography, and something every mirrorless camera should offer.

The ergonomics of the GFX50S are not great. The body is thick like a DSLR. The buttons, dials, and joystick could be a bit bigger. Still, we like the analog ISO and shutter controls and the top screen is attractive. Overall, we like holding it much better than the smaller GFX50R.

GFX50R ($3,500 new, $3,000 used)

The GFX50R is marketed as a compact camera for activities like street and travel photography, and that's how we tested it. Unfortunately, it really failed at those tasks.

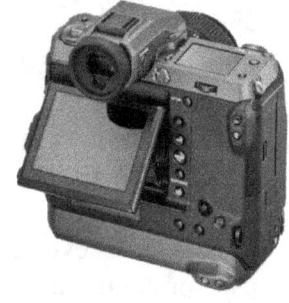

While smaller than the other GFX cameras, the GFX50R is still huge. Big enough that people in the street you don't know stop and say things like, "Woah, big camera." It's a little heavy to carry on a strap. It's the opposite of discreet and definitely made me feel like a target for theft. I found myself much happier using my Fujifilm X-T4 for street photography.

For other types of travel photography, it's OK! I never enjoyed using it though; the ergonomics are rather poor. The autofocus system, like the GFX50S, is slow and unreliable, even for still subjects. It has eye-detect autofocus, but when shooting portraits of my daughter, it missed every photo with shallow depth-of-field. Ultimately, I would have gotten better results with just about any other similarly priced camera, regardless of sensor size.

The 3.6 million-dot electronic viewfinder doesn't compare with similarly priced full-frame cameras, which offer 2 or 3 times the resolution, nor does it offer the tilting viewfinder of the other GFX cameras. However, I like the rangefinder placement of the viewfinder, which makes it more comfortable for me to hold against my face. If you use your left eye, it would be uncomfortable.

If you can't tell, I didn't love my time with the GFX50R. The lenses and autofocus just weren't as powerful as my Canon or Sony mirrorless cameras.

GFX100 ($10,000)

The GFX100 is the highest-priced camera in this guide, and for good reason. It has twice the resolution, 100 megapixels, of its closest competitors. It has sensor stabilization (unlike the other two GFX cameras) and an improved autofocus system (though it still doesn't keep up with cameras like the Canon R5). Like the GFX50S, you can attach a standard viewfinder, a tilting attachment, or completely remove the viewfinder.

It's huge and heavy. The vertical grip isn't removable. You can hand-hold it no problem at all, but everyone who sees it will think, "Woah, that's a huge camera." Many will feel the need to point it out to you, as if you didn't already know. If that sounds appealing, you'll love it. If you like portable, discreet cameras, this isn't the camera for you... though if you're shopping for a $10,000 camera, you probably have multiple cameras for different purposes, anyway.

Here's how camerasize.com compares the GFX100 to the 60-megapixel Sony a7R IV, probably its closest competitor.

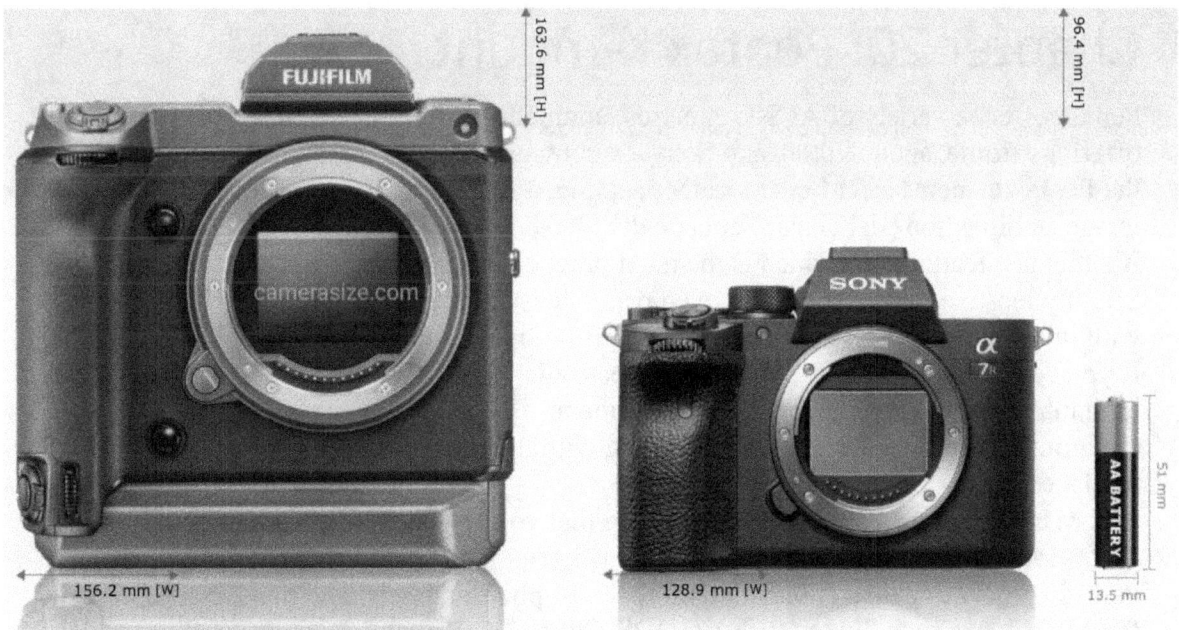

You can see from that image that the Sony is about half the size and weight. The cost and size do come with real, tangible benefits: unmatched resolution and dynamic range. The GFX100 has the best technical image quality of any 4-figure camera (it technically costs $9,999). And, though the ergonomics and autofocus aren't the best, it actually is very usable for such a serious camera.

Will you see those benefits in your images? In the lab, taking pictures of test charts, yes. The dynamic range is better. Pictures are sharper. In the real world, where the less-than-perfect autofocus system often misses focus, and atmospheric conditions and subject movement start to serious limit sharpness beyond about 40 megapixels, you won't see much difference.

But when you shoot with the GFX100, you know you're getting the world's best image quality. You'll never feel like your shot could have been a little sharper. You will probably never have gear envy. If you're an emotional buyer, that might be enough.

GFX100S ($6,000)

The GFX100S is a less-expensive and less modular version of the GFX100. It promises a much more portable experience with similar image quality, and should generally be a better choice over the older camera.

We've asked Fujifilm to borrow a camera to review and will update the buying guide when we've spent more time with it.

Chapter 20: Pentax K-mount

Pentax offers a variety of APS-C cameras, competing with the entry-level to mid-range offerings from Canon, Nikon, and Sony. In comparison to the more popular competitors, the Pentax cameras offer better weather sealing, a reputation for durability, and in-body image stabilization (IBIS) that Pentax calls Shake Reduction (SR).

It's this last feature, SR, which is the most practical benefit of the Pentax cameras. With SR, any lens is image stabilized—even prime lenses, such as the 50mm f/1.8. Fast primes with image stabilization provide unbeatable low-light handholding capabilities.

If you're interested in IBIS but want an electronic viewfinder and a wider variety of lenses, look into the Olympus mirrorless camera lineup. However, the smaller sensor of the Olympus Micro Four-thirds cameras, along with the higher Olympus base ISO (200), produces noisier images.

Before choosing the Pentax system, be sure that you're happy with the Pentax DA and DA* variety of lenses. Canon and Nikon each offer a much wider variety of lenses, but Pentax offers more than enough lenses for most photographers, and they're all optimized for APS-C. Additionally, my favorite APS-C lens, the Pentax 18-35mm f/1.8, is available for the Pentax format, and that might be the only lens you need.

While you can use the Pentax FA lenses, they were designed for film cameras, and as a result they won't be quite as fast or sharp as the more modern lenses.

Unfortunately, all Pentax cameras lack modern niceties that are becoming common in other cameras of this price range, including touch screens, Wi-Fi, NFC, and GPS. The autofocus capabilities don't match those of similarly priced cameras from Canon and Nikon, either.

In short, most photographers looking into an APS-C DSLR should seriously consider Canon or Nikon cameras instead. Choose Pentax only if:

- You plan to handhold the camera with fast prime lenses in low light (thus taking advantage of the IBIS).

- You really need the improved weather sealing.

- You want to take pictures rapidly of still subjects (thus taking advantage of the higher frame rates, but acknowledging the slower autofocus performance).

- You plan to be deeply immersed in countries where electricity might not be available, but from where you can purchase AA batteries (thus taking advantage of the fact that Pentax cameras can use standard batteries).

- You want a camera in wild color combinations.

The sections that follow provide an overview of the Pentax DSLR lineup.

Model	Price (new)	Price (used)	Megapixels	Max shutter	Frames per second
K-500 (kit)		$250	16	1/6000	6
K-50		$275	16	1/6000	6
K-70	$600	$525	24	1/6000	6
K-5II *http://amzn.to/10Zcnyo*		$300	16	1/8000	7
K-5IIs *http://amzn.to/TNY3n0*		$400	16	1/8000	7
K-S1		$230	20	1/6000	5.4
K-S2		$450	20	1/6000	5.4
K-3 *http://amzn.to/TNY6z8*		$400	24	1/8000	8.3
K-3 II	$800	$650	24	1/8000	8.5
KP	$900		24	1/24000	8.3
K-1	$1,700	$1,400	36	1/8000	4.4
K-1 II	$2,000	$1,400	36	1/8000	4.4

K-500 ($250 used)

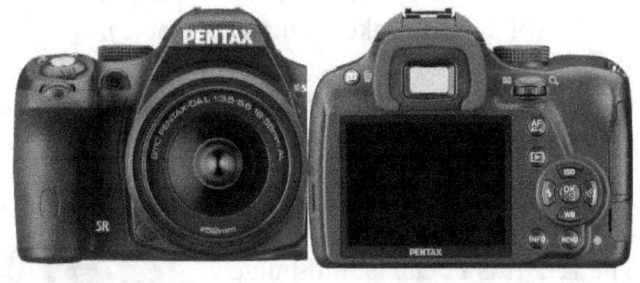

The K-500 is Pentax's entry-level body. Compared to similarly priced Canon and Nikon competitors, the faster shutter speed and higher frame rate should make it better for action. However, Pentax cameras have a bad reputation for tracking action in low light, so if you need a sports camera for indoor basketball, volleyball, or other sports, you might instead look for a used Canon 7D or Nikon D5200.

It's hard to recommend the K-500 over the K-50. Pentax prices the K-500 at $600, compared to $780 for the K-50. However, at the moment, the K-50 is actually less expensive at most online stores. Therefore, I recommend upgrading to the K-500 instead to take advantage of the improved weather proofing and electronic level.

K-50 ($275 used)

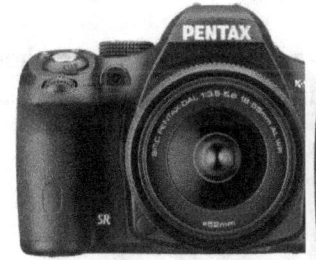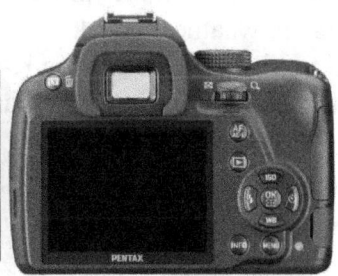

One step up from the base model, the K-50 is a great value for the price. Compared to the Canon T5 or Nikon D3200 (similar models at this price point), the K-50 offers weather sealing and in-body image stabilization. If you plan to use fast prime lenses, this makes the K-50 a better choice for handholding pictures in low light environments.

The K-50 is available in a variety of colors. In fact, you can custom order your K-50 in 120 different color combinations. Buy one to match every outfit!

K-70 ($600 new, $525 used)

At this price point, the K-70 blows away competing Canon and Nikon DSLRs in several ways:

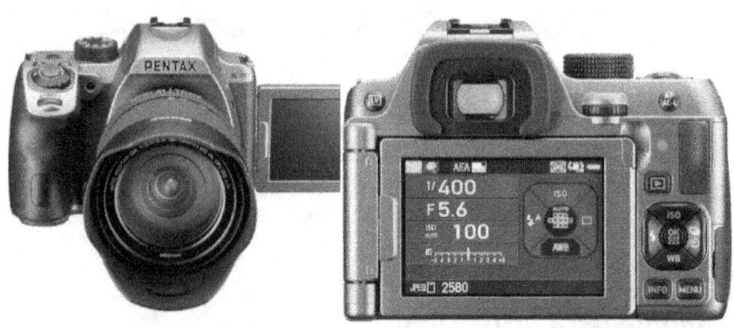

- Shake Reduction stabilizes hand-held shots with any lens, allowing you to use fast primes handheld in low light. Effectively, you can produce sharper, cleaner images than you could with any unstabilized full-frame camera (when shooting handheld in low light).

- Pixel Shift resolution produces cleaner, sharper images when photographing still subjects on a tripod, such as for product photography. However, any movement in the frame can result in strange artifacts that will require editing in post-processing (or simply make pictures unusable).

Video quality and autofocus lags behind the competition, however. At this price point, sports and wildlife photographers should consider a used Canon 7D or 70D. For video, a Panasonic G7 will do a much better job than the K-70.

K-5 II ($300 used)

The K-5 II is Pentax's mid-range body, and though it's a couple of years old, it's still a capable APS-C camera at a great price. The 16-megapixel sensor won't yield as much detail as the 24-megapixel APS-C sensors from Nikon and

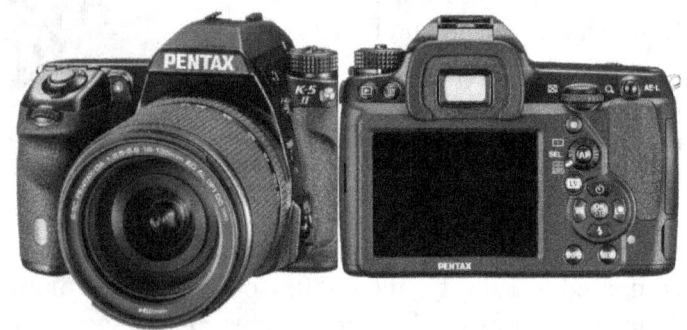

Sony, but you'll never notice the difference with the kit lens, anyway.

Unlike the similarly priced competitors from Canon, Nikon, and Sony, the K-5 II and K-5 IIs are weather sealed.

If you're considering the K-5 II, I'd recommend choosing the K-5 IIs instead. The price is the same, and you'll get sharper images.

K-5 IIs ($400 used)

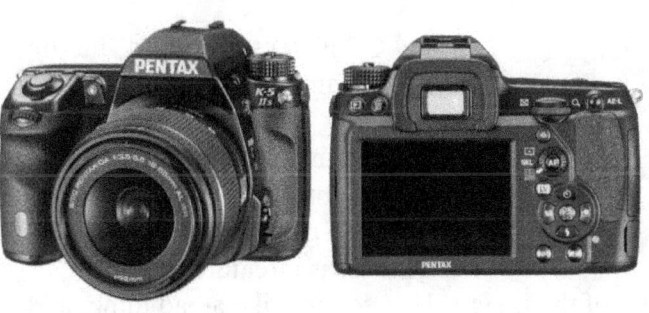

The K-5IIs is identical to the K-5 II in every way except that Pentax has removed the anti-aliasing (AA) filter, allowing sharper images. In 2012, when the cameras were released, a camera without an AA filter was an important distinction. However, most cameras released in the past year have had the AA filter permanently removed.

Given that the K-5 IIs is available for the same price as the standard K-5 II, I recommend choosing this camera over the K-5 II. You'll get sharper images with almost no drawbacks.

K-S1 ($230 used)

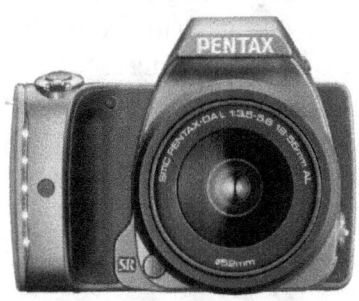 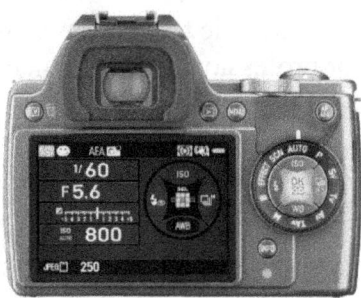

The K-S1 is a small, cool, and capable APS-C DSLR. Pentax is appealing to the fashion-conscious photographer with a wide range of camera colors, as shown in the following image. It's a welcome relief from the standard black DSLRs.

If you like the small size of the K-S1 but want access to a wider variety of lenses, check out the Canon SL-1, which is a bit smaller but not as colorful.

K-S2 ($450 used)

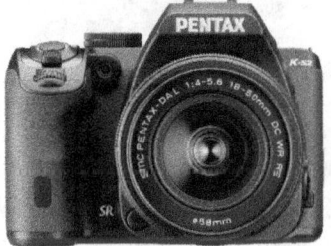 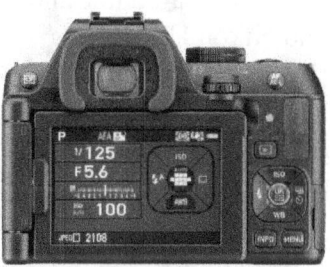

The K-S2 is Pentax's mid-level DSLR, and it's a better value overall than the Nikon D5500 and Canon T6i, both $750, which are the closest alternatives.

The controls are rather chunky compared to the competition, but the camera feels sturdy. The image quality is similar to other cameras of this price range, but the focusing is excellent. Pentax's Wi-Fi app, however, is terrible.

> Video: K-S2 Review
> **23:14** - *sdp.io/T6iReview*

K-3 II ($800 new, $650 used)

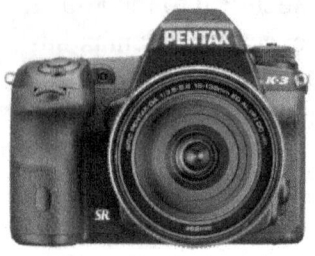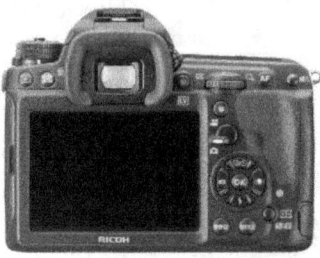

The K-3 is Pentax's top-of-the-line APS-C camera. For its price, the incredibly fast 8.3 frames per second simply can't be beat. Another interesting feature is the selectable anti-aliasing filter. The AA filter is a feature of many digital cameras that reduces the sharpness of the image. That sounds like a bad thing, and it usually is. However, the AA filter also reduces some artifacts, such as moiré. The K-3 is the only camera that allows you to turn the AA filter on or off. However, most people should simply choose to leave it off, and I've never missed having an AA filter on my cameras that lack it.

Like many Pentax cameras, the K-3 is weather sealed, which won't matter to most people. All cameras can handle a certain amount of weather. If you know that you might get caught in a serious downpour but not be able to put your camera in your bag (for example, if you're doing photojournalism or professional sports), then the weatherproofing is meaningful.

The K-3 has 25 cross-type autofocus points, making it the best camera in the Pentax lineup for action. However, it still doesn't autofocus as well as similarly priced competitors from Canon and Nikon, such as the Canon 70D or Nikon D7100.

KP ($900 new)

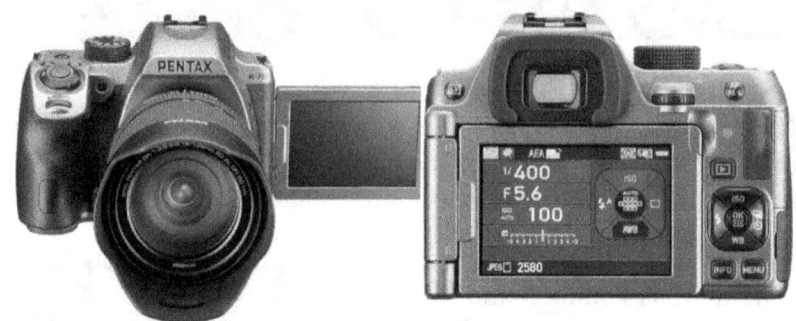

The KP is Pentax's top-end APS-C camera and represents an amazing value compared to Canon and Nikon DSLRs. In particular, the sensor stabilization gives it a huge advantage in low-light, because you can handhold fast primes at slow shutter speeds. In that scenario, the KP can produce cleaner, sharper images than a vastly more expensive Canon 5DS-R or Nikon D810.

When photographing a completely still subject with a tripod, the Pixel Shift Resolution feature lets you produce cleaner, sharper images than any other 24 megapixel DSLR, regardless of sensor size. However, our tests have shown that the technology doesn't handle water or leaves blowing in the wind as perfectly as it should, resulting in long, frustrating editing sessions. Pentax claims that their software detects movement between frames and handles it, and sometimes it works well, but other times it completely wrecks your picture. As a result, we primarily use the technology for product photography.

Another unique feature of the KP (when compared to other DSLRs) is the electronic shutter, which allows shutter speeds of up to 1/24,000 of a second. However, the rolling shutter effect is absolutely awful, making shots with any moving subjects unusable… and

256

the only time you'd need a shutter of 1/24,000 of a second is when shooting a fast-moving subject. The electronic shutter does allow quieter shutter speeds, but it's not silent.
The KP is not an ideal sports camera. The raw buffer fills up after 10 seconds, and the focus tracking doesn't keep up with the comparable Canon 80D or Nikon D5600. However, if you want to spend $1,000 on a DSLR, the KP is overall the best camera available. Pentax's APS-C lens selection is excellent, though we recommend pairing the KP with the Sigma 18-35 f/1.8 for the best quality.

K-1 ($1,700 new, $1,400 used)

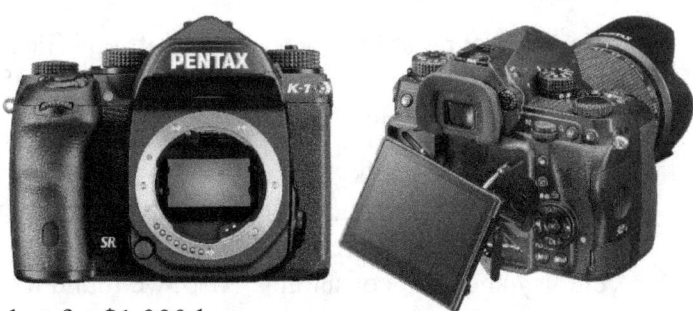

The K-1 will be Pentax's first full-frame DSLR, and once again it looks like Pentax will provide the best value. With 36 megapixels and no anti-aliasing filter, the K-1 produce images similar to those of the Nikon D810, but for $1,000 less.
A fairly unique feature for full-frame DSLRs is a stabilized sensor. The stabilized sensor will allow photographers to hand-hold pictures with slow shutter speeds, allowing them to use unstabilized lenses (like fast primes) in low light conditions. Because of this, the K-1 should be the best low-light DSLR ever made.
Another great features include:

- Dual SD cards slots to protect your images in case of a memory card failure.

- GPS and a compass to record the location of every picture.

- Wi-Fi to allow you to transfer images to your mobile device.

The unique tilting and twisting screen isn't any more useful than a standard tilting screen, however, it worked well, and the Nikon D810 and Canon 5DS-R lack this feature entirely. The only other full-frame camera with a tilt screen is the Nikon D750, despite the fact that they're extremely useful for shooting on a tripod, or for shooting a high or low angles.
It has some serious drawbacks, however.
The most significant drawback is a lack of proper full-frame lenses. Compared to Nikon and Canon, the selection is extremely slim.
In fact, as of June 2015, Pentax offers a total

Video: K-1 Preview
21:06 - *sdp.io/K1Preview*

of five full-frame lenses. However, that number is sure to increase, and Pentax might have every full-frame lens that you need.
The 15-30mm and 24-70mm f/2.8 lenses are rebranded Tamron lenses. We really liked the 15-30mm f/2.8 Tamron lens, and happily use it on our Canon and Nikon bodies. Therefore, if that focal length is important to you, you're not giving up anything significant.
However, we found the Tamron (and now Pentax) 24-70 f/2.8 lens to be lacking compared to the Canon and Nikon varieties. While the sharpness matches the Nikon, it is much less sharp than the Canon. It also has a lower quality build than the Canon and Nikon versions,

and when focusing at close range, it has *severe* focus breathing. While we don't normally complain about bad bokeh, this lens' bokeh is rather disturbing.

The PixelShift feature moves the sensor 1 pixel in each direction, and then automatically stacks those four images. Effectively, this reduces noise by 400% and noticeably improves the sharpness of the image. Unfortunately, PixelShift in its current form has some severe limitations:

- You must use a tripod.

- You cannot reliably process PixelShift raw files in Adobe Lightroom. While Lightroom seems to correctly process still PixelShift images, any images with movement (including moving leaves or water) create strange artifacts.

- You can use the SilkyPix software to process the PixelShift raw files. It does a better job of detecting movement between the four frames. However, SilkyPix is simply awful compared to Adobe Lightroom, and switching between applications is clumsy and time-consuming. Also, we found that SilkyPix still regularly left ugly artifacts in moving water. As a result, we could never trust that PixelShift would reliably get the shot, and always needed to shoot an important scene with both PixelShift and without.

Because of the limited Pentax full-frame lens selection, and the current limitations of PixelShift, we still prefer the images we get from the Canon 5DS-R for serious landscape photography. However, the 5DS-R body and canon 24-70 f/2.8 L II cost about 30% more than the Pentax.

Unfortunately, the slow frame rate, slow SD memory card format, and small buffer make the Pentax K-1 almost unusable for action photography, even when shooting JPG images.

K-1 Mark II ($2,000 new)

The K-1 Mark II is almost exactly the same as the original K-1. In fact, for a limited time after the release of the K-1 Mark II, owners of the original K-1 could send their cameras in and have them upgraded.

Most buyers will see the best value by buying a used copy of the original K-1; they're durable cameras that can easily survive multiple owners, and the $600 you save is better spent on buying lenses. However, the K-1 Mark II does offer these benefits:

- Pixel Shift while handheld. The Pixel Shift feature in both the K-1 and K-1 Mark II stacks four pictures together, shifting the image by one pixel for each photo. This produces a sharper, cleaner image. Unfortunately, we found the feature to be useful only when shooting still life photography, such as product photography. Even landscapes weren't still enough, because clouds, trees, and water moves too much, and Pixel Shift often ruined the resulting image. Handheld Pixel Shift works completely differently; it simply uses your natural handheld movement to move the sensor, and stacks the images in software. It's an in-camera implementation of super resolution, which I teach (for any camera) here at *sdp.io/SR*. In the sample files I reviewed, I saw that handheld Pixel Shift did eliminate moire, but it did not seem to increase sharpness.

- Better focusing. We'll have to test this to comment on whether the focusing is substantially improved. We already know that the K-1 Mark II has the exact same focusing system as the original K-1, however, which means the only possible changes are tweaks to the software and perhaps faster processing of data from the focusing system. All the focusing points are still clustered in the middle, making it a challenge to use even for landscapes and portraits. The low frame rate and small buffer still make the camera unacceptable for action.

Pentax 645Z ($5,000)

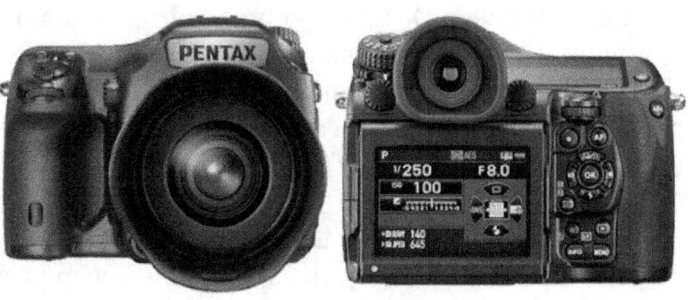

The 645Z is Pentax's medium format camera, supporting lenses from the company's long history of 4x5 medium format film cameras. However, the 645Z's digital sensor isn't as large as their film cameras. It's still quite a bit larger than full-frame 35mm digital cameras, with a 0.79X crop factor.

The Sony-made 51 megapixel sensor is clearly the star of this heavy, expensive body. If you absolutely must have more pixels than a Nikon D810, the Pentax 645Z is your only option under $10,000.

Medium format cameras offer a certain prestige because of their history in commercial photography, but even if you have the budget, I'd still recommend the D810 over the 645Z to most photographers. Nikon offers a much wider variety of lenses than Pentax, and those lenses are less expensive than the Pentax equivalents. The larger size of the 645Z (especially with a few lenses) significantly hampers usability in all but commercial studio scenarios.

If you want to understand the potential benefits, read the *Sensor Size And Crop Factor* section of Chapter 3. Let's explore a few examples from the Pentax lens lineup:

> Video: Pentax 645Z Preview
> **14:39 -**
> *sdp.io/645ZPreviewhttp://sdp.io/d810preview*

- The Pentax-D 55mm f/2.8 ($1,000) is equivalent to a full-frame 43.5mm f/2.2. Therefore, despite the larger sensor, Pentax's normal prime is slower than the full-frame 50mm f/1.4 standard (about $400) and offers less background blur.

- The Pentax 28-45mm f/4.5 ($4,900) is equivalent to a full-frame 22-35mm f/3.5. Again, despite the larger size, it gathers less light and offers less background blur than a full-frame Tamron 24-70 f/2.8 ($1,100) lens.

- The Pentax 80-160mm f/4.5 ($1,935) is equivalent to a full-frame 63-126mm /f3.5. In the full-frame world, a 70-200 f/2.8 offers a wider zoom range and gathers more total light.

Because the lenses aren't particularly fast, the 645Z can't fulfill the promises of larger sensors: gathering more light and providing shallower depth-of-field. However, the higher megapixel count can allow extracting more detail, given proper technique.

Another consideration is whether you can achieve cleaner images with lower noise in ideal conditions, such as tripod-mounted landscapes or studio photography. The 645Z's minimum ISO of 100 means that it can gather more total light than a full-frame 35mm camera at ISO 100—about two-thirds more light, in fact. That should translate to two-thirds cleaner images from the 645Z than a full-frame 35mm camera at ISO 100.

However, the Nikon D810 supports an unusually low ISO of 64, which gathers almost exactly the same total amount of light as the 645Z will at ISO 100.

Obviously, anyone considering the 645Z should seriously consider the Nikon D810 instead. However, those with a total budget over $15,000, no concern about camera size or weight, and an obsession with capturing as much detail as possible should evaluate the 645Z.

If you're truly obsessed with detail and low noise, you might also consider the Mamiya 645DF Credo 80 ($38,000), which offers 80 megapixels and ISOs as low as 35. With a bigger sensor than the 645Z (it has a crop factor of about 0.62) and lower minimum ISO, the Mamiya can gather about two stops more light than the 645Z, or about three stops more light than a full-frame camera at ISO 100.

Chapter 21: Micro Four-Thirds (Panasonic/Olympus)

Micro Four-Thirds (MFT) is a standard that allows cameras and lenses from different manufacturers to work together. For example, with MFT, you can use an Olympus lens on a Panasonic camera, or vice-versa. This is quite remarkable considering that every other camera manufacturer, including Canon, Nikon, Sony, Pentax, Samsung, Fuji, and Leica, created a proprietary system that prevents you from using equipment from other manufacturers.

MFT uses a relatively small sensor—about ¼ the size of a full-frame DSLR, with roughly a 2x crop factor. That sensor is still much larger than the sensor in most smartphones and point-and-shoot cameras.

If you have the right lenses, smaller sensors are theoretically capable of taking images that are similar in quality to those taken by larger sensors. Fortunately, the MFT system has an amazing variety of lenses. To get the best image quality from your MFT camera, I highly recommend using fast prime lenses and a body that supports In-Body Image Stabilization (IBIS), especially:

- Panasonic 20mm f/1.7
- Panasonic 20mm f/1.4
- Olympus 45mm f/1.8
- Olympus 75mm f/1.8

In addition to Olympus and Panasonic, BlackMagic Design also supports MFT lenses on their Pocket Cinema Camera, an appropriately named small, professional video camera. The Pocket Cinema Camera has an even smaller sensor than MFT, with a 2.9x crop factor compared to full-frame cameras.

Like with all mirrorless cameras, you can use adapters to attach SLR lenses to your camera (though you might lose sharpness, autofocus, and aperture control, depending on the lens and adapter). Non-professionals should generally avoid the use of adapters with MFT because the severe crop factor drastically changes the lens focal length and quality. Professional videographers frequently use adapters with the Pocket Cinema Camera and GH2/GH3/GH4 cameras.

To partially offset the effect of the crop factor, MetaBones has created a series of Speed Booster adapters. These adapters work exactly like a teleconverter, except opposite. Whereas a teleconverter spreads the light from just the center of the lens over the camera sensor, a speed booster focuses the light from the lens onto a smaller sensor.

For example, if you connect a 50mm f/1.4 lens to a MFT camera with a traditional adapter, it would behave like a 100mm f/2.8 lens on a full-frame camera. With the Nikon G to Micro Four Thirds Speed Booster (*http://sdp.io/nmftsb*, $400), the lens would behave like a 75mm f/2 lens.

This table summarizes the current Micro Four-Thirds Panasonic and Olympus cameras. They all have similar 16 megapixel sensors.

Model	Price (new)	Price (used)	Touch screen	Tilt or Articulating Display	Wi-Fi	EVF	FPS	IBIS	Video
Olympus PEN E-PM2	$290	$210	*				8	3-axis	1080i
E-PL5 *http://amzn.to/WqTjSH*	$300	$275	*	Tilt	*		8	3-axis	1080i
E-PL7	$500	$390	*	Selfie	*		8	3-axis	1080p
E-P5 *http://amzn.to/Wb90fT*	$660	$400	*	Tilt	*		9	5-axis	1080p
F	$1,200	$900	*	Articulating	*	*	10		
E-M10 *http://amzn.to/133Bxvc*	$550	$450	*	Tilt	*	*	8	3-axis	1080p
E-M10 II	$700	$650	*	Tilt	*	*	8.5	5-axis	1080/60p
E-M5 *http://amzn.to/12Bwt2h*	$500	$300	*	Tilt	*	*	9	5-axis	1080i
E-M5 II	$1,000	$750	*	Articulating	*	*	10	5-axis	1080/60p
E-M1	$1,000	$650	*	Tilt	*	*	10	5-axis	1080p
Panasonic Lumix GF6	$360	$350	*	Tilt	*		4		1080p
GM1	$400	$300	*		*		5		1080p
GM5	$500 kit	$380			*	*	5.8		1080/60p
G6	$540	$460		Articulating			7		1080/60p
GF-7	$510	$450	*	Selfie	*		5.8		1080p
G7	$600	$500	*	Articulating	*	*	6		4k
GX7	$725	$400	*	Tilt	*	*	5	2-axis	1080/60p
GX8	$1,000	$800	*	Articulating	*	*	6	2-axis	4k
GX85	$800		*	Tilt	*	*	8	5-axis	4k
GH2		$350	*	Articulating		*			1080p
GH3	$900	$550	*	Articulating	*	*	6		1080p
GH4	$1,300	$1,100	*	Articulating	*	*	12		4k
GH5	$2,000		*	Articulating	*	*	9-30	5-axis	4k/60p
GH5S	$2,000		*	Articulating	*	*	9-30		4k/60p

Unfortunately, in 2020, I can no longer recommend new buyers invest in Micro Four-thirds equipment. While I'm sure Panasonic will release a GH6, the system hasn't been able to keep up with the competition in areas of image quality and autofocus. Olympus has completely existed the camera business, and Panasonic is investing more heavily in their full-frame L-mount.

Olympus

Olympus, a Japanese company, made its first film cameras in 1936, and has been making cameras and optics ever since. In 2008, they joined Panasonic to launch the Micro-Four Thirds (MFT) system for digital cameras.

Olympus is known for two series of mirrorless cameras:

- **PEN**. Compact, inexpensive, and capable compact cameras designed for casual photographers. Olympus made film cameras using the Pen name from 1959 until the early 1980s, and resurrected the name in 2009 for digital cameras. The PEN cameras do not include a viewfinder, so you'll rely on the back of the camera to frame your shots, which works fine for casual photography. Though you can add an optional viewfinder (such as the Olympus VF-1, VF-2, VF-3, or VF-4), if you want a viewfinder, you should probably just choose an OM-D model.

- **OM-D**. Rugged, retro, and DSLR-like cameras designed for more serious photographers. Olympus made film cameras using the OM name from 1972 until 2002, and resurrected the name in 2012 for digital cameras. Olympus seems to have indicated that they're focusing all their future efforts on the OM-D cameras and dropping further PEN development.

In 2018, I warned the photography community about the upcoming death of the Micro Four-thirds platform as a result of the continuing decline of the camera industry and the increased competition in the mirrorless segment as Canon and Nikon launched their own systems. As Panasonic continues to put more engineering and development into their full-frame L-mount platform, it seemed inevitable that Olympus would have to leave the camera industry.

In 2020, that prediction came true as Olympus announced the sale of their camera department. After 84 years, Olympus was leaving the camera business entirely. The new owners, Japan Industrial Partners, are primarily known for helping Japanese companies sell assets and lay off employees (tasks that are almost impossible in Japan). They aren't in the photography industry and are known primarily for selling assets and creating low-cost, low-quality electronics.

There is some possibility you will be able to buy Olympus cameras in the future; JIP is committed to updating the system. I now believe they will release some new camera in 2021, however, I doubt it will be the compelling tech-leader that so many Olympus buyers chose the system for.

If you already own Olympus gear, it will continue to work. Enjoy it! There's no reason to think resale value is going to drop. In fact, resale value might even increase when no more cameras or lenses are manufactured.

If you're thinking about buying an Olympus camera for the first time, maybe look elsewhere. There's no guarantee that you'll be able to upgrade as camera technology continues to improve. You might never see a new Olympus lens after 2020. Instead, I would recommend checking out the Fujifilm X-mount, Canon RF, or Sony E-mount cameras. Fujifilm provides Olympus' great looks and small size. Canon RF, specifically the $900 Canon RP, provides small, light, and fun cameras, and the Canon 600mm f/11 provides a compact wildlife lens like those Olympus shooters have grown to love.

PEN E-PM2 ($290 new, $210 used)

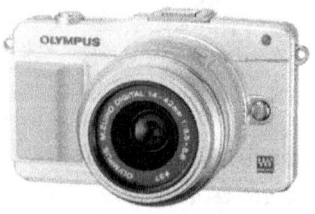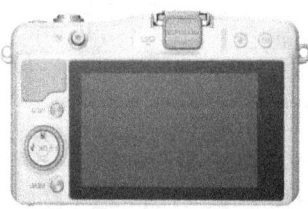

Olympus' entry-level camera is an amazing value. It's small, light, and inexpensive, but produces the exact same images as every other Micro Four-Thirds camera.

For the sake of argument, let's compare the $290 E-PM2 to the $1,350 E-M1. The E-M1 has a glorious viewfinder that's better than real life, it takes 8 frames per second, and you can shower with it. But if you chose the E-PM2 instead, you'd have an extra $1,050 to spend on lenses, and that could buy you:

- Panasonic 20mm f/1.7: $350
- Olympus 45mm f/1.8: $330
- YongNuo YN-560 II flash: $75
- Manfrotto MKC3-H01 Tripod: $45
- Flashpoint 180 Monolight: $200

That's not to say the E-M1 isn't a better camera than the E-PM2; it definitely is. But the E-PM2 and $1,000 worth of extra lenses, a flash, and a tripod will definitely take better pictures than an E-M1. In other words, don't spend your whole budget on your camera body, and don't underestimate entry-level cameras.

PEN E-PL5 ($300 new, $275 used)

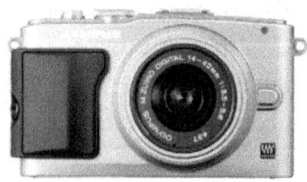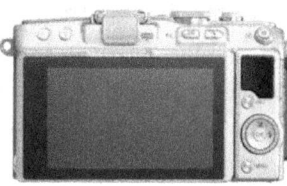

Compared to the base model E-PM2, the E-PL5 adds an articulating screen, which makes it easier to shoot from high or low angles and to take selfies. I love an articulating screen and use it regularly; however, it does add some weight and bulk to the camera.

This model also adds built-in Wi-Fi, so you can more easily transfer photos to your smartphone. With the E-PM2, you need to use a special SD card to access pictures wirelessly.

PEN E-PL7 ($500 new, $400 used)

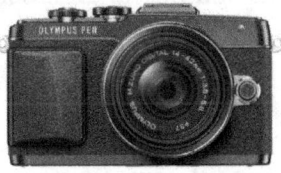 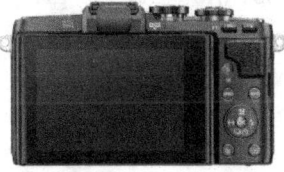

The PEN E-PL7 is very similar to the older E-PL5. The video quality has been improved by supporting 1080p instead of 1080i. The biggest difference, however, is that the tilting screen has been upgraded to a selfie screen, as shown by the following product image. Olympus has also improved autofocusing.

PEN E-P5 ($660 new, $400 used)

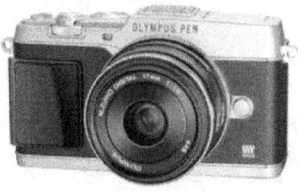 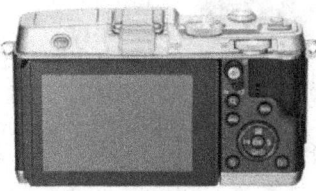

Compared to the less expensive PEN cameras, the E-P5 upgrades to 5-axis image stabilization over the standard 3-axis image stabilization. According to Olympus, you'll only notice the difference when shooting close-up macro photos. While it's difficult to test (since testing it requires handholding the camera), we've found that 5-axis stabilization does give us more usable shots at slow shutter speeds than 3-axis stabilization. It's a minor difference, however.

Perhaps more importantly, the E-P5 adds more dials for manual control. These dials are critical to any serious photographer because they allow you to adjust the aperture, shutter speed, or exposure compensation quickly. With candid photography, they are often the difference between getting and missing the shot. Because of those dials, I'll recommend the E-P5 over the other PEN cameras for anyone who is even slightly serious about photography.

The E-P5 also provides access to ISO 100 and a maximum shutter speed of 1/8000th. Though you'll probably never need either, serious photographers working in bright light with professional lenses will appreciate the upgrade. The E-P5 also looks and feels better than the less expensive PEN cameras.

PEN-F ($1,100 new)

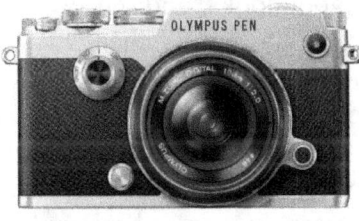 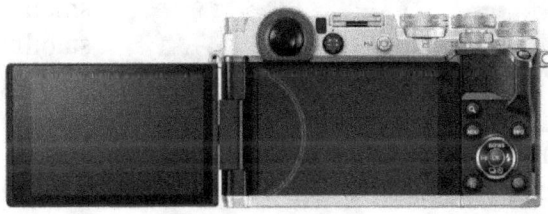

The PEN-F is a beautiful, stylish, and full-featured rangefinder-style camera. Like the E-P5, the PEN-F includes 5-axis sensor stabilization, so handheld shots are stabilized with any lens. The PEN-F's touchscreen flips out from the side, allowing selfies. The PEN-F also adds 50 megapixel high res shot mode, providing cleaner, sharper images than you can take with a full-frame camera like the Canon 5DS-R… assuming you're shooting a still subject and using a tripod. Therefore, that mode is mostly useful for product photography.

Unfortunately, the video is limited to 1080p. At this price point, you'll get superior video from a Panasonic G7 or GH4... but you'd lose the PEN-F's great looks and the rangefinder styling, which I prefer.

OM-D E-M10 ($500 new, $450 used)

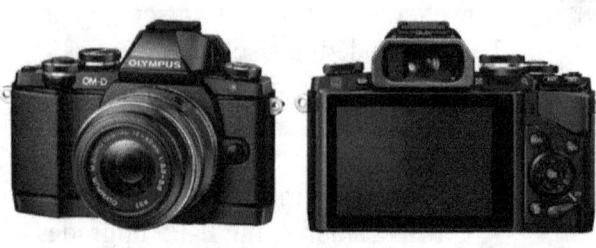

The E-M10 is the baby of the OM-D series of cameras, and it's my single favorite camera for travel and casual photography. It's stylish, light, capable, inexpensive, and fun. The Panasonic 20mm f/1.7 pancake lens is simply

perfect with it; it's sharp enough to crop, and the in-body image stabilization and fast aperture make the combination amazing for low light.

It feels a bit plasticky, and the E-M10 lacks the durability of its bigger brothers, the E-M5 and E-M1. However, I've dragged it thousands of miles, and through storms and blizzards, and never used a bag or lens cap. I abuse my gear, and it's durable enough for me.

Video: E-M10 Review
20:49 - *sdp.io/em10review*

Video: E-M10 Travel Review (& E-M1 Comparison)
12:33 - *sdp.io/em10travel*

Like with all other current Micro Four-Thirds cameras, the autofocusing can be frustrating in low light. At times, I wish it had a couple more customizable buttons. The viewfinder isn't as fast or beautiful as the E-M1. However, those drawbacks aren't enough to stop us from bringing this camera everywhere, and we've owned almost every modern camera.

OM-D E-M10 II ($700 new, $650 used)

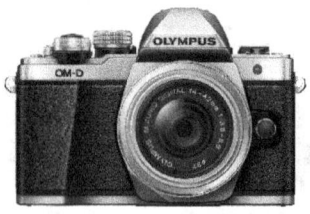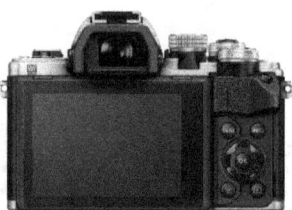

The updated E-M10 is still one of our favorite travel cameras because of its small size, good looks, and sensor stabilization. It has hardly changed compared to the original E-M10, so refer to the previous section for

detailed information.

Compared to the original E-M10, this version adds 60 fps video (for smoother video) and better sensor stabilization. If you don't care about those two features, find a used original E-M10 and save yourself a few hundred dollars.

OMD-D E-M10 IV ($700)

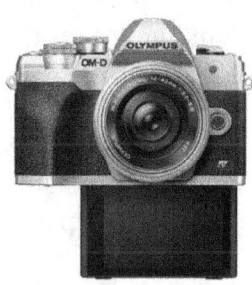

The latest update to the E-M10 entry-level camera upgrades the sensor to 20 megapixels. It also adds a flip-down screen for handheld selfies (though it would be blocked by a tripod unless you also added a bracket). The autofocus is improved, too. It continues to be a great first camera, though if you think photography might be a hobby you stick with for a long time, you should consider the fact that buyer of Olympus' camera business might not continue to manufacture high-quality cameras for many years into the future.

OM-D E-M5 ($500 new, $300 used)

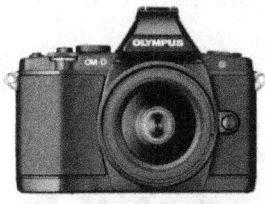 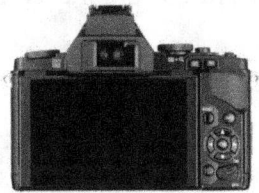

Olympus' mid-range OM-D camera, the E-M5, has essentially been replaced by the newer and more fully featured E-M10. Compared to the E-M5, the E-M10:

- Is a bit less expensive new
- Is a bit lighter
- Has built-in Wi-Fi
- Has HDR, an intervalometer, and other software-based features

Compared to the E-M10, the E-m5 is more weather sealed and sturdier. If you already have an E-M5, it's probably not worth it to upgrade. But if you're buying new, I'd recommend the E-M10 instead.

OM-D E-M5 II ($1,000 new, $750 used)

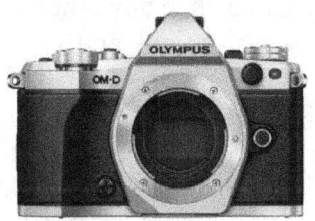 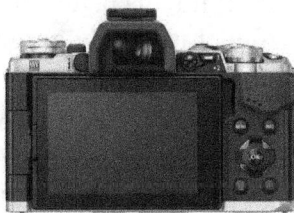

Olympus' newest DSLR-like mirrorless camera is also our favorite (especially in the two-tone silver and black combination). It's like a fashion accessory that also takes great pictures.

Technically, it sits between the top-end E-M1 and entry-level E-M10. Compared to the E-M1, the grip is smaller and there are fewer buttons, making it less of a "serious" camera. The bigger E-M1 was never particularly good at being a serious camera, anyway, especially compared to DSLRs of a similar price and size.

The E-M5 II has one unique feature: a high-res mode that shifts the sensor by half a pixel, allowing the camera to take 40 megapixel low-noise pictures with a 16 megapixel sensor.

> Video: E-M5 II High-Res Mode
> **41:03** - *sdp.io/em5hrii*

With a good prime lens like the Olympus 75mm f/1.8, the camera produced far better looking images than our 36-megapixel Nikon D810 DSLR, at a lower total price, and with a much smaller size. You absolutely have to use a tripod, however, and we couldn't get high-res mode to create acceptable results for landscape photography. The only scenario it seemed useful in was still product photography.

For travel and all handheld work, many will be happier with the smaller and lighter E-M10 instead. Compared to the E-M10, the E-M5 II has a screen that flips out sideways, making it easier to take selfies. The build quality is slightly better; it looks more expensive (which it actually is). The image quality is slightly better, but you probably won't notice. The E-M5 II offers 5-axis in-body image stabilization, which might allow you to get a higher percentage of sharp shots with slow shutter speeds, though we didn't notice a difference. Of course, The E-M5 II is the only Micro Four-Thirds camera with the amazing high-res mode.

OM-D E-M1 ($1,000 new, $650 used)

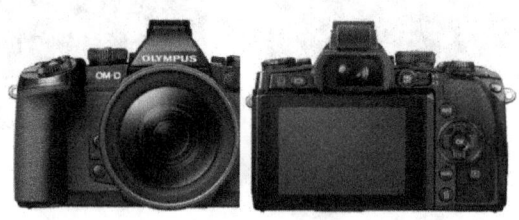

Olympus' flagship mirrorless camera is also the ultimate Micro Four-Thirds body. The OM-D E-M1 is clearly designed to steal market from DSLR camera bodies by shattering the perception that mirrorless cameras are flimsy and not suitable for professional use.

The E-M1 is certainly not a delicate camera. It feels like it was carved from a solid piece of steel. It's advertised as beyond weatherproof (when used with the right lenses); Olympus' own ads show it being drenched in the water.

The E-M1 has more than enough buttons, dials, and switches for any photographer, and they can all be customized to your needs. Like the other OM-D cameras, it has a tiltable touchscreen. The

Video: E-M1 vs. X-T1
41:03 - *sdp.io/em1vxt1*

look and feel of the E-M1 are beyond reproach; it feels better in my hands than any other mirrorless camera I've ever used. The viewfinder is simply gorgeous: fast, bright, and detailed.

But the E-M1 is not the right camera for most people. The single biggest benefit of the Micro Four-Thirds architecture is the small size, and this camera is as heavy as an APS-C DSLR. In the DSLR world, bigger and more expensive cameras also advertise bigger sensors and better image quality, but the E-M1 has about the same image quality as all Micro Four-Thirds cameras, even those costing less than $500.

The E-M1's focusing system works well for still subjects, and it does a decent job of tracking a walking subject. It's not able to accurately track a running person, however.

Though Olympus advertises this camera as having phase detection autofocus, that system only works with adapted Four-Thirds lenses. When you use it with native Micro Four-Thirds lenses (which most will do), it will use

Video: E-M1 vs. X-T1 AF
9:05 - *sdp.io/em1af*

old-fashioned contrast-based focusing, which is frustrating in many scenarios.

If you like the design of the E-M1 but you'd prefer a lighter camera, or you want to save some money for lenses, consider the E-M10. The E-M10 doesn't have as many buttons, making it a bit clumsier for advanced users, and it has only three-way In-Body Image Stabilization instead of five-way, but I never noticed much difference. For my video review and comparison of the two cameras, visit *http://sdp.io/em1v10*.

If you like the weight, size, and controls of the E-M1 and you're not yet invested in Micro Four-Thirds, you should also consider the Fuji X-T1. The X-T1 can track running subjects better (with some tweaks applied to the focusing system) and the image quality is far better than that of the E-M1. However, the E-M1 has far better usability than the X-T1. These two videos will help you decide which you prefer:

- *Fujifilm X-T1 vs Olympus O-MD E-M1: http://sdp.io/xt1vem1*

- *Fujifilm X-T1 vs Olympus O-MD E-M1 autofocus rematch*: http://sdp.io/xt1vem1af

OM-D E-M1 Mk II ($2,000 new, $1,600 used)

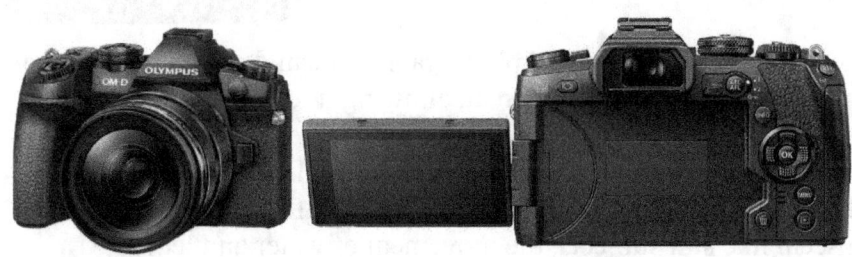

Olympus' top-end camera is perhaps the greatest all-around camera ever made. It's a jack-of-all-trades, master-of-none camera. The still images are the best of any micro four thirds camera, but won't match larger sensors in normal conditions. 4K video quality is great, though not quite as versatile as a Panasonic GH5. Focusing is also great, though not as good as the Sony a9.

The E-M1 Mk II is the master of one thing: still image quality on a tripod. Like other recent Olympus cameras, it moves the sensor to stack multiple images into a higher-resolution picture. The latest version is more sophisticated than earlier versions, and it can compile a massive 80 megapixel RAW file. In our testing, it produced sharper, cleaner images than the full-frame Canon 5DS-R. We plan to test it against 50-megapixel medium format cameras from Fujifilm and Hasselblad, but based on the images I've seen so far, I expect it to soundly beat them soundly.

Like Pentax, Olympus claims high-res mode automatically fixes moving subjects in the picture. In real-world usage using it as a landscape camera, I found that it didn't work nearly well enough. Therefore, I can only recommend people rely on high-res mode for product photography. Some landscapes without blowing wind or moving water will work OK with high-res mode, too.

Without high-res mode, the 20 megapixel images look better than any other Micro Four-Thirds camera we've ever tested. Especially compared to the GH5 with its heavy AA filter, the E-M1 Mk II is far sharper with the same lens. However, 20 megapixels is still a really low resolution sensor for a $2,000 camera. For example, the Canon 6D Mk II has 30% more pixels on a full-frame sensor.

But I'm not recommending the Canon 6D Mark II over the E-M1 Mk II. Once you use the Olympus, you won't ever want to go back to an optical viewfinder. Nor will you want to give up sensor stabilization or 4K video. Plus, DSLRs feel fat and plasticky by comparison.

If you're already invested in Micro Four Thirds lenses and you're primarily a stills shooter, the E-M1 Mk II will produce the best possible images. If this is going to be your first Micro Four Thirds camera, consider the Fujifilm X-T2, which has better image quality in ideal conditions, but lacks an articulating screen, a touch screen, and sensor stabilization.

OM-D E-M1X ($2,400)

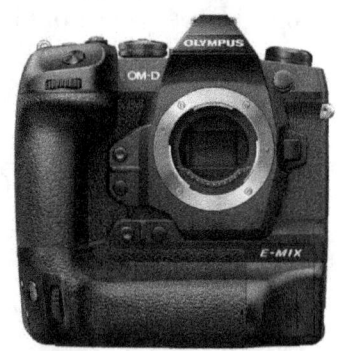

The 20 megapixel E-M1X was Olympus' biggest and most expensive camera. With its built-in vertical grip, it was positioned to compete against cameras like the Canon 1DX II and Nikon D5 (both about $6,000). Unfortunately, our testing found that the image quality and autofocus system didn't compare against those more expensive cameras... nor does it compare against full-frame cameras like the Sony a7 III ($2,000) or Canon R6 ($2,600).

It's primary area of innovation was computational photography. It can do in-camera image averaging, allowing long exposures. That worked great, allowing it to create clean, handheld, low-light images with better quality than similar full-frame cameras. You could shoot waterfalls with smooth motion without needing to attach an ND filter.

With moving subjects, that feature simply didn't work very well, unfortunately. Even with landscape photos, which seem like still subjects, the movement of water and leaves confused the algorithms and ruined photos.

In hindsight, the E-M1X was Olympus' last-ditch effort to save their camera business. Obviously, since Olympus has decided to exit the camera business, that last-ditch effort failed. Still, for Micro Four-Thirds owners, it represents the best camera they can buy – and possibly the best camera they will ever be able to buy.

OM-D E-M1 Mk III ($1,700)

The E-M1 Mark III is very similar to the E-M1X, but smaller. I personally prefer it over the E-M1X, but if you like the feel of a more substantial camera, you should choose the E-M1X instead. Or, better yet, get one of both and use them for different purposes.

Panasonic Lumix

Panasonic doesn't have Olympus' long history of making cameras (they released the first Lumix camera in 2001), but their Lumix cameras all take the same lenses and flashes as the Olympus cameras. Therefore, you can switch between the brands without selling all your gear, and if you have a Panasonic camera and find an Olympus lens that you like, you can use it without worry.

Panasonic cameras differ from Olympus cameras in a couple of ways:

- While all the Olympus cameras have in-body image stabilization (IBIS), only the Panasonic GX7 has IBIS. IBIS allows you to handhold non-stabilized lenses (such as most prime lenses) for longer.

- Panasonic lenses tend to have image stabilization built-in (because the feature is lacking in the camera body), while Olympus lenses do not have image stabilization.

- Panasonic lenses come with lens hoods, while you'll usually need to buy a separate lens hood for an Olympus lens.

- Panasonic prioritizes video capabilities, and many of their cameras have useful video features that other brands lack, such as the ability to carefully control video color and codecs.

GF6 ($360 new)

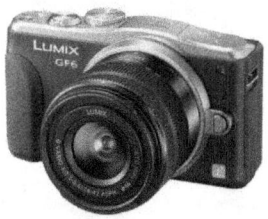 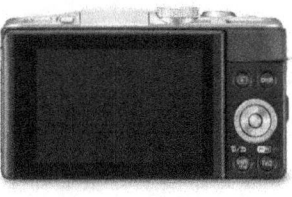

The baby on the Panasonic lineup, the GF6 offers the same still image quality as the rest of the Micro Four-Thirds lineup, making it an excellent value. The tilt screen offers versatility and even flips 180 degrees to make selfies easier. In fact, it offers many benefits to the selfie generation, including easy connections to Android smartphones that support NFC and to Wi-Fi networks.

If you want to control your aperture, shutter speed, and exposure compensation, you'll want to look to the rest of the lineup for more powerful controls. For the casual photographer stepping up from a smartphone or point-and-shoot, the GF6 is an excellent choice. You might consider upgrading to the G6 or GX7 if you want to use flash, because the GF6 lacks a hot shoe. Upgrading to the GX7 also adds in-body image stabilization, which is only useful if you plan to use prime or Olympus lenses without stabilization.

GM1 ($400 new, $300 used)

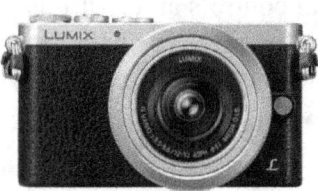 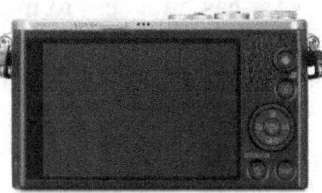

The GM1 is a small, lightweight, retro-styled camera perfect for travel and casual photography on the go. The diminutive size makes it the ideal take-everywhere camera, but it feels more solid and serious than the GF6. However, the size can also make the buttons more challenging to use. It relies on the touch screen for common tasks such as changing the aperture or shutter speed, or even manually focusing with the kit lens, which will frustrate more serious photographers.

In fact, it's so small that many smaller MFT lenses, such as the popular Panasonic 20mm f/1.7, are taller than the camera itself, preventing it from sitting flat on a table. This camera also lacks a grip, making it less pleasant to handhold for extended periods of time. You can, however, buy an optional grip that adds size and weight to the GM1, helping it balance many lenses.

Despite its size, the GM1 is capable of taking the same high quality pictures as the rest of the Micro Four-Thirds Lineup. If you plan to use prime lenses that don't have image stabilization, you might consider one of the Olympus PEN cameras instead because those bodies have image stabilization built in. With the GM1, you will need to choose lenses that have image stabilization.

The GM1 includes both a mechanical shutter and an electronic shutter. Thanks to the electronic shutter, the maximum shutter speed is an outrageous 1/16,000[th]. That's rarely useful, though; you'd need an incredible amount of light for such a fast shutter speed. Unfortunately, this novel design limits the sync speed (the maximum shutter speed you can use with a flash) to 1/50th of a second. That can be a problem if you plan to use fill flash outdoors for portraits.

If you're considering the GM1, you might consider upgrading to the larger GM5, G6, or GX7 if you want to use flash, because the GM1 lacks a hot shoe. Upgrading to the GX7 also adds in-body image stabilization, which is only useful if you plan to use prime or Olympus lenses without stabilization.

GM5 ($500 new kit, $380 used)

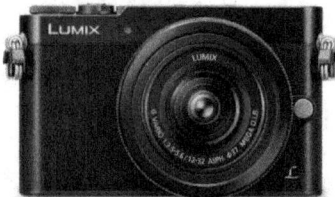 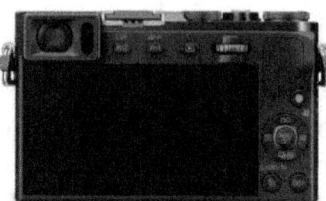

The GM5 is the smallest Micro Four-thirds camera with an electronic viewfinder. The small size can make it difficult to place sufficient buttons and dials, but there is a single main click dial easily accessible with your thumb, allowing you to adjust the aperture or shutter speed easily. More serious manual settings will require multiple presses and turns.

Unlike the GM1, the GM5 has a hot shoe so you can use an external flash. Another nice feature of this camera is silent mode, which uses an electronic shutter to take a picture without making any noise.

You might consider upgrading to a GX7 for easier access to manual controls and a tilting touch screen. If you prefer to use your left eye with the viewfinder, you might be happier with a G6 or GH4, which have a central viewfinder aligned with the lens. If you prefer to use the display on the back of the camera instead of the electronic viewfinder, you'll probably be happy with the less expensive GM1. The GM1 also fits more easily in your pockets because the viewfinder doesn't protrude.

G6 ($540 new, $460 used)

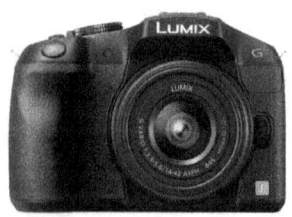 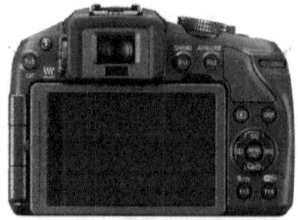

The G6 is Panasonic's mid-range DSLR-like camera. By DSLR-like, I mean that it physically looks like a traditional DSLR, with a deep grip, rugged looks, and a viewfinder centered over the lens rather than placed in the upper-left corner.

The centered placement of the viewfinder is a requirement for SLRs because of the mirror that physically bounces the light from the lens through the viewfinder. However, mirrorless cameras like the G6 have an electronic viewfinder that can be placed anywhere on the body. The centered viewfinder causes the back of the camera to press against your nose, making it less comfortable for most photographers. However, photographers who use their left eye to shoot might find the centered viewfinder more comfortable.

The G6 has excellent controls, with both a main dial and secondary dial that are easy to access. This allows you to rapidly change the shutter speed and aperture.

Compared to the less-expensive Panasonic cameras, the G6 is less portable but much more flexible. If you'd like most of that flexibility in a smaller package, consider the GX7. I would only upgrade to the GH4 if you want to record 4k video (which is overkill for most amateurs).

GF7 ($510 new, $450 used)

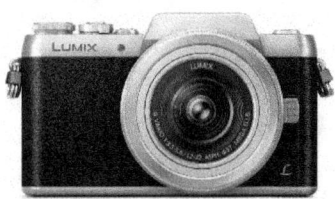 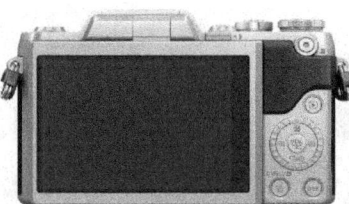

Panasonic's GF7 is a classy, portable, and capable camera for casual photography. I say *casual* photography because, while it's capable of taking the same images as any Micro Four-Thirds cameras, the lack of an electronic viewfinder and limited buttons and dials can frustrate more serious photographers.

If you're looking for a camera that you won't notice while you walk around a city, or something to bring on a night with friends that will create better selfies than your smartphone, the GF7 with its standard kit lens is a perfect choice.

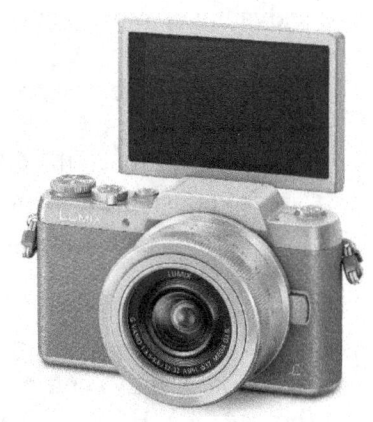

If you don't care about taking selfies, you might prefer the GM-1 instead. If you hope to use prime lenses in low-light, you might consider the Olympus PL7, which has a selfie screen but also offers in-body image stabilization (IBIS) that will let you use lower ISOs for cleaner images when handholding in low light.

Oh, it's also available in pink.

G7 ($600 new, $500 used)

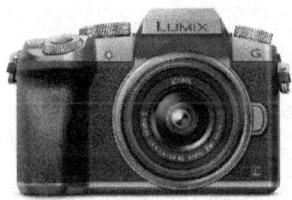 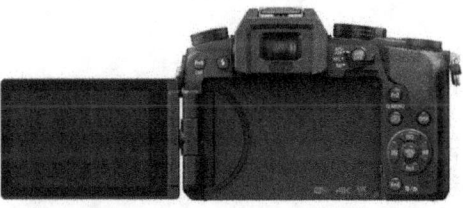

The G7 was one of the most exciting cameras announced in 2015, but not because it offered any new exciting features. Indeed, the G7's only surprise was that it record full 4k video, with interchangeable lenses, at the amazing $800 price point. The GH4 does have more powerful professional-oriented videography tools, such as V-Log that increases the dynamic range of a video (but requires you to grade it in photo editing software), and a headphone jack so you can record audio internally. Another important difference: the G7 is limited to a 30 minute maximum recording time (like most cameras). The GH4 can record for an unlimited amount of time without stopping, making it a better camera for unmanned shooting.

However, if you've been longing after a GH4 for 4k video, the G7 is the better choice for those on smaller budgets. At barely more than half the price, the G7 leaves more budget for lenses, lighting, sound, and post-processing tools, and those aspects of videography will make more of a difference than the upgrade from the G7 to the GH4. Be sure to set aside some budget for batteries, though: the G7's battery life is less than half that of the GH4. You should also look at the Samsung NX500 ($700 with lens), which also records 4k video, but is a bit less refined and offers fewer native lenses. If you don't need interchangeable lenses, the $700 Panasonic DMC-FZ1000 is a fixed-lens 4k camera with a decent zoom. The image and video quality isn't as good as the G7, but you won't have to buy a separate lens.

G85 ($1,000 new)

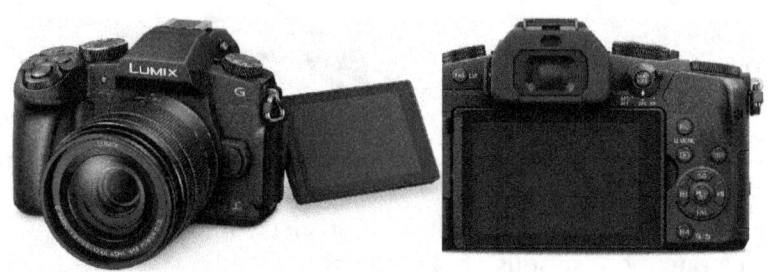

The G85 is a step up from the G7 if you plan to shoot stills, because it removes the G7's anti-aliasing filter. If you plan to shoot video, that might actually be a disadvantage. While your video will look sharper, it can also introduce unpleasant moire effects when shooting patterns such as clothing.

The G85 also offers 5-axis sensor stabilization, which will help you shoot handheld stills and video. Like the G7, it supports 4K/30p video.

These features make the G85 the best video/stills hybrid shooter at the $1,000 price point. If you're considering stepping up to the $2,000 GH5, you'll get better controls, 4K/60p video, and a headphone jack.

G9 ($1,700 new, $1,550 used)

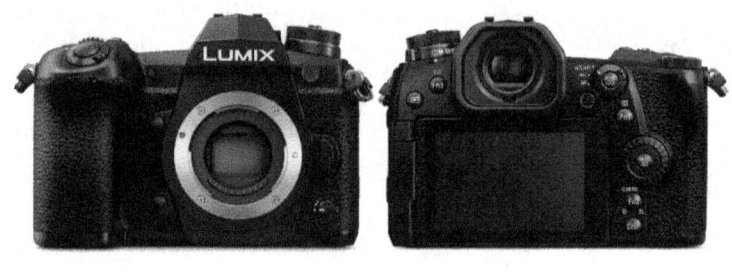

The G9 is Panasonic's top-end stills photography centric camera. Whereas the GH5 is primarily focused on video (but also does a great job with stills), the G9 is primarily focused on stills (but also does a great job with video).

Compared to its predecessor, the G85, the G9 adds dual card slots, high-resolution mode, a top LCD screen, improved sensor stabilization, Bluetooth (to make connecting easier from a smartphone app) and about twice the weight.

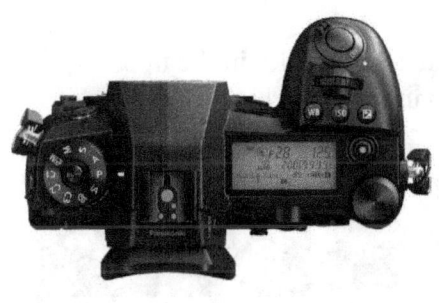

One feature that separates it from the GH5 is the LCD top display, something that will feel very familiar to Canon or Nikon DSLR users. This is simply a holdover from film SLRs that completely lacked a rear display. I've never missed it on the GH5 or any other digital camera because it's easy to look at the rear screen as you adjust settings.

Like the GH5, it offers sensor stabilization, allowing you to handheld photos with any lens. It also has dual card slots so a failed SD card won't result in lost images.

Like the Olympus E-M5 Mark II and E-M1 Mark II, the G9 offers a high-resolution tripod mode that stacks 8 separate photos, shifting the sensor by one pixel between each photo. This does effectively increase resolution and reduce noise, creating higher-quality images that you could get with a single photo from a full-frame Canon or Nikon DSLR. However, it only works on completely still photos, such as product photography. If you plan to use the technology for landscape photography, be prepared to remove artifacts in Photoshop later.

GX7 ($725 new, $400 used)

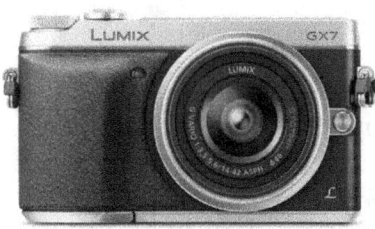 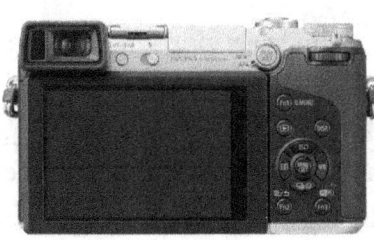

While still very compact, the GX7 offers many features serious photographers will appreciate, including a main dial for setting the aperture and shutter speed, and a touch screen that simplifies selecting a focusing point (even when your eyes is to the tiltable viewfinder). The grip feels good in the hand, too.

The GX7 also includes in-body image stabilization (IBIS), allowing you to handhold lenses that don't have image stabilization built in. This makes primes such as the Panasonic 20mm f/1.7 and the Olympus 45mm f/1.8 much more usable in low light conditions. While the IBIS isn't as robust as that included with the Olympus OM-D E-M5 or E-M1, you'll probably never notice the difference. You will, however, notice that it lacks the viewfinder hump in the middle of the body, making it much easier to carry.

GX85 ($800 new)

Virtually identical to the GX7, the GX85 provides IBIS, an articulating screen, and a 16-megapixel sensor. This newer model increases sharpness for both stills and 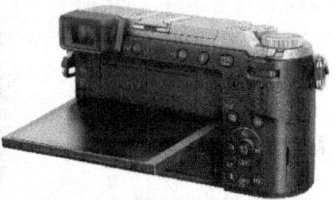 video by removing the anti-aliasing filter and adding 4k capabilities.

GX8 ($1,000 new, $800 used)

Like the GX7, the GX8 is defined by attractive styling and a viewfinder in the upper-left corner, which is more comfortable for photographers who use their right eye, because it doesn't require you to poke your nose into the back of the camera.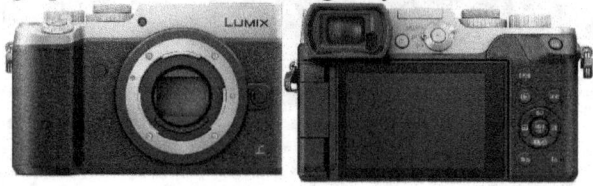

The GX8 has a 20 megapixel sensor, which is 25% increase from the 16 megapixel sensor built into almost every other Micro Four-Thirds camera. Typically, you can expect to see an extra 10-15% more detail in your photos, making bigger prints sharper and allowing heavier cropping.

The GX8 is almost a perfect camera for YouTubers:

- Small size provides fantastic portability

- Micro Four-thirds lenses are numerous, inexpensive, and small

- 4K video gives extra detail for cropping in post

- In-body image stabilization to steady the shot, no matter which lens is being used (though it doesn't work with 4K video)

- Tilting viewfinder for composing video or stills at a low angle in bright light

- The fully articulating screen can be flipped forward for still and video selfies

Be aware, however, that 4K video recording crops the image by about 1.25X. Therefore, you will need wider-angle lenses for video than you will for stills. While the camera has a mic jack, it lacks a headphone jack, so you won't be able to monitor your sound.

GX9 ($1,000 new)

The Panasonic GX9 is very similar to the GX8, except that it's a little bit smaller. That can be a good thing, except that the grip is less deep, too. You definitely wouldn't want to use any larger lenses with the GX9. Though the GX9 can shoot at 1/16000 with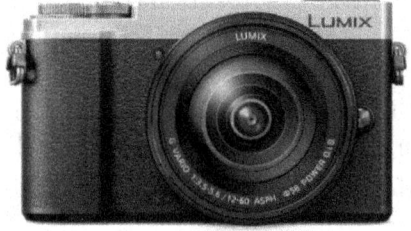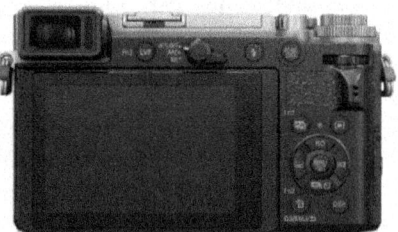
the electronic shutter, the mechanical shutter is limited to 1/4000, while the GX8 has a minimum shutter speed of 1/8000. The GX9 also offers Bluetooth, which can make connecting to your smartphone app a little quicker.

The GX9 also includes several technologies to improve sharpness:

- An improved shutter mechanism (also on the GX85) which improves sharpness at some shutter speeds.

- Removal of the AA filter on the GX9.

- Dual-IS for better handholding when using stabilized Panasonic lenses.

GH2 ($350-$450 used)

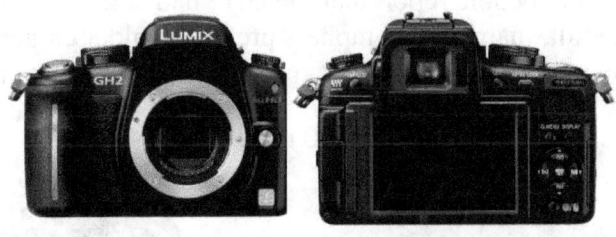

After Canon released the 5D Mark II with its 1080p support, everyone wanted to jump into DSLR video to take advantage of the large sensors and high-quality lenses that had been designed for still photography. The GH1 and more popular GH2 were Panasonic's answer to that—cameras designed to look and feel like a DSLR, but that act more like a video camera (while still being a capable still photography camera).

In many ways, they're superior to full-frame DSLR video:

- Smaller size

- Lower cost

- More depth-of-field

- Clean HDMI out

- Electronic viewfinder visible while recording

- Touch screens

- In the US, no 30-minute recording limit

The GH2 revolutionized videography and helped secure the entire MFT format. Many videographers, including the author, switched from Canon to Panasonic for videography. While the GH2 is no longer made, it still deserves mention because used models are available for less than $500, and with the right lens, they provide incredible sharpness and video quality. Getting the color right is a challenge, however; the GH2 is known for having a green color cast in video.

The GH2 (and GH3) cameras can be hacked, which replaces their firmware with a version that's not officially supported by Panasonic but can improve video recording quality. These hacks can lead to unreliable operation, though, so use them only if you really need them. In my personal experience buying used GH2s, I purchased some that had been hacked and some that had not been hacked, and the hacked models would randomly stop recording.

GH3 ($900 new, $550 used)

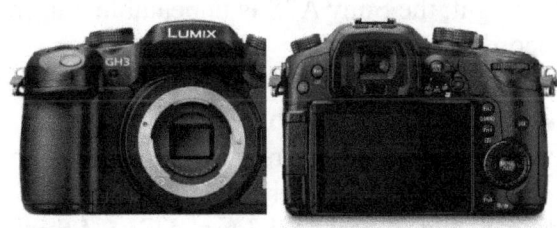

The follow-up to the GH2, the GH3 adds the ability to send sound over the HDMI cable. If you plan to record the on-camera audio through HDMI, you'll need that feature. The GH3 also increases the size of the body and improves the controls, allowing you to more quickly select a shutter speed, aperture, exposure compensation, and focusing point.

Many people report that the GH2 had slightly better video quality than the GH3, and as a result, many videographers prefer the older camera. Overall, the GH3's bigger size and better controls make it a superior camera, albeit at a slightly higher price used.

GH4 ($1,300 new, $1,100 used)

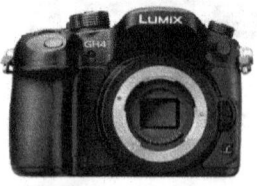 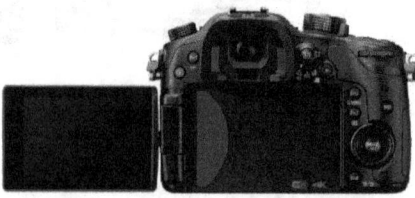

Panasonic's latest video-oriented camera is physically very similar to the GH3, but it offers a substantial improvement: 4k video. Panasonic offers an optional interface unit for the GH4 featuring XLR audio inputs, allowing you to record audio directly to the GH4 and improve your audio quality without using a separate recorder.

If you're interested in serious 4k video, the GH4 is currently your best value. There are less expensive 4K cameras (such as the G7) but they lack the higher-end capabilities of the GH4.

The GH4 is a remarkable video camera. We have two, and use them for most of our filming. Our GH4s have been through six countries, climbed mountains, and been thoroughly soaked in flash flooding. They continue to perform wonderfully.

| Video: GH4 Review
 40:58 - *sdp.io/GH4review* |  |

As a video camera, the GH4's advantages go beyond just 4k video. The articulating touch screen is wonderful. For candid filming, you can hold the viewfinder to your eye and film, which is impossible with a DSLR. The light weight makes it much easier to hand-hold steadily. Focus peaking makes manual focus fast and reliable.

Our favorite lens for filming is the Panasonic 14-140mm f/3.5-5.6 ($630 new). 4k video is just as sharp as with the Panasonic 12-35 f/2.8, and we prefer having the extra range at the telephoto end.

You might also consider the $3,000 BlackMagic Design Production Camera 4k (*http://sdp.io/bm4k*), which supports full-frame Canon lenses. If you need to film in very low light, the Sony A7S is unbeatable, but requires an expensive and bulky external recorder for 4k.

Mildly interesting trivia: Panasonic has usually avoided using the number "4" in its product names because of superstition in some Asian markets. For example, it skipped from the G3 to the G5, and from the GF3 to the GF5. I think it

| Video: a7S vs NX1 vs GH4
 24:10 - *sdp.io/a7sNX1GH4* | |

included the 4 in the GH4 name simply because it coincides perfectly with the 4k video feature.

GH5 ($2,000 new, $1,825 used)

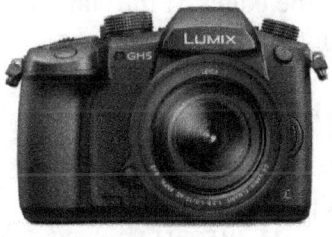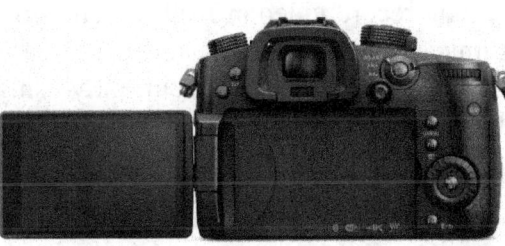

The GH5 is the best DSLR-style video camera we've ever used. It's not perfect, but the flexibility to shoot 4k at 60 frames per second has noticeably improved the quality of our videos. Either we publish at 60p for silky-smooth, lifelike quality, or we slow it down to 30p and take advantage of being able to stretch video for dramatic effect to make B-roll fill more space.

It's not a perfect video camera, though. Its biggest flaw is that autofocus isn't anywhere near as good as Canon's Dual Pixel AF or the Sony a9, especially at tracking moving subjects.

But, none of the Canon Dual Pixel cameras currently offer 4K video in a codec that we consider to be practical. The Sony a9 is a great video camera, but the screen doesn't flip forward, it's more than twice the price, and it doesn't record in 4K/60p.

Video: GH5 Review
40:58 - *sdp.io/GH5review*

The biggest competitor to the GH5 is probably the GH4. The GH4 is still an excellent camera, and you can almost buy two used GH4's for the price of a single GH5. The GH5 is definitely a better camera, however. The key improvements are the addition of 4k/60p, 1080/180p (for 6X slow motion in HD), better color and video quality, and utilizing the full-width of the sensor.

For stills shooters, the GH5 offers a good-quality 20 megapixel sensor. However, the Olympus E-M1 Mark II has better focusing, sharper images (thanks to not having an AA filter) and an amazing High-Res Mode. If you're not already invested in micro four-thirds lenses, you could also buy a full-frame Sony a7 III for $2,000.

GH5S ($2,500 new)

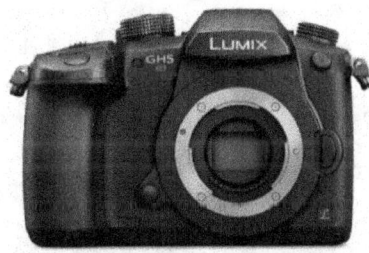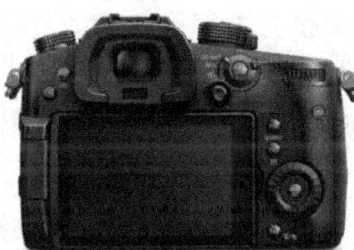

The Panasonic GH5 is a well-rounded but video-centric camera. The GH5S is a variant of the GH5 that's even more specialized; it's designed exclusively as a low-light video camera for use on a tripod or gimbal, but not handheld.

Panasonic introduced a few key changes in the GH5S when compared to the original GH5:

- The sensor is 10 megapixels instead of 20. Yes, it's more expensive and lower resolution. This low resolution means it's not at all optimal for stills photography.

- The sensor is a bit larger. Oddly, it's larger than a standard micro four-thirds sensor, allowing you to record video in 16:9 and still go to the edges of the standard micro four-thirds image circle. If you're using standard micro four-thirds lenses, you'll appreciate this difference. If you're currently using a GH5 with a speedbooster and adapted lenses, you might already have the optimal image circle, depending on the strength of your speedbooster.

- The low-light image quality is vastly batter; perhaps by as much as two stops. If you're recording at ISO 6,400 or above, the GH5S simply looks MUCH better than the GH5 or any other camera, including full-frame cameras like the a7S II. The GH5S, especially with fast lenses, is a fantastic choice for low-light video.

- It lacks sensor stabilization. This is a deal-breaker for many, especially vloggers who often shoot handheld. Of course, you can still use a stabilized lens, but your handheld shots still won't be as stable as the GH5 which can use both lens and sensor stabilization simultaneously. As a result, the GH5S' should primarily be used on a tripod, gimbal, or other support.

Most videographers should choose the GH5 over the GH5S; it's more versatile. However, if you need a primary and a backup body, as we do for filming our YouTube videos, consider getting one of each. Use the GH5 when you need the extra stabilization and use the GH5S when you need to film in low-light conditions.

GH5 II ($1,500)

The GH5 II is a good vlogging camera, and that's not something I could say about the original GH5. For those of us handholding a camera and relying on autofocus to keep us sharp, the GH5 II is a really good option (especially at this price point). Like the original GH5, it's also very good at traditional filmmaking, where you might be manually focusing and carefully controlling your exposure. However, traditional filmmakers will produce similar results with the original GH5 and have quite a few dollars left in their pockets.

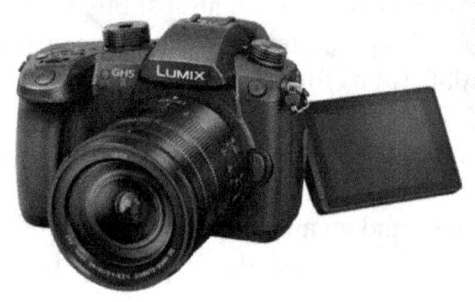

There are a few new features everyone might like in the updated camera:

- Built in log (which you had to buy for the original GH5)

- 4K/60 with 10-bits for better post-processing and dynamic range

- Like streaming to places like YouTube without a computer (though setup is cumbersome and even as a YouTuber I was never tempted to use it)

- Better image stabilization

- USB-C power and charging

For existing Micro Four-Thirds users, the GH5 II is a good upgrade. If you're buying your first camera, however, it's hard for me to recommend a Micro Four-Thirds camera in 2021 when there are so many great full-frame cameras (including those also made by Panasonic).

GH6 ($2,200)

The Panasonic GH6 is *so close* to being the greatest vlogging camera ever made. No close, and yet so far.

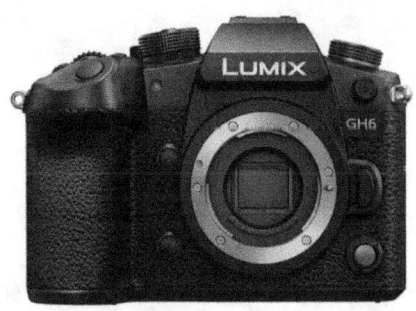

Full-width 4k/60 video is absolutely gorgeous and it'll record it almost all day without overheating. It feels great in the hands, and the new Panasonic f/1.7 zooms produce truly full-frame results.

The sensor stabilization is shocking; video is so rock-solid that walking shots look like they were filmed on a gimbal. That's the biggest reason I want the GH6 to be my main video camera; a gimbal is clumsy, heavy, and the opposite of discreet.

However, the GH6 has a fatal flaw that prevents me from using it as a video camera: autofocus. Autofocus simply isn't reliable; it'll hunt in and out randomly. Even when it does lock onto a subject, the contrast-detect focus pulses in and out, creating distracting bokeh. If I shot video in manual focus, I would absolutely use this as my main camera, but the shallow depth-of-field look we use requires good autofocus tracking even for static shots.